# Lecture Notes in Geoinformation and Cartography

Series Editors: William Cartwright, Georg Gartner, Liqiu Meng, Michael P. Peterson

Peter van Oosterom • Sisi Zlatanova •
Friso Penninga • Elfriede Fendel
(Eds.)

# Advances in 3D Geoinformation Systems

With 235 Figures

Editors:

Peter van Oosterom
Jaffalaan 9
2628 BX, Delft
The Netherlands
E-mail: p.j.m.vanoosterom@tudelft.nl

Friso Penninga
Jaffalaan 9
2628 BX, Delft
The Netherlands
E-mail: f.penninga@tudelft.nl

Sisi Zlatanova
Jaffalaan 9
2628 BX, Delft
The Netherlands
E-mail: s.zlatanova@tudelft.nl

Elfriede M. Fendel
Jaffalaan 9
2628 BX, Delft
The Netherlands
E-mail: e.m.fendel@tudelft.nl

ISBN  978-3-540-72134-5 Springer Berlin Heidelberg New York

Library of Congress Control Number: 2007939908

This work is subject to copyright. All rights are reserved, whether the whole or part of the material is concerned, specifically the rights of translation, reprinting, reuse of illustrations, recitation, broadcasting, reproduction on microfilm or in any other way, and storage in data banks. Duplication of this publication or parts thereof is permitted only under the provisions of the German Copyright Law of September 9, 1965, in its current version, and permission for use must always be obtained from Springer-Verlag. Violations are liable to prosecution under the German Copyright Law.

Springer is a part of Springer Science+Business Media
springeronline.com
© Springer-Verlag Berlin Heidelberg 2008

The use of general descriptive names, registered names, trademarks, etc. in this publication does not imply, even in the absence of a specific statement, that such names are exempt from the relevant protective laws and regulations and therefore free for general use.

Cover design: deblik, Berlin
Production: A. Oelschläger
Typesetting: Camera-ready by the Editors
Printed on acid-free paper   30/2132/AO 54321

# Preface

Society is expecting and demanding more 3D support since users have experienced the added value in emerging visualisation applications such as 3D globe based interfaces, navigation systems presenting a 3D perspective, etc. Due to the rapid developments in sensor techniques more 3D data have become available. Effective algorithms for (semi) automatic object reconstruction are required. Integration of existing 2D objects with height data is a non-trivial process, and further research is needed. The resulting 3D models can be maintained in several forms: TEN (Tetrahedral Network), constructive solid geometry (CSG) models, regular polytopes, TIN boundary representation and 3D volume quad edge structure, layered/topology models, voxel based models, 3D models used in urban planning/polyhedrons, and n-dimensional models including time. 3D analysis and 3D simulation techniques explore and extend the possibilities of spatial applications. In such a dynamic scientific environment, it is very important to have high quality and an open exchange of ideas on these new developments. It is also very important to carefully review and document the progress that is made. This book and the associated 3D GeoInfo workshop are an attempt to achieve this goal. The workshop is the second in a series on 3D geo-information. The previous event took place in Kuala Lumpur, Malaysia, on 7-8 August 2006 (http://www.fksg.utm.my/3dgeoinfo2006). Selected papers from the first workshop were published in 'Innovations in 3D Geo Information Systems', Springer-Verlag, 2006. The current (2007) workshop was held in Delft, the Netherlands, while future discussions on 3D issues are expected to be held in Seoul, South Korea, on 12-14 November 2008. The chapters in this book are the result of the '2nd International Workshop on 3D Geo-Information: Requirements, Acquisition, Modelling, Analysis, Visualisation' (12-14 December 2007, Delft, the Netherlands). The workshop's website contains many details, including the programme of the event (http://www.3d-geoinfo-07.nl). The five themes – mentioned in the sub-title – give a good indication of the thematic scope and the chapters in this book, which have been organised accordingly. The chapters have been selected based on a full-paper submission and were thoroughly reviewed by three members of the international programme committee. The authors of the best and most original submissions were asked to submit revised versions based on these comments. Additionally, this book contains two chapters that are related to the invited key-notes, both with a Geo-ICT industry origin (TeleAtlas: 'Maps Get Real, Digital Maps evolving from mathematical line graphs to virtual reality models' and Oracle: 'On Valid and Invalid Three-Dimensional Geometries'). These chapters together make up the main part of the book. During the workshop there were also working group sessions organised according to each of the specific themes: Requirements & Applications, Acquisition, Modelling, Analysis, and Visualisation. All of the working group sessions followed a given format: current problems to be solved, potential solutions, and recommendations by the working group. The discussions started with a position paper that was usually prepared by the chairs of the working groups. These position papers are also included in the last part of this book. The discussion sessions were coordinated by the chair, and the concluding summaries of the results were presented at the closing plenary session.

This series of workshops is an excellent opportunity for the exchange of ideas on 3D requirements and the comparison of the different techniques of 3D acquisition, modeling and simulation. The 3D GeoInfo workshops aim to bring together international state-of-the-art research in the field of 3D geo-information. They offer an interdisciplinary forum to researchers in the closely related fields of 3D data collection, modelling, management, data analysis, and visualisation. We hope that this series will become a very interesting yearly event with many sparkling discussions on all aspects of handling 3D geo-information!

The editors of this book would like to thank the co-organisers (Eveline Vogels, Marc van Kreveld and George Vosselman) for the pleasant cooperation from the first initial idea to organise the workshop through the final preparations. Further, we are grateful to all of the authors for their original contributions (also to the authors of contributions that were not selected). Special thanks to the members of the programme committee; they had the difficult task of critically reviewing the contributions and providing constructive comments, thus enhancing the quality of the chapters included in this book. The editors are also grateful to the support provided by the Advanced Gaming and Simulation (AGS) research centre and the two projects RGI-011 '3D topography' and RGI-013 'Virtual reality for urban planning and security', funded by the Dutch Program 'Space for Geo-information'.

Delft,
October 2007

*Peter van Oosterom*
*Sisi Zlatanova*
*Friso Penninga*
*Elfriede M. Fendel*

# Organisation

Programme chair | Peter van Oosterom,
Delft University of Technology

**Local organising committee**
Peter van Oosterom | Delft University of Technology
Sisi Zlatanova | Delft University of Technology
Friso Penninga | Delft University of Technology
Elfriede Fendel | Delft University of Technology
Eveline Vogels | Delft University of Technology
George Vosselman | ITC Enschede
Marc van Kreveld | Utrecht University

**Programme committee**
Alias Abdul-Rahman | University of Technology Malaysia (Malaysia)
Bart Beers | Cyclomedia (the Netherlands)
Tim Bevan | 1Spatial (United Kingdom)
Roland Billen | University of Liege (Belgium)
Lars Bodum | Aalborg University (Denmark)
Arnold Bregt | Wageningen University and Research Centre (the Netherlands)
Styli Camateros | Bentley (Canada)
Volker Coors | University of Applied Sciences Stuttgart (Germany)
Andrew Frank | TU Wien (Austria)
Georg Gartner | TU Wien (Austria)
Christopher Gold | University of Glamorgan (United Kingdom)
Cecil Goodwin | TeleAtlas (USA)

| | |
|---|---|
| Armin Gruen | ETH Zürich (Switzerland) |
| Norbert Haala | University of Stuttgart (Germany) |
| Muki Haklay | University College London (United Kingdom) |
| John Herring | Oracle Corporation (USA) |
| Daniel Holweg | Fraunhofer Institute Darmstadt (Germany) |
| Thomas Kolbe | Technical University Berlin (Germany) |
| Marc van Kreveld | Utrecht University (the Netherlands) |
| Hugo Ledoux | Delft University of Technology (the Netherlands) |
| Jiyeong Lee | University of North Caroline at Charlotte (USA) |
| Paul Longley | University College London (United Kingdom) |
| Twan Maintz | Utrecht University (the Netherlands) |
| Paul van der Molen | FIG/ITC Enschede (the Netherlands) |
| Martien Molenaar | ITC Enschede (the Netherlands) |
| Stephan Nebiker | Fachhochschule Nordwestschweiz (Switzerland) |
| András Osskó | FIG/Budapest Land Office (Hungary) |
| Peter van Oosterom | Delft University of Technology (the Netherlands) |
| Chris Parker | Ordnance Survey (United Kingdom) |
| Wanning Peng | ESRI (USA) |
| Friso Penninga | Delft University of Technology (the Netherlands) |
| Norbert Pfeifer | TU Wien (Austria) |
| Clemens Portele | Interactive Instruments (Germany) |
| Jonathan Raper | City University London (United Kingdom) |
| Carl Reed | Open Geospatial Consortium (USA) |
| Massimo Rumor | University of Padova (Italy) |
| Aidan Slingsby | City University London (United Kingdom) |
| Jantien Stoter | ITC Enschede (the Netherlands) |
| Rod Thompson | Queensland Government (Australia) |
| George Vosselman | ITC Enschede (the Netherlands) |
| Peter Widmayer | ETH Zürich (Switzerland) |
| Peter Woodsford | 1Spatial / Snowflake (United Kingdom) |
| Alexander Zipf | University of Applied Sciences FH Mainz (Germany) |
| Sisi Zlatanova | Delft University of Technology (the Netherlands) |

# List of Contributors

Alias Abdul Rahman
Department of Geoinformatics, Faculty of Geoinformation Science and Engineering, Universiti Teknologi Malaysia, Malaysia e-mail: alias@fksg.utm.my

Thierry Badard
Centre for Research in Geomatics and Geomatics Department, Laval University, Quebec, Canada

Jens Basanow
University of Bonn (Cartography), Germany, e-mail: basanow@geographie.uni-bonn.de

Karine Bédard
Centre for Research in Geomatics and Geomatics Department, Laval University, Quebec, Canada

Ahmad Biniaz
Department of Computer Science and Engineering, Shiraz University, Iran e-mail: biniaz@cse.shirazu.ac.ir

Martin Breunig
Institute for Geoinformatics and Remote Sensing, University of Osnabrück, Germany, e-mail: martin.breunig@uni-osnabrueck.de

Björn Broscheit
Institute for Geoinformatics and Remote Sensing, University of Osnabrück, Germany, e-mail: bjoern.broscheit@uni-osnabrueck.de

Arno Bücken
Institute of Man-Machine-Interaction, RWTH Aachen, Germany, e-mail: buecken@mmi.rwth-aachen.de

Edgar Butwilowski
Institute for Geoinformatics and Remote Sensing, University of Osnabrück,

Germany, e-mail: edgar.butwilowski@uni-osnabrueck.de

Eddie Curtis
Snowflake Software, United Kingdom, e-mail: eddie.curtis@snowflakesoftware.co.uk

Gholamhossein Dastghaibyfard
Department of Computer Science and Engineering, Shiraz University, Iran
e-mail: dstghaib@shirazu.ac.ir

Etienne Desgagné
Centre for Research in Geomatics and Geomatics Department, Laval University, Quebec, Canada

Jürgen Döllner
Hasso-Plattner-Institute at the University of Potsdam, Germany, e-mail: doellner@hpi.uni-potsdam.de

Ludvig Emgård
SWECO Position AB, Sweden, e-mail: ludvig.emgard@sweco.se

Rob van Essen
Tele Atlas NV, 's-Hertogenbosch, the Netherlands e-mail: rob.vanessen@teleatlas.com

Andrew U. Frank
Department of Geoinformation and Cartography, Technical University Vienna, Austria e-mail: frank@geoinfo.tuwien.ac.at

Tassilo Glander
Hasso-Plattner-Institute at the University of Potsdam, Germany, e-mail: tassilo.glander@hpi.uni-potsdam.de

Christopher Gold
University of Glamorgan, Pontypridd, Wales, United Kingdom, e-mail: cmgold@glam.ac.uk

Baris M. Kazar
Oracle USA, Inc., USA, e-mail: Baris.Kazar@Oracle.comu

Dave Kidner
University of Glamorgan, Pontypridd, Wales, United Kingdom, e-mail: dbkinder@glam.ac.uk

Henk de Kluijver
dBvision, Utrecht, the Netherlands, e-mail: henk.dekluijver@dBvision.nl

Ravi Kothuri
Oracle USA, Inc., USA, e-mail: Ravi.Kothuri@Oracle.com

Marc van Kreveld
Department of Information and Computing Sciences, Utrecht University, the Netherlands, e-mail: marc@cs.uu.nl

List of Contributors

Tobias Krüger
Leibniz Institute of Ecological and Regional Development (IOER), Dresden, Germany, e-mail: t.krueger@ioer.de

Vinaykumar Kurakula
L.I.G 'B' 543, Hyderabad, India, e-mail: kurakulavinay@rediffmail.com

Hugo Ledoux
section GIS Technology, OTB, Delft University of Technology, the Netherlands, e-mail: h.ledoux@tudelft.nl

Jiyeong Lee
Department of Geoinformatics, University of Seoul, 90 Jeonnong-dong, Dongdaemun-gu. Seoul, 130-743, South Korea, e-mail: jlee@uos.ac.kr

Gotthard Meinel
Leibniz Institute of Ecological and Regional Development (IOER), Dresden, Germany, e-mail: g.meinel@ioer.de

Shyamalee Mukherji
Centre of Studies in Resources Engineering, Indian Institute of Technology, Bombay, India e-mail: shyamali@csre.iitb.ac.in

Pascal Neis
University of Bonn (Cartography), Germany, e-mail: neis@geographie.uni-bonn.de

Steffen Neubauer
University of Bonn (Cartography), Germany, e-mail: neubauer@geographie.uni-bonn.de

Peter van Oosterom
section GIS Technology, OTB, Delft University of Technology, the Netherlands, e-mail: oosterom@tudelft.nl

Sander Oude Elberink
International Institute for Geo-Information Science and Earth Observation (ITC), Enschede, the Netherlands e-mail: oudeelberink@itc.nl

Friso Penninga
section GIS Technology, OTB, Delft University of Technology, the Netherlands, e-mail: f.penninga@tudelft.nl

Jacynthe Pouliot
Centre for Research in Geomatics and Geomatics Department, Laval University, Quebec, Canada e-mail: jacynthe.pouliot@scg.ulaval.ca

Shi Pu
International Institute for Geo-information Science and Earth Observation, Enschede, the Netherlands, e-mail: spu@itc.nl

Jonathan Raper

The giCentre, Department of Information Science, City University, London, United Kingdom e-mail: raper@soi.city.ac.uk

Siva Ravada
Oracle USA, Inc., USA, e-mail: Siva.Ravada@Oracle.com

Paul Richmond
University of Sheffield, United Kingdom, e-mail: paul@dcs.shef.ac.uk

Daniela Romano
University of Sheffield, United Kingdom e-mail: d.romano@dcs.shef.ac.uk

Jürgen Rossman
Institute of Man-Machine-Interaction, RWTH Aachen, Germany e-mail: rossmann@mmi.rwth-aachen.de

Arne Schilling
University of Bonn (Cartography), e-mail: schilling@geographie.uni-bonn.de

Aidan Slingsby
The giCentre, Department of Information Science, City University, London, United Kingdom e-mail: a.slingsby@city.ac.uk

Jantien Stoter
ITC Enschede, Department of Geo Information Processing, the Netherlands, e-mail: stoter@itc.nl

Chen Tet Khuan
Department of Geoinformatics, Faculty of Geoinformation Science and Engineering, Universiti Teknologi Malaysia, Malaysia e-mail: kenchen@fksg.utm.my

Vincent Thomas
Centre for Research in Geomatics and Geomatics Department, Laval University, Quebec, Canada

Andreas Thomsen
Institute for Geoinformatics and Remote Sensing, University of Osnabrück, Germany, e-mail: andreas.thomsen@uni-osnabrueck.de

Rodney Thompson
Department of Natural Resources and Water, Queensland, Australia, e-mail: Rod.Thompson@nrw.qld.gov.au

Rebecca Tse
University of Glamorgan, Pontypridd, Wales, United Kingdom, e-mail: rtse@glam.ac.uk

Alexander Zipf
University of Bonn (Cartography), Germany, e-mail: zipf@geographie.uni-bonn.de

Sisi Zlatanova
section GIS Technology, OTB, Delft University of Technology, the Netherlands, e-mail: s.zlatanova@tudelft.nl

# Contents

**Part I Keynotes**

1  Maps Get Real: Digital Maps evolving from mathematical line graphs to virtual reality models .................... 3
   Rob van Essen

2  On Valid and Invalid Three-Dimensional Geometries ..... 19
   Baris M. Kazar[1], Ravi Kothuri[1], Peter van Oosterom[2] and Siva Ravada[1]

**Part II Papers**

**Theme I: Requirements & Applications**

3  Navigable Space in 3D City Models for Pedestrians ....... 49
   Aidan Slingsby and Jonathan Raper

4  Towards 3D Spatial Data Infrastructures (3D-SDI) based on open standards – experiences, results and future issues  65
   Jens Basanow, Pascal Neis, Steffen Neubauer, Arne Schilling, and Alexander Zipf

5  Re-using laser scanner data in applications for 3D topography ................................................. 87
   Sander Oude Elberink

6  Using Raster DTM for Dike Modelling .................... 101
   Tobias Krüger and Gotthard Meinel

7  Development of a Web Geological Feature Server (WGFS) for sharing and querying of 3D objects .......... 115

xvii

Jacynthe Pouliot, Thierry Badard, Etienne Desgagné, Karine Bédard, and Vincent Thomas

**Theme II: Acquisition**

8  Using 3D-Laser-Scanners and Image-Recognition for Volume-Based Single-Tree-Delineation and -Parameterization for 3D-GIS-Applications .............. 131
Jürgen Rossmann and Arno Bücken

9  Automatic building modeling from terrestrial laser scanning ................................................... 147
Shi Pu

10  3D City Modelling from LIDAR Data .................... 161
Rebecca (O.C.) Tse, Christopher Gold, and Dave Kidner

**Theme III: Modelling**

11  First implementation results and open issues on the Poincaré-TEN data structure ........................... 177
Friso Penninga and Peter van Oosterom

12  Drainage reality in terrains with higher-order Delaunay triangulations ........................................... 199
Ahmad Biniaz and Gholamhossein Dastghaibyfard

13  Surface Reconstruction from Contour Lines or LIDAR elevations by Least Squared-error Approximation using Tensor-Product Cubic B-splines ......................... 213
Shyamalee Mukherji

14  Modelling and Managing Topology in 3D Geoinformation Systems[1] ............................................... 229
Andreas Thomsen, Martin Breunig, Edgar Butwilowski, and Björn Broscheit

15  Mathematically provable correct implementation of integrated 2D and 3D representations .................... 247
Rodney Thompson[1,2] and Peter van Oosterom[1]

16  3D Solids and Their Management In DBMS .............. 279
Chen Tet Khuan, Alias Abdul-Rahman, and Sisi Zlatanova

17  Implementation alternatives for an integrated 3D Information Model ...................................... 313
Ludvig Emgård[1,2] and Sisi Zlatanova[1]

## Theme IV: Analysis

**18  Serving CityGML via Web Feature Services in the OGC Web Services - Phase 4 Testbed** .......................... 331
Eddie Curtis

**19  Towards 3D environmental impact studies: example of noise** ........................................................... 341
Jantien Stoter[1], Henk de Kluijver[2], and Vinaykumar Kurakula[3]

**20  The Kinetic 3D Voronoi Diagram: A Tool for Simulating Environmental Processes** ................................. 361
Hugo Ledoux

## Theme V: Visualisation

**21  Techniques for Generalizing Building Geometry of Complex Virtual 3D City Models** ......................... 381
Tassilo Glander and Jürgen Döllner

**22  Automatic Generation of Residential Areas using Geo-Demographics** ........................................ 401
Paul Richmond and Daniela Romano

## Part III Position papers

**23**  *Working Group I – Requirements and Applications – Position Paper:*
**Requirements for 3D in Geographic Information Systems Applications** ............................................. 419
Andrew U. Frank

**24**  *Working Group II – Acquisition – Position Paper:*
**Data collection and 3D reconstruction** ..................... 425
Sisi Zlatanova

**25**  *Working Group III – Modelling – Position Paper:*
**Modelling 3D Geo-Information** ........................... 429
Christopher Gold

**26**  *Working Group IV – Analysis – Position Paper:*
**Spatial Data Analysis in 3D GIS** .......................... 435
Jiyeong Lee

**27**  *Working Group V – Visualization – Position Paper:*
**3D Geo-Visualization** ...................................... 439
Marc van Kreveld

# Part I
# Keynotes

# Chapter 1
# Maps Get Real: Digital Maps evolving from mathematical line graphs to virtual reality models

Rob van Essen

**Abstract**

This paper describes the production process of 3D maps at Tele Atlas. Starting from a 2D City map which contains relevant features by their outline, 3D City maps are created by adding height to the outlines of buildings and other objects. Further enhancements are made by adding roof representations and facade textures. These textures can either be an exact representation of the facade or be composed of predefined elements from a library which leads to significant data size reductions. In the production process, Mobile Mapping a technology developed at Tele Atlas plays an important role. It serves to collect building heights and facade textures and components. To date this has resulted in the availability of 3D City maps of 12 cities with important extension planned for 2008. The paper ends by stating that whereas the original reason to develop 3D maps stems from a desire to improve the user interface of navigation and other Intelligent Transport applications, it is likely that 3D models will have growing significance in future also for in-car safety systems.

## 1.1 Introduction

Map-making appears to predate written language by several millennia. One of the oldest surviving maps is painted on a wall of the Catal Huyuk settlement in south-central Anatolia (now Turkey); it dates from about 6200 BC. Cartography as a science is generally considered to have started with

Vice-President Strategic Research and Development
Tele Atlas NV,
Reitscheweg 7f,
POBox 420, NL-5201AK, 's-Hertogenbosch,
e-mail: rob.vanessen@teleatlas.com

Eratosthenes (276BC-194BC), a Greek mathematician and geographer born in today's Lybia. Eratosthenes developed a system of latitude and longitude and was the first to calculate the circumference of the earth. As such he was at the roots of modern GPS. As a practical result he published what is generally considered the first world map (see figure 1.1) [1]. Maps remained relatively

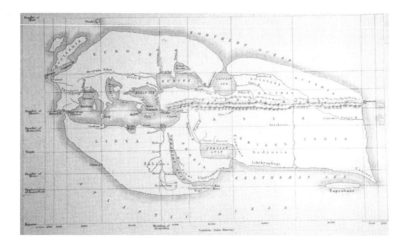

**Fig. 1.1** Erastosthenes' world map [1]

unchanged until around 1965 when the US Census bureau, with the help of GDT (Geographic Data Technologies Inc.) founder Donald Cooke, started to design a topological model for map information called Dual Independent Map Encoding (DIME) for the US bureau of the Census which together with the USGS (US Geological Survey) Digital Line Graphs in 1990 evolved into the US TIGER (Topologically Integrated Geographic Encoding and Referencing) files, published by the Census Bureau. Maps were no longer two-dimensional images on a physical sheet but now also models in a computer. Whereas the TIGER files were created mainly for Census purposes, i.e. the registration of address information, soon people started to think about other application areas. Despite the change in the map concept, these new applications were similar to the applications of the traditional map, the main one being Navigation, ie. finding locations and guiding the user to it. In the beginning of the eighties ETAK started to create the navigable map of the US and a few years later Tele Atlas started the same job in Europe. In 2000 ETAK and Tele Atlas joined forces. In 2004 Tele Atlas and GDT joined forces. Today the complete North-American and Western-European territory is covered by Tele Atlas maps and extensive coverage increases take place in Eastern-Europe, China, the Far East and in South America. It can be expected that all the streets on the earth surface will have been captured before long.

The first navigation systems hit the European market around 1995. These were closely tight to the car and therefore called in-car navigation systems. The tight bond with the car was caused by the fact that the GPS signal which enabled the car to position itself was distorted for use by non-US military purposes by the US Military. As a consequence navigation was very similar to traditional nautical navigation. After the initial (inaccurate) (GPS) fix, a process called dead reckoning was started which on basis of direction and distance sensors estimated the new positions of the car. As soon as it was possible to match the position of the car with a position on the map (e.g. when the car turned into a side street), the estimated position was corrected. This process is called map matching. Dead reckoning and map matching tied navigation to the car because the sensors which these processes required, (gyro-) compasses, odometers, wheel sensors etc. were generally not part of the navigation system but external components built in the car. These navigation systems typically had one of two types of user interface. First there was the low cost interface where the user was guided by icons on a small (monochrome) display. The other type combined the icon based guidance with an image of the map around the vehicles position on a (color) display. Both interfaces were generally combined with audible instructions.

In 2000, US President Clinton 'turned off Selective Availability' or, in other words, he removed the distortion from the GPS signal. As a result the GPS signal increased its accuracy to such an extent that positioning without the help of external sensors became possible and navigation was no longer tied to the car. This initiated what can be called the biggest change in modern land navigation to date. Navigation was leaving the car, devices dramatically dropped in price and navigation became main stream and mass market. This change gave rise to a series of very important changes in digital maps and in the process of digital mapping. One of these is the subject of the paper: The adaptation of the displayed image of the map to the requirements of the mass-market. The general public had only limited appreciation for the very strong technical connotation of the digital line graphs displayed on the early navigation systems. The first step to respond to these requirements was to turn back to the 'traditional' paper cartography and to apply the map display principles common there. In paper maps, roads were no lines but linear areas and buildings were rectangles and not points. I.e. objects were represented by their outlines rather then by a mathematical abstraction. The result of applying these 'traditional' map principles onto the digital map is the 2D City map.

Still, significant portions of the population have problems reading such a map. At the same time, the gaming industry was showing the way: Computers made it possible to display virtual reality in consumer applications not as a model but rather as it is, as virtual reality, without the actual correspondence between a map image at a certain location and the view at that location in reality though. The map was leaving its traditional two dimensional shelter and became three dimensional. The 3D City Map was born.

Clearly, 3D computer models of buildings had been developed before, in the gaming industry but also by academia and government researchers [3]. The described development can be considered novel because these 3D models were created in an accurate geographic reference frame (thus justifying the term map) in an industrial production environment focused on large scale production for mass-market applications. Tele Atlas has been leading the way in the process of adapting the map to the requirements of the mass-market. It was the first to publish a 2D City map for navigation applications and the first corresponding 3D City Map products also came from Tele Atlas. This paper describes the process how Tele Atlas realized these new products.

Essential in these processes is the Tele Atlas' Mobile Mapping technology. The paper will first give a description of Mobile Mapping. Secondly it will shortly describe the production process of 2D City maps after which the 3D City map process is described. Finally some outlook on the future will be given where it is likely that these virtual reality models not only appeal to the user but also fulfill requirement of a new generation of in-car systems which focus on increasing safety and reducing environmental consequences.

## 1.2 Capturing Reality: Mobile Mapping

### 1.2.1 The early days

Mobile mapping is a technology developed by Tele Atlas. It was first launched in 1990 as a result of a Eureka project [2]. At that time, geometrically correct paper map material often was the basic source for digital maps. Clearly, the information which could be derived from these paper sources was not sufficient to create a navigable map. Therefore, also a wide-variety of other sources had to be used ranging form postal files, tourist maps and construction plans of new road geometry. In addition it was necessary to actually go to the spot and check the information on correctness and currentness and fill in the information which was not available on any other source like one-way traffic regulations or prohibited turn information. The thought behind Mobile Mapping was that instead of sending a surveyor who had to manually capture the information to send a vehicle equipped with camera's which record reality and afterwards interpret the images in an optimally tuned office environment. The project resulted in thousands of video tapes. Difficulties of handling these large amounts of media and accessing individual images sequences finally made the project less successful than originally anticipated and the project was stopped.

1 Maps Get Real

**Fig. 1.2** First generation Mobile Mapping van (1990)

## 1.2.2 Mobile Mapping re-invented

In 2004 technology had made big leaps compared to 1990. Positioning and camera technology had improved and completely new, random access storage media had become available at considerable lower prices. As a result also big advances in Mobile Mapping technology had been achieved (see also [4]). This was for Tele Atlas the signal to revisit the concept of Mobile Mapping. The result was a highly accurately positioned vehicle equipped with up-to six digital camera's of which the two forward facing were stereo camera's allowing geometrically accurate measurements in the images. The positional accuracy achieved is sub-half-meter. The pictures have a resolution of 1300x1000 pixels. The novel aspect of this new generation of mobile mapping was not so much the technical components but more the support for industrial production processes. Aspects of this include quasi-continuous operation by none expert people at typical traffic speeds positioned highly accurately and seamlessly in different countries. The main thought behind the Mobile Mapping concept in 2005 was similar to that of 1987. Instead of sending surveyors to check and complete the content of the database it was better to record reality and to extract the information in an optimally tuned office environment. In this respect had the need for Mobile Mapping considerably increased compared to 1990. This was the consequence of the ever increasing content of a digital map database. Next to the needs of simple navigation applications now also the need of advanced safety applications requiring the inclusion of information like lanes, traffic signs, curve information, traffic lights etc. had to be accommodated [5]. In addition, it became apparent that Mobile Mapping also can fulfill the need of improved map display requirements. This

**Fig. 1.3** Second generation Mobile Mapping vehicle (2005)

will be the further detailed in the following sections. Currently Mobile Mapping is extensively used by Tele Atlas. World wide, 50 vehicles are driving the streets resulting 250 TB of imagery yearly. These images are interpreted by highly trained database experts and the necessary content is extracted. When customer complaints come in via the Tele Atlas' Customer Feedback system 'MapInsight' the images corresponding to the complaint position are interpreted which in a large number of cases leads to a resolution of the complaint.

### 1.2.3 Mobile Mapping: the future

The potential of Mobile Mapping is by far not exploited fully yet. Contents requirements are still increasing. And a lot of this new contents can be captured using Mobile Mapping. 3D information is one example which will be dealt with later. Another example is content which can be captured using additional sensors on the Mobile Mapping van. A typical example of this is the slope of roads which is required to optimize the automatic gearbox of trucks resulting in significant fuel consumption decreases. In current prototyping activities, slope is being captured with Mobile Mapping by a 3D gyro installed in the vehicle. Other sensors include laser scanners which enable collection of banking (transverse slope) and street lay out. Novel aspects of these developments mainly relate to the use of different sensors in combination and exploiting the synergies, also referred to as sensor fusion. Another field where

# 1 Maps Get Real

further developments are expected is the automatic interpretation of images. Currently research on automatic object recognition techniques is undertaken at Tele Atlas and the first results have been adopted in production. The expectation is that in future more and more of the information present in the Mobile Mapping images can be extracted automatically.

## 1.3 Advanced Map Display

### 1.3.1 2D City Maps

Traditionally, the Japanese navigation industry has put much more emphasis on map display than their European counterparts. The 2D City Map therefore originates from Japan. Tele Atlas started producing 2D City maps in cooperation with its Japanese partner IPC [6] in 1999. As indicated above 2D City Maps typically contain features described by their outline. Streets, including traffic islands and side walks are included as well as buildings. Also walkways (through e.g. parks) are included. Railway lines and water areas

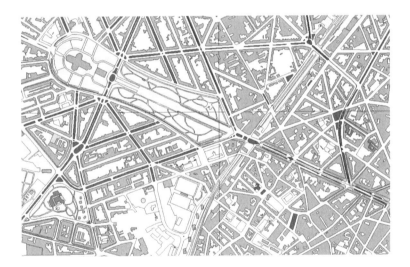

**Fig. 1.4** An extract of the 2D City Map of Brussels

are included with greater detail. 2D City Maps are mainly produced in a traditional way. From a source like a paper (cadastral) map or areal photograph, the outlines of the objects are extracted, usually via a manual digitizing process. The accuracy requirements for 2D City Maps are 5m RMS which is higher than the accuracy requirements for the traditional in-car navigation road network. Because the two need to fit, there is a final production step involving the possible adaptation of the original navigation network geometry to the geometry of the 2D City Map.

## *1.3.2 3D City Maps*

Basically, a 3D City Map is a 2D City map in which a height attribute has been attached to the building outlines and possibly to the sidewalks. The buildings can have been extended with a roof and the facades of the buildings can have been assigned a certain texture. In addition standard 3D representation of road furniture objects like traffic signs, traffic lights, trees, fences etc. can have been added. Also 3D Landmarks should be mentioned here as potential components, which are separately description further down. From all the 'can's' in the above description it becomes clear that a 3D City Model can take on very different shapes. Tele Atlas decided to go for a model in which roofs and facade textures are included. Variation occurs in the type of texture which is used. This will be elaborated further down. Despite all the variance possible, there is one very important fixed factor and that is that 3D Building representations are defined as extended 2D City maps. The consequence of this is that they are perfectly aligned with the navigation road network and that they thus can be deployed in navigation systems.

Building height is derived from externally acquired photogrammetric elevation models. By subtracting the terrain height from normal digital elevation models, building heights are acquired and attached to the corresponding building outlines. This leads to good results in the majority of the cases. Where this is not the case building heights are derived from the stereoscopic camera imagery of the Mobile Mapping vehicles. Facade textures are also derived from Mobile Mapping images. The approach taken here depends on the 'commercial' value of the facade. The most basic approach makes use of standard components from a library [7]. This library contains typical facade components like windows, doors and wall textures. The library has been built during a manual pre-process on basis of mobile mapping imagery. Depending on what the 3D City map operator sees on the Mobile Mapping images he will compose the facade using different components from the library. This is a semi-automatic process in which technology interprets the image and proposes a limited set of components to the operator. It is clear that such an approach facilitates the re-use of components which will greatly reduce the storage requirements for these types of models. In a lot of cases, facades have a repetitive character. The model has been designed such that this repetitive character also is explicitly supported thus further reducing storage and data size consequences. The component based model has been successfully proposed for standardization to ISO/TC204 SWG3.1: GDF. The next version of GDF4.0 [8] , GDF5.0 will contain the model described. Figure 1.8

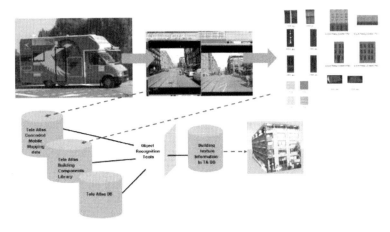

**Fig. 1.5** Adding textures via a library based approach

gives the UML model. Please note that this model is more extensive than the described products. Pending the finalization of this standard, 3D models are made available in the Shape format [9]. An alternative approach takes

**Fig. 1.6** The result of repetition of components

entire facades (or parts of facades) from Mobile Mapping images and attaches them to the blocks. Experience has shown that the repetitive component approach does not work for characteristic facades like the ground floor level of shopping streets with the typical and individual window lay-out with dominant lettering and logo's. Here, only a complete image of the shop front will have sufficient similarity. A typical problem with this approach is that the Mobile Mapping images are 'polluted' with trees, parked cars and accidental pedestrians. Special technology facilitates the semi-automatic removal of this pollution.

**Fig. 1.7** A Mobile Mapping images before and after clean-up

It is important to note that both the technology and the GDF model allow that both approaches are deployed simultaneously. In other words, the image based approach can be deployed for one building in a street and for the ground floor of four shops further down the street, while the rest of the street is modelled with the component based approach. In such a way an optimal combination of data size requirements and quality requirements can be adopted. Currently, 3D City Maps of 2 Cities are available fully textured (Berlin and London) and 10 cities as blocks models with roofs. In 2008 24 cities will be available fully textured and 52 as block models.

# 1 Maps Get Real

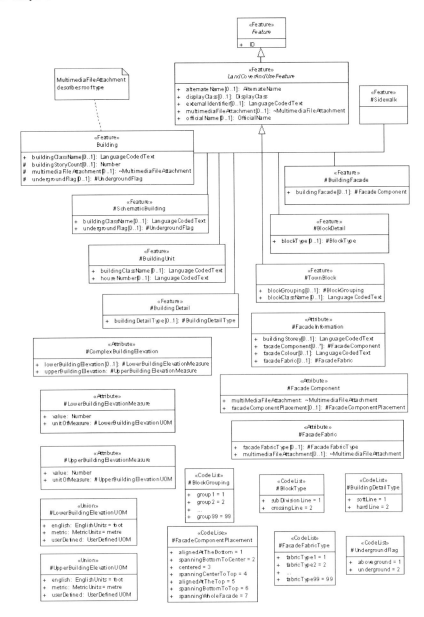

**Fig. 1.8** The UML Model for 3D building as proposed for GDF5.0

## 1.4 3D Landmarks

Despite its flexibility, it is clear that the 3D City model approach only applies to cubically shaped buildings with facades which consist of vertical planes

**Fig. 1.9** An example of a 3D City Model of the city of Graz

only. Consequently not all building can be modelled with it. Notable examples where this will be the case include the Arc de Triomphe in Paris or the Gurkin Building in London. These buildings or rather "constructions" can be modelled as 3D Landmarks.. 3D Landmarks are created by photographing the object from all sides with an accurately positioned digital camera. With special software the images are interpreted and converted in a so called wireframe model. Planes are identified in this wire frame model and the corresponding picture parts are identified and transformed and added to the planes. This results in a high quality 3D model of the object. 3D Landmarks are treated as an object external to the digital map. A proper connection between the two is a pointer to a location which is in fact an absolute WGS '84 coordinate. A second horizontal coordinate assures that aspects like direction and scaling are properly taken care of. The high quality of the 3D Landmark also makes the methodology fit to represent buildings which do have a cubic shape but which because of their prominence need to be modelled with higher quality than the 3D City models.

## 1.5 Data Volumes

Data volumes are important criteria for on- and off-board navigation applications. The GDF model of the component based approach facilitates very efficient data size requirements. Research has shown that a standard non compressed Tele Atlas 2D City Map needs about 200 kb/km2. For the basic 3D block models an additional 75 kb/km2 is required. For the production of complete 3D block models with roof types and generic facade information 330kb for the complete library has to be added. Those libraries can be

**Fig. 1.10** 3D Land mark of the Arc de Triomphe

**Fig. 1.11** 3D landmarks

used for a complete country. If buildings need to be presented by individual images, an extra of 10kb per processed facade picture has to be added. An extra of about 100 kb per km2 has to be added in order to link the map data to those libraries. The volume of standard VRML 3D models is high, but can significantly be reduced using compression technology. They have a data volume between 70 to 150 kb per 3D landmark. In Tele Atlas products the footprint of the 3D components is significantly higher. This is because Tele

Atlas customers have expressed the desire to make an own trade-off between data size and quality and apply corresponding data compression techniques.

## 1.6 Storage of 3D data

In the recent past, an interesting trend has become visible: The use of off the shelf relational database technology for storing and managing geographic data and even delivering geospatial applications. More and more geographic features are natively supported by relational database products. With respect to 3D the recent version of the Oracle database product contains interesting new features. This Oracle 11g database has as one of the key features support for 3D. Oracle targeted its software development in supporting 3-dimensional data toward specific applications such as:

1. GIS for city planning and property rights
2. City modelling and adopting features to support CityGML guidelines
3. Business Intelligence for real estate and advertising
4. Virtual-Reality solutions

Specifically, the 3D data types that are supported include points, lines, polygons and solids, as well as multi-points, multi-lines and multi-surfaces. It follows Geography Markup Language (GML) 3.1.1 [10] and ISO19107 [11] specifications. Support for simple solids includes closed surfaces such as a cubes or pyramids. 3D support for arcs or parametric surfaces is not included. These enhancements will support large, high density and volume 3D city models. Additional new data types support the massive volumes of point data, such as point clouds, obtained from laser scanners and LiDAR, as well as triangulated irregular networks. Part of Oracle 11g are also many spatial functions to manage and operate on 3D data. Tele Atlas contributed to the Oracle 11g beta testing effort. As a consequence, the 3D models described in the previous sections are compliant to Oracle 11g 3D model and easy upload of the Tele Atlas 3D models in Oracle 11g is provided. This trend has also influenced the GDF standardisation where in the upcoming version 5.0 also a database neutral relational storage model is supported, which is compliant to the SQL/MM standard [12]. In version 5.0, a syntactic model will be provided, modelling the entities of the GDF conceptual model one to one in relations, e.g. node, edge, face, point, feature, line feature, area feature etc. In future, it is likely that GDF will also provide semantic model, by splitting the generic table further according to feature themes. Also, the current 2.5D model can be extended towards a 3D model, by supporting corresponding geometric primitives such as solids. Thus, the GDF relational extensions are to be seen complementary to the commercial off-the-shelf solutions from Oracle and others.

## 1.7 The future: 3D Maps beyond map display

The relevance of 3D modelling has in this paper mainly explained in the context of the need for more user-friendly map display. As a consequence (digital) maps have evolved from digital line graphs to virtual reality models. Maps are getting real. An industrial production process has been outlined with which these maps are produced. Mobile Mapping has been presented as an important data collection technology in this process. Mobile Mapping was developed as an efficient capturing mechanism for the ever growing content of the navigable database which mainly relates to the upcoming use of in-car safety systems. These safety systems are also referred to as Advanced Driver Assistance Systems (ADAS). They typically work on basis of sensors which constantly monitor the surroundings of the car. The map is considered one of the sensors which allow the system to look ahead. Map information is either used as a primary sensor (e.g. to detect the sharp curve ahead and warn the driver) or as a secondary sensor where it is used to validate information from other sensors (e.g. a lane departure system working with video as a primary sensor in the case that lane markings are missing). The sensor equipment of the car will further grow in future and there will be a growing need to use a map to validate the information coming from these sensors. 3D representations of objects in the map are expected play an important role here. In future not only people will need 3D maps but also cars.

## References

[1] http://en.wikipedia.org/wiki/Eratosthenes
[2] http://www.eureka.be/inaction/AcShowProjectOutline.do
[3] City of Helsinki converts to 3D Mapping: http://www.bentley.com/en-US/Markets/Geospatial/User+Stories/3D+Mapping.htm
[4] Tao, Vincent, Jonathan Li (Eds.): Advances in Mobile Mapping Technology. Taylor&Francis, London, 2007
[5] T'Siobbel, Stephen: Mobile Digital Mapping: Field data collection Matters!. Proceedings ITS World 2004, Nagoya 2004
[6] http://www.incrementp.co.jp/english/business1.html
[7] Vande Velde, Linde: Tele Atlas 3D City Models. International Workshop Next generation 3D city mod-els, Bonn, 21-22 June 2005
[8] ISO14285:2003 Intelligent Transport Systems - Geographic Data Files - Overall Data Specifications (colloquially referred to as GDF4.0), ISO/TC204 Intelligent Transport Systems..
[9] http://www.esri.com/library/whitepapers/pdfs/shapefile.pdf
[10] http://www.opengeospatial.org/standards/gml
[11] ISO19107: http://www.iso.org/iso/en/CatalogueDetailPage.CatalogueDetail?CSNUMBER=26012

[12] ISO/IEC 13249-3 Information technology - Data-base languages - SQL Multimedia and Application Packages - Part 3: Spatial, May 15, 2006.

# Chapter 2
# On Valid and Invalid Three-Dimensional Geometries

Baris M. Kazar[1], Ravi Kothuri[1], Peter van Oosterom[2] and Siva Ravada[1]

**Abstract**

Advances in storage management and visualization tools have expanded the frontiers of traditional 2D domains like GIS to 3Dimensions. Recent proposals such as CityGML and associated gateways bridge a long-standing gap between the terrestrial models from the GIS and the CAD/CAM worlds and shift the focus from 2D to 3D. As a result, efficient and scalable techniques for storage, validation and query of 3D models will become a key to terrestrial data management. In this paper, we focus on the problem of validation of 3D geometries. First we present Oracle's data model for storing 3D geometries (following the general OGC/ISO GML3 specifications). Then, we define more specific and refined rules for valid geometries in this model. We show that the solid representation is simpler and easier to validate than the GML model but still retains the representative power. Finally, we present explicit examples of valid and invalid geometries. This work should make it to easy to conceptualize valid and invalid 3D geometries.

## 2.1 Introduction

The combination of civil engineering surveying equipment, GPS, laser scanning and aerial and satellite remote sensing coupled with strong spatial research has enabled the growth and success of the GIS industry. These industries typically cater to a wide variety of application-specific databases includ-

---

[1]Oracle USA, Inc.
One Oracle Drive, Nashua, NH 03062, USA
[2]Delft University of Technology, OTB, section GIS Technology,
Jaffalaan 9, 2628 BX the Netherlands
Baris.Kazar, Ravi.Kothuri, Siva.Ravada@Oracle.com, oosterom@tudelft.nl

ing cadastral databases for property/land management, and utility management in addition to popular location-based searching [G84, BKSS90, BDA04, MTP04, MA05] and services [J04, YS05]. Cadastral and Census databases typically maintain exact geometry boundaries for personal properties or administrative boundaries. For example, most countries in the European Union including the Netherlands maintain exact boundaries of properties and use that information for better visualization, property tax calculation etc. The United States Census Bureau maintains the database of exact geometrical shapes for various entities in different levels of the administrative hierarchy. The hierarchy consists of census blocks, block-groups, cities, counties, states etc. In most of these GIS applications, the data is represented using 2Dimensional geometries and stored inside a spatial database. The databases can be quite huge storing up to tens and hundreds of millions of spatial geometries. These geometries are typically stored, managed and queried using efficient processing techniques developed in the spatial and database research community [A94, BKSS90, BKS94, DE04, G84, S89].

Till recently, 3D research has been primarily confined to the graphics, CAD/CAM, gaming, virtual reality and computational geometry communities [FD97, SE02]. The rapid strides in scanner equipment and the declining costs of the same in recent years has generated renewed enthusiasm in 3D data management in various fields like GIS, CAD/CAM, medical and city modeling. Recent proposals such as CityGML [KGP05, OGC06C] bridge a long-standing gap between the terrestrial models from the GIS and the CAD/CAM worlds and shift the focus from 2D to 3D. As a consequence, more and more city departments are planning to utilize 3D models for storing building footprint information for city records databases. Figure 1 shows one such example of a 3D model for the city of Berlin.

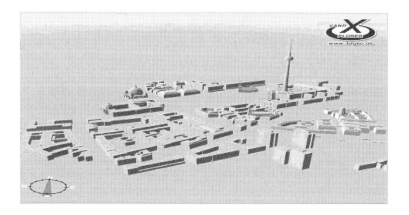

**Fig. 2.1** 3D Model for the Buildings part in the City of Berlin (courtesy data from the www.citygml.org and LandXplorer visualization tool)

The renewed interest on the acquisition and storage of 3D data in the GIS industry and city departments is likely to result in large amounts of 3D information in the near future. As a result, efficient and scalable techniques for storage and querying of 3D models will be essential to scalable terrestrial data management. To address this need, Oracle Spatial [KAE04] enhanced its SDO_GEOMETRY data type to store 3D data and created additional functionality for the efficient storage, query and management of such 3D data inside the database. One important piece of successful data models is "validation" which verifies whether or not the data generated by third parties, say, a city department, conforms to the data model supported. Validation is an essential and important component of 3D data modeling and enables subsequent operations on valid data to run correctly and efficiently (without the need to check for validity). Standardization organizations such as ISO and OGC have tried to give unambiguous and complete definitions of valid geometric primitives. But as it was already pointed out in [OT03] and [OT04] it turns out that the standards are not unambiguous and complete even in the 2D case of polygons. It should further be noted that ISO [ISO03] provide abstract descriptions, but not a detailed implementation specification enabling system interoperability. This aspect has been provided by OGC/ISO in [OGC99, OGC06a, OGC06b]. It is interesting to note that OGC did improve their definition [OGC06a] of a valid polygon in 2D by adding the condition that the outer boundary must have a counter clockwise orientation compared to the previous definition [OGC99]. What is still missing is a good treatment of tolerance values (also noted in [OT04]), which somehow is implied in the used terms such as spikes and also needed in a final digital machine to decide if two rings of a polygon do touch. Further different vendors of geo-DBMSs did also have different interpretations of a valid polygon, both compared to each other and to the ISO and OGC standards (again see [OT04]). However it must be stated that in the meantime also the producers did take notice of this issue and significant improvements have been noticed (though a systematic benchmark has not been repeated). For 3D geometric primitives there is an abstract ISO specification (again [ISO03]), but this is not the needed implementation specification. In [ASO05] a proposal was made to extend Oracle Spatial with a polyhedron primitive. Despite the fact that much attention was paid to the validation of this primitive, this paper did have some limitations. For example, it was stated that every edge in a valid polyhedron must exactly be used two times, but in this paper we will show that also valid cases exits where an edge is used four (or higher even number) times. Also the paper did focus on a single polyhedron (called in ISO terms 'simple solid'), and did not discuss composite and multi solids. In this paper a more complete set of validation rules are given for the complete set of 3D geometric primitives. Also this paper gives more illustrations of valid and invalid polyhedrons (simple solids), illustrating that quite complicated configurations are possible for which it is not easy to decide whether the primitive is valid.

One could wonder if the subject of valid or invalid solids has not been treated within CAD, with its long-standing 3D basis. There one has to distinguish between the different type of CAD models: voxels, CSG (constructive solid geometry) and boundary representation using (curved) surfaces [M97]. Only the boundary representations are comparable to the discussion of valid solids. However, within CAD the focus is more on the shape of the (curved) boundary, than on the validness of the solid object. Most CAD systems work with CSG and form complex objects from primitive objects. Validity of the resulting objects is typically assumed (due to the underlying nature of the primitives). In practical GIS applications, Boundary representation is more natural. The OGC's GML defines solids in a boundary-representation format [GML03]. GML also allows composite solids to mimic CAD world's convenience of making complex man-made objects. Oracle's model closely follows OGC's specifications in both aspects. In this paper, we illustrate how the Oracle 3D model compares with the GML definitions and present explicit validation rules for checking for validity in a practical implementation. This paper will also help in shedding more light on the valid geometry definitions of GML by providing explicit examples of valid and invalid geometries.

Section 2 defines the data model for storing 3D geometries in Oracle. Section 3 describes rules for determining 'structurally-valid' polygons and surfaces. Note that the definition of polygons is quite similar to those in GML. However, the contribution of that section is Lemma 1 (which can be converted to an algorithm) to prove that polygons with inner rings can be converted to composite surfaces without inner rings. Besides, this section also forms the basis for solid modeling. Section 4 describes solid modeling in Oracle using either 'simple solids' or 'composite solids' and how these differ from the standard definitions in GML. This section also establishes that simpler representations using simple solids without inner rings in faces (or the equivalent composite solids) are sufficient to model 3D geometries (as long as they don't have arcs and parametric curves). By adopting this paradigm (of simple solids) for storing solids, much complexity is avoided and validation algorithms are simplified. Section 5 describes validation rules and examples for collection geometries. The final section discusses the implementation details of these rules (with pointers to the full report) and concludes the paper with pointers for future research.

## 2.2 3D Geometry in Oracle Spatial

The SDO_Geometry data type for storing 3D data in Oracle has the class structure depicted in Figure 2. The SDO_GEOMETRY is represented by an array of one or more elements. The element array can be used to represent a point, linestring, surface, solid, or collection (heterogeneous collection or homogenous collection such as multi-point, multi-curve, multi-surface, or

2 On Valid and Invalid Three-Dimensional Geometries 23

multi-solid). Two or more points form a line-string. A closed line-string is called a ring. One or more rings (1 outer and 0 or more inner rings) within the same plane form a polygon [OT04]. One or more polygons form a surface (called a composite if consisting of more than 1 polygon and the surface is connected via shared parts of edges). Two or more outer rings of a composite surface can be on the same or different planes. One outer surface and 0 or more inner surfaces (represented as 1 or more in Figure 2) surfaces form a simple solid (Ssolid in Figure 2) if all the involved surfaces bound a closed volume. One or more adjacent (sharing at least a 2D part of a face) simple solids form a composite solid (referred to as CSolid in Figure 2). As we will show in the next sections, any composite solid can be represented as simple solid by removing the shared polygons. Composite solid is however very convenient from the user's perspective and part of the ISO standard [ISO03]. The collection types are formed as one or more elements of the appropriate type (e.g., one or more points form a multi-point, one or more solids form a multi-solid etc.). Note that the elements of the multi-X (X=surface, or solid) are either disjoint or touching via lower dimensional shared part of their boundary.

Buildings and other architectural elements of city models will likely be represented using simple solids (Ssolids in Figure 2), composite solids (CSolids in Figure 2), multi-solids or collection-type geometries. In this paper, we mainly focus on these types of 'complex' geometries (and do not discuss the rest of the simpler types as they are simple extensions of the 2Dimensional counterparts). The above SDO_GEOMETRY is realized as an implementation object type called SDO_GEOMETRY in the Oracle database. This object has attributes such as SDO_GTYPE, SDO_ELEM_INFO and SDO_ORDINATES. The SDO_GTYPE specifies the type for the geometry. The SDO_ORDINATES stores the ordinates of the vertices of the geometry. The SDO_ELEM_INFO is an array of element descriptors each describing how to connect the ordinates stored in the SDO_ORDINATES field. The following SQL shows the constructor for a Composite surface geometry composed of two (axis-aligned) rectangle polygons. More examples of the SDO_GEOMETRY type with additional examples can be found in Appendix B.

```
SDO_GEOMETRY (
   3003,    -- SDO_GTYPE: 3Dimensional surface type geometry
   NULL, NULL, -- SDO_SRID for coordinate system ID
SDO_ELEM_INFO_ARRAY( -- SDO_ELEM_INFO constructor
   1,       -- starting offset for element
   1006,    -- Element type is a COMPOSITE SURFACE
   2,       -- Number of primitive polygon elements making up
            -- the composite surface
   1,       -- starting offset of first polygon of Composite
            -- surface
   1003,    -- Polygon element type
   3,       -- Axis-aligned rectangle specified by two corners
   7,       -- starting offset of second polygon of Composite
```

```
                -- surface
      1003,     -- Polygon element type
      3         -- Axis-aligned rectangle specified by two corners,
    )
    SDO_ORDINATE_ARRAY ( -- Constructor for SDO_ORDINATES:
      -- Store the actual ordinates of the vertices
      2,0,2, -- 1st corner or 1st axis-aligned rect. polygon
      4,2,2, -- 2nd corner of 1st axis-aligned rect. polygon
      2,0,2, -- 1st corner of 2nd axis-aligned rect. polygon
      4,0,4  -- 2nd corner of 2nd axis-aligned rect. polygon
    )
)
```

Note that the class structure of Figure 2 closely follows the class structure for 3D geometries in GML 3.1.1 [GML03] but with a notable restriction: arcs and parametric curves are not currently supported in the SDO_Geometry. The validation rules and algorithms described in this paper work even when arcs are incorporated into the data model.

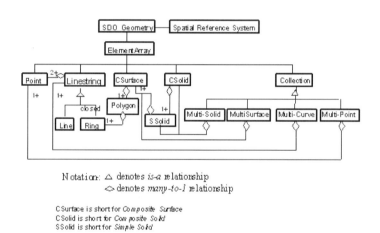

**Fig. 2.2** The Class structure for the SDO_GEOMETRY data type

In the next sections, we describe how surfaces, solids and collections are defined in Oracle and present specific validation rules with appropriate examples. We skip points and lines as for these types the validation rules are no different than their 2D counterparts, they are relatively trivial, and there is not much ambiguity in the OGC/ISO standards (no repeated notes, no self intersection of edges). The only tricky part is the treatment of tolerance values (needed to decide if there is indeed an intersection or equal node), but this will be discussed for surfaces.

## 2.3 Surfaces

In this section surfaces and their validation rules are discussed, as the closed surfaces are the building block for defining solids (outer and possible inner boundaries). First the ring is discussed, the simple surfaces (polygon) and finally the composite surface. This section does not present new validation rules but it presents the concepts and proves an important lemma needed for the next section on solid modeling.

### 2.3.1 Rings

Based on our interpretation of the ring definition in OGC/ISO GML3, a ring can be defined as a closed string of connected non-intersecting lines that lie in the same plane. A ring R is typically specified as the sequence of n+1 vertices $R = <V_1, V_2, \ldots, V_n, V_1>$ where the first and the $(n+1)^{th}$ vertex are the same (to make the ring closed). All other vertices are distinct. Each pair $<V_i, V_{i+1}>$ represents a directed edge connecting the vertex $V_i$ to $V_{i+1}$. Also note that no two edges of the ring can intersect (no self intersection). The only exception is the first edge $<V_1, V_2>$ and the last edge $<V_n, V_1>$ can touch at the vertex $V_1$.

**Validation Rules for a Ring:**

- **Closedness Test**: The first and last vertices of the ring are identical.
- **Planarity Test**: All vertices of the ring are on the same plane (within a planarity-tolerance error).
- **Non-intersection of edges**: If edge $e_i$ connects vertex $<V_i, V_{i+1}>$ and edge $e_j$ connects $<e_j, e_{j+1}>$, $e_i$ and $e_j$ have the following properties:
  - If ($j = i+1 \mod n$), then $e_i$ and $e_j$ only touch at vertex $V_j$
  - Otherwise, $e_i$ and $e_j$ do not intersect.
- **Distinct Vertex Test**: Adjacent vertices Vi, Vi+1 should not represent the same point in space. Vi, Vi+1 are considered to duplicates of the same point if the distance between Vi, and Vi+1 is less than a tolerance error.

Note that the planarity tolerance discussed in bullet 2 and tolerance in bullet 4 are different. These tolerance values ensure that spikes and other degenerate cases are invalidated. Note that due to the tolerance also spikes to the inside and outside are not allowed (in [OT04] spikes to the inside were allowed, but not to the outside, which was a bit asymmetric).

## 2.3.2 Polygon in GML

In GML [GML03], a Polygon is a planar Surface defined by one exterior boundary and zero or more interior boundaries. Each interior boundary defines a hole in the Polygon. GML has the following assertions for Polygons (the rules that define valid Polygons):

a) Polygons are topologically closed.
b) The boundary of a Polygon consists of a set of LinearRings that make up its exterior and interior boundaries.
c) No two Rings in the boundary cross and the Rings in the boundary of a Polygon may intersect at a Point but only as a tangent.
d) Polygon may not have cut lines, spikes or punctures.
e) The interior of every Polygon is a connected point set.

The exterior of a Polygon with one or more holes is not connected. Each hole defines a connected component of the exterior.

### Simple Surface, Polygon in Oracle

A polygon $P$ in Oracle strictly adheres to the definition in GML (nothing new here). It is defined as a single contiguous area in a planar space bounded by one outer (or exterior) ring $PR_e$ as the exterior boundary and zero or more (interior) rings $PR_{i1},\ldots,PR_{ik}$ which are interior to the outer ring and non-overlapping with one another. The inner rings should be oriented in opposite direction as the outer ring. The outer ring itself can always be oriented in any manner. In addition to the mentioned assertions for polygons in GML, we also add the implicitly mentioned co-planarity of points.

### Validation Rules for a Polygon:

The rules for polygons can be listed as follows.

- **Validity of Rings**: The rings in a polygon are valid (satisfy closedness, planarity, No Self-intersection tests, and distinct vertex).
- **Co-planarity of Rings**: Since the polygon defines an area in a plane, all rings are on the same plane (within tolerance).

- **Proper orientation**: The inner rings (if any) must have the opposite orientation compared to the outer ring.
- **Single Contiguous Area**: Together the outer ring and the interior rings define a single area. This means the inner rings cannot partition the polygon into disjoint areas.
- **Non-overlapping inner rings**: Inner rings cannot overlap (tolerance) with each other but may touch at a point (under the single contiguous area condition).
- **Inner-outer disjointedness**: Every inner-ring must be inside outer-ring and can only touch (tolerance) at a single point (under the single contiguous area condition).

Note that for 2Dimensional data, Oracle required that the vertices of the outer ring be specified in counterclockwise direction and those of the interior rings in clockwise direction. Such orientation restrictions are not needed for polygons and surface geometries (only the fact that the inner boundaries have opposite orientation compared to the outer boundary). Orientation becomes more important when these polygons/surfaces become components in a solid geometry. Modeling the polygon on the 3D ellipsoid is difficult (co-planarity may not be enforced as points on ellipsoidal surface are not on the same plane) and is not discussed here.

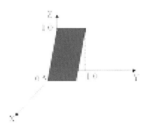

**Fig. 2.3** Example of a 3D polygon

## 2.3.3 Composite Surface

A composite surface is a contiguous area formed as a composition of M non-overlapping polygons. Note that the polygons may or may not be in the same plane. GML does not give any explicit rules here.

**Validation Rules for Composite Surfaces:**

The validation rules that we propose for composite surface are as follows.

- **Validity of Polygons**: Each of the M polygons has to be a valid polygon.
- **Non-overlapping but edge-sharing nature**: Any two polygons $P_i$ and $P_j$ should not overlap, i.e. if $P_i$ and $P_j$ are in the same plane, the area of intersection of the two polygons has to be zero. However, two polygons may touch (tolerance) in a (part of a) line/edge.
- **Contiguous area**: Every polygon in the composite should be reachable from any other polygon by appropriate tracing of the shared (parts of) edges.

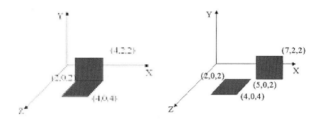

**Fig. 2.4** Left: Valid composite-surface. Right: Invalid composite-surface: Not a single contiguous area. Right can be modeled as a homogenous (multi-surface) or heterogeneous collection

## Decomposition of a Polygon with inner rings into a Composite Surface

**Lemma 1**: Any polygon $P$ with an outer ring $P_o$ and (non-overlapping) inner rings $P_{i1}, \ldots, P_{in}$ can always be decomposed into a composite surface $S$ where each polygon has no inner rings with the following characteristics:

- Every edge/vertex in $P$ is an edge in one of the polygons of $S$.
- Area($P$) = Union of Area($Q$) for all $Q$ in $S$.
- No polygon of $S$ has an inner ring
- Every edge in $S$
    - Either belongs to $P$ and is traversed only once in $S$
    - Or is an edge inside the outer ring $P_o$ and is traversed twice.

**Proof**: see Appendix A.

## 2.4 Solids

GML and ISO define specific representations for solids. Oracle's definition of the solid is quite equivalent except that it does not allow arcs and parametric curves in the solid specification. The GML definition of a solid is as follows [GML03]:

*The extent of a solid is defined by the boundary surfaces (shells). A shell is represented by a composite surface, where every shell is used to represent a single connected component of the boundary of a solid. It consists of a composite surface (a list of orientable surfaces) connected in a topological cycle (an object whose boundary is empty). Unlike a Ring, a shell's elements have no natural sort order. Like Rings, shells are simple. The element 'exterior' specifies the outer boundary of the solid. Boundaries of solids are similar to surface boundaries. In normal 3Dimensional Euclidean space, one (composite) surface is distinguished as the exterior. In the more general case, this is not always possible. The element 'interior' specifies the inner boundary of the solid. Boundaries of solids are similar to surface boundaries.*

In this paper, we will only focus on solid modeling in Euclidean spaces and restrict our attention to solids without arcs and parametric curves. Oracle supports two types of solids: a simple solid, and a composite solid. Both representations are equivalent but a composite solid may be more convenient to form for the user. Further composite solids are also required by the OGC/ISO standards.

One problem with this definition is: it allows inner rings in a polygon of a composite surface. Consider a solid such as that in the following figure that has an inner ring on its top surface. Is this solid, valid or invalid? The answer depends on whether the inner ring in the top surface is complemented with inner walls too as shown in the subsequent figure.

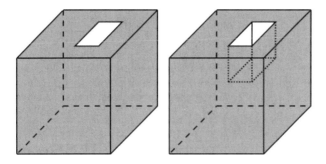

**Fig. 2.5** Is the white inner ring an empty surface-patch on the solid, not matched with other faces as shown on the left side? (In that case, the solid is not valid.) Or does it connect to other 'inner' faces in which case it could become valid as shown on the right side

Oracle offers a simpler representation: all polygons only have an outer ring but no inner ring (thus avoiding unnecessary computation). It turns out that this simpler representation does not lose any power (i.e. can represent any solid with polygons that have inner rings). Oracle supports two variants, simple solids and composite solids, that both exhibit this property.

## 2.4.1 Simple Solids in Oracle

In Oracle, a simple solid is defined as a 'Single Volume' bounded on the exterior by one exterior composite surface and on the interior by zero or more interior composite surfaces. To demarcate the interior of the solid from the exterior, the polygons of the boundary are oriented such that their normal vector always point 'outward' from the solid. In addition, each polygon of the composite surfaces has only an outer ring but no inner ring. (This is a restriction compared to the GML definitions, but without loosing any expression power).

**Validation Rules for Simple Solids:**

Based on these above definitions, we can define the rules/tests for validation of solids (again all operations are using tolerance values):

- **Single Volume check**: The volume should be contiguous.
  - **Closedness test**: The boundary has to be closed. [Z00] show that the vector sum of the edges in the boundary traversal should be zero (i.e. every edge on the boundary needs to be traversed even number of times: note that some implementations check for just 2 times but that may disallow some valid solids as shown in Figure 9). Necessary condition but not sufficient (Figure 11 left, Figure 12 left, Figure 13 left are invalid)
  - **Connectedness test**: For sufficiency, volume has to be connected. (Figure 11 right, Figure 12 right, Figure 13 right are valid). This means each component (surface, solid) of the solid should be reachable from any other component.
- **Inner-outer check**:
  - Every surface marked as an inner boundary should be 'inside' the solid defined by the exterior boundary.
  - Inner boundaries may never intersect, but only touch under the condition that the solid remains connected (see above)
- **Orientation check**: The polygons in the surfaces are always oriented such that the normals of the polygons point outward from the solid that

they bound. Normal of a planar surface is defined by the right-hand thumb rule (if the fingers of the right hand curl in the direction of the sequence of the vertices, the thumb points in the direction of the normal). The volume bounded by exterior boundary is computed as positive value if every face is oriented such that each normal is pointing away from the solid due to the Green's Theorem. Similarly, the volume bounded by interior boundary is computed as negative value. If each exterior and interior boundary obeys this rule and they pass connectedness test as well, then this check is passed.
- **Element-check**: Every specified surface is a valid surface.
- **No-inner-ring in polygons**: In the composite surfaces of a solid, no inner rings are allowed.

A solid cannot have polygons that overlap. Due to the use of tolerances some very thin volume parts could collapse to spikes (dangling faces or edges). However, it is not possible to have spikes (either linear or areal shaped) as it is not allowed to have the vector sum of edges unequal to 0.

**Theorem 1**: Any valid solid S where polygons have inner rings can also be represented as a simple solid without inner rings in the faces.

**Proof**: Consider a polygon $P$ that has interior rings in $S$. During a traversal of $P$, every edge of $P$ is traversed just once. Each of these edges is traversed a second time in the traversal of the rest of the polygons that close the solid. Replace polygon $P$ (that has interior rings) in $S$ by its equivalent composite surface consisting of the no-interior-polygons $P_1, \ldots, P_k$ as in Lemma 1. Since every edge in $P$ is traversed only once during a traversal of $P_1, \ldots, P_k$, the boundary is preserved. All edges that are in $P_1, \ldots, P_k$ but not in $P$ are traversed exactly twice in opposite directions and are cancelled out in the traversal. Thus preserving the solid-closedness properties. Other properties are also likewise preserved.

In Figure 6, simple solid has an outer boundary represented by a closed composite surface and a hole (Hole) represented by an inner composite surface. These two solids share a face. This solid is invalid since it violates the inner-outer disjointedness rule of a simple solid. However, this solid can be modeled as a valid simple solid represented by a single outer composite surface without any inner surfaces (solid with dent).

The composite-solid in Figure 7 is composed of Solid 1 and Solid 2. Even though solids have common (shared) area, Solid 2 is inside Solid 1, which violates the No-volume-intersection of composite-solids and hence invalid.

The examples in Figures 8-13 give a number of valid and invalid simple solids. The captions of the figures do explain why. From these figures it becomes clear that the validation includes a non-trivial topological analysis of the inner and outer boundary elements. Let us consider another example. The geometry in Figure 15 is designated as a composite-solid geometry consisting of simple solids: Solid 1 and Solid 2. Solid 2 has an outer boundary and an additional surface patch intersecting (overlapping) one of its faces. This violates the No-surface-patch rule of simple solids and hence is an in-

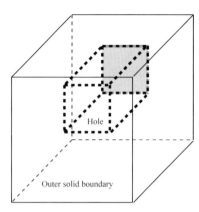

**Fig. 2.6** Simple Solid: invalid if modeled as outer, inner surfaces. Note that the back face (i.e. shaded area) of the inner solid boundary is partly shared with the back face of the outer solid boundary

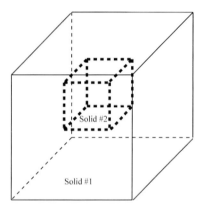

**Fig. 2.7** Invalid composite solid because solid elements cannot be inside the other. If modeled as a simple solid, the object will be invalid as it has two outer boundaries (which are disconnected)

valid geometry. The geometry, however, can be modeled as a (heterogeneous) collection.

## 2.4.2 Composite Solid in Oracle

In addition to a simple solid, Oracle (and GML too) also allows the specification of a 'composite solid'. In Oracle, a composite solid is a combination of n simple solids. Compared to the simple solid definition in Section 4.1, what this allows is the overlap of the polygons of different simple-solids but the

## 2 On Valid and Invalid Three-Dimensional Geometries

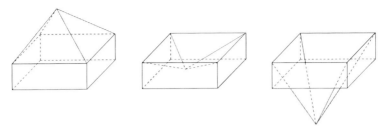

**Fig. 2.8** The importance of checking the intersection between faces: all 3 simple solids have the same node-edge connectivity, but the last (rightmost) one has intersecting faces and is therefore invalid

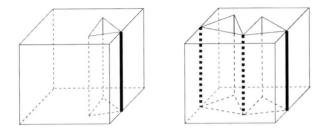

**Fig. 2.9** Simple solid with 'inner' boundary touching the outer boundary in one line (fat edges are used 4 times). The left solid is valid, while the right simple solid is invalid (together the two 'inner' boundaries separate the solid into two parts). Note that both solids do not really have inner boundaries (it is a more complex single outer boundary causing through holes touching the outer boundary in other places)

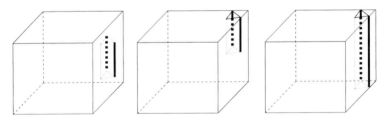

**Fig. 2.10** Simple solid with inner boundary touching the outer boundary in two lines (fat edges are used 4 times). The left solid is the only solid with a true inner boundary (touching the outer boundary in two lines: right outer boundary in the fat line and the back outer boundary in the fat dashed line). The left is a valid simple solid. The middle solid is also valid (because inner boundary does not continue through the whole), while the right simple solid is invalid (as inner boundary does separate the solid into two parts). Again note that the middle and right solids do not really have inner boundaries (it is a more complex single outer boundary)

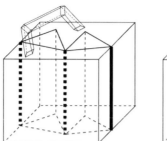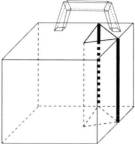

**Fig. 2.11** Invalid simple solids of previous figures becoming valid via adding an additional handle making it possible to travel from one part to another part of the object (completely via the interior). Note: where handle touches the face, a part of the faces is removed (that is an interior ring is added within the exiting face to create the open connection). So, all faces have always (and everywhere) on one side the object and on the other side something else (outside, where the normal is pointing to)

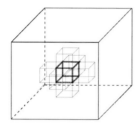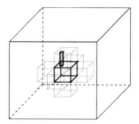

**Fig. 2.12** Left: simple solid with 6 internal (cube-shaped) boundaries separating the big cube into two parts (the internal one draw with fat lines is implied by the 6 boundaries of the 6 smaller cube-shaped holes). Therefore the left simple solid is invalid (note that removing one of the 6 holes, makes it valid again). Right: Invalid simple solids of previous figures becoming valid via adding an additional handle making it possible to travel from one part to another part of the object (completely via the interior). Right: the two parts are connected via a 'pipe' making it a valid simple solid again

boundary is still closed. Note that this does not allow overlapping polygons in the same simple solid but only across multiple simple solids.

Following theorem shows the equivalence of a 'simple solid' and a 'composite solid'. Composite solids are defined for convenience: it is easier to combine two or more simple solids and make a composite. For the same reason, they are also included in GML specification too. However, the composite solids in GML do allow inner rings in polygons whereas the Oracle model does not but still equivalent (from Theorem 1 and Theorem 2).

**Theorem 2**: Every valid composite solid can also be represented as a simple solid.

**Proof**: Composite solid always has solids attached to each other via partially or fully shared faces. Having detected these shared faces, one can get rid

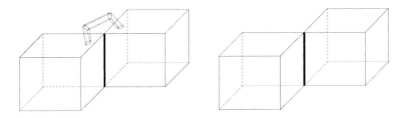

**Fig. 2.13** Left: valid simple solid (fat edge still used 4 times), but handle is added through which it is possible to travel from one part to the other part via the interior only, Right: invalid simple solid with one edge being used four times (fat line)

of these faces and redefine the solid without these shared areas. Figure 14 illustrates an example.

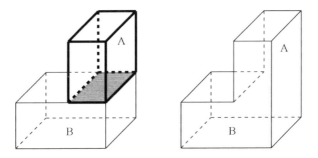

**Fig. 2.14** (a) Composite Solid consisting of two simple solids A and B. Shared portion of bottom face of A (i.e. shaded area) in both A and B. (b) Equivalent Simple Solid. Shared portion of bottom face of A should not be included in the boundary.

**Validation Rules for Composite Solids**

- **Component Validity**: Each component simple solid of a composite is valid.
- **Shared-face but no-volume intersection**: Intersection of two simple solid components of a composite solid has to be a zero volume (can be non-zero area).

- **Connectedness**: The volume of the composite is contiguous, i.e. we can go from any point in one component to any other component without going out of the composite solid.

Since composite solids are equivalent to a simple solid, an alternate way is simply convert the composite to the equivalent simple solid and then validate the resulting solid.

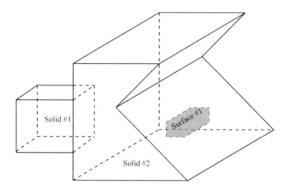

**Fig. 2.15** Invalid Composite Solid: Cannot have surface patches on a solid

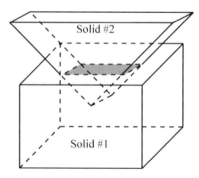

**Fig. 2.16** If modeled as a Composite Solid, the object is invalid due to intersection of solid elements. If modeled as a simple solid, the object is invalid due to 'overlapping polygons' of composite surface

Consider a third example on composite-solids in Figure 16, which is composed of a cube and a triangular prism where a prism goes into the cube through its top face. This is an invalid geometry because it violates the No-volume intersection rule of composite-solids.

## 2.5 Collections

Following GML [GML03] specifications, collections in Oracle can be either homogenous or heterogeneous. Let us look at each of these in turn.

### 2.5.1 Homogenous Collections

A homogenous collection is a collection of elements, where all elements are of the same type. A homogenous collection can be either a multi-point, multi-line, multi-surface, or multi-solid corresponding to the element type point, line, surface and solid. For example, in a multi-solid, all elements have to be (valid) solids.

**Validation Rules for Homogenous Collections:**

- All elements of same type and conform to the homogenous collection type (multi-point, multi-line, multi-surface, multi-solid).
- Each element should be valid.

In addition to the above rules, Oracle also adds a specific rule for determining disjointedness of elements in a homogeneous collection in validation procedures (this is a deviation from the GML specification). The reason to add the disjointedness is to distinguish between composite solids and multi-solids. If the user does not want this restriction, he can model it as a heterogeneous collection.

The geometry in Figure 16 is also invalid as a multi-solid because it violates the disjointedness rule for multi-solids. Note that the disjointedness in multi-solids not only implies no-volume intersection but also no-area intersection. The latter is represented by the geometry in Figure 17, which is invalid due to the disjointedness (no-area intersection) rule of multi-solids.

### 2.5.2 Heterogeneous Collections

In a heterogeneous collection[1], the elements can be a mixture of different types. For example, a building (simple solid) with the windows/doors (surfaces) can be modeled as a heterogeneous collection.

---

[1] In Oracle documentation, a heterogeneous collection is referred to simply as a 'collection'. All homogenous collections are referred to by their names (e.g. multi-line, multi-point, multi-surface, multi-solid).

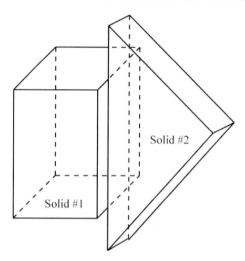

**Fig. 2.17** Invalid multi-solid because solid elements are non-disjoint (but a valid composite solid

**Validation Rules for Heterogeneous Collections:**

- Each element of the collection should be valid.

Figure 18 shows an example of a building modeled as a heterogeneous collection: The building shape is modeled as a solid whereas the windows and doors are modeled as surfaces.

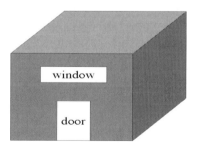

**Fig. 2.18** Building modeled as a Heterogeneous collection

## 2.6 Discussions and future work

In this paper, we presented a data model for storing 3D objects that occur in the rapidly evolving area of city modeling. We examined different types and described rules for validating these types of geometries based on OGC/ISO standards [ISO03, OGC6a, OGC6b, OGC6c]. We then developed and presented in this paper the more detailed, unambiguous validation rules and applied them to different examples to determine validity or otherwise. Such a data model and validation rules can form the backbone of 3D data models needed for proper interoperability (having exactly the same definition of the geometric primitive during the exchange of 3D data). Our specific contributions include: simpler representation for solids (by eliminating the need for inner rings in polygons) thereby simplifying the validation of solids, explicit rules for validation of each geometry type, and concrete examples for valid and invalid 3D geometries that bring out the concepts of the validation model. This model and validation rules have been implemented in the Oracle 11g database (the implementation details have not been included due to space restrictions). The rules that are more involved to implement are the Closedness test and the Reachability/Connectedness tests. Full details on our implementation of these tests are available in [Oracle07]. Future work could concentrate on performance evaluation of the 3D data model, validation and query on large-scale city models. Specific optimizations based on 3D query window sizes [KR01] could be investigated. Besides, the 3D data modeling using SDO_GEOMETRY can be compared with other models such as Tetrahedron-based models [POK06] in terms of functionality and performance. Since the area of 3D modeling in databases is a fledgling topic, we need to investigate on easy approaches for deriving 3D data by extruding 2D footprints. By first converting arbitrary 2D polygons to a composite surface, via Lemma1, and then extruding them to 3D may be a good approach. Fast algorithms for doing this conversion may also be investigated (there exist more efficient algorithms than the recursive construction in Lemma 1). Another important aspect in 3D modeling is visualization. Popular tools like Google Sketchup work well with 3D surfaces but do not understand 3D solids. Other tools such as Landxplorer and Aristoteles visualize all types of 3D GML geometries. Commercial products from Autodesk work with their own proprietary 3D formats and are planning to publish to GML geometries. The Xj3D tools that are popular in the gaming community are also interoperating with the GML forums. Using appropriate functions, the 3D data stored as SDO_GEOMETRY can be converted to GML thereby opening it further to other interoperable formats. Future work can be devoted to developing fast rendering of the native 3D geometries.

# Acknowledgements

We would like to thank Friso Penninga for proof reading an earlier version of this paper. The contribution of Peter van Oosterom to this publication is the result of the research programme 'Sustainable Urban Areas' (SUA) carried out by Delft University of Technology and the Bsik RGI project 3D Topography.

# References

[Oracle07]. 'Validation Rules and Algorithms for 3D Geoemtries in Oracle Spatial', 2007.

[ASO05] Calin Arens, Jantien Stoter and Peter van Oosterom, Modelling 3D spatial objects in a geo-DBMS using a 3D primitive, In: Computers & Geosciences, Volume 31, 2, 2005, pp. 165-177.

[A94] Lars Arge, Mark de Berg, Herman J. Haverkort, Ke Yi: The Priority R-Tree: A Practically Efficient and Worst-Case Optimal R-Tree. SIGMOD Conference 2004: 347-358.

[BKSS90] Beckmann, N., Kriegel, H., Schneider, R. and Seeger, B., The R* tree: An efficient and robust access method for points and rectangles. In Proc. ACM SIGMOD Int. Conf. on Management of Data, pages 322-331, 1990.

[BKS94] Brinkhof, T., Kriegel, H., and Seeger, B., Efficient processing of spatial joins using R-trees. In Proc. ACM SIGMOD Int. Conf. on Management of Data, pages 237-246, 1994.

[BDA04] Nagender Bandi, Chengyu Sun, Amr El Abbadi, Divyakant Agrawal: Hardware Acceleration in Commercial Databases: A Case Study of Spatial Operations. VLDB 2004: 1021-1032.

[DE94] Dimitris Papadias, Yannis Theodoridis, Timos K. Sellis, Max J. Egenhofer: Topological Relations in the World of Minimum Bounding Rectangles: A Study with R-trees. SIGMOD Conference 1995: 92-103.

[FD97] Foley, van Dam, Feiner, Hughes. Computer Graphics: Principles and Practice, The Systems Programming Series, 1997.

[G84] A. Guttman. R-trees: A dynamic index structure for spatial searching. Proc. ACM SIGMOD Int. Conf. on Management of Data, pages 4757, 1984.

[GML03] The Geographic Markup Language Specification. Version 3.1.1, http://www.opengeospatial.org.

[GSAM04] Thanaa M. Ghanem, Rahul Shah, Mohamed F. Mokbel, Walid G. Aref, Jeffrey Scott Vitter: Bulk Operations for Space-Partitioning Trees. ICDE 2004.

[HKV02] Marios Hadjieleftheriou, George Kollios, Vassilis J. Tsotras, Dimitrios Gunopulos: Efficient Indexing of Spatiotemporal Objects. EDBT 2002: 251-268.

[ISO03] ISO/TC 211/WG 2, ISO/CD 19107, Geographic information - Spatial schema, 2003.

[J04] Christian S. Jensen: Database Aspects of Location-Based Services. Location-Based Services 2004: 115-148.

[KGP05] Th. H. Kolbe, G. Gröger, L. Plümer. CityGML: Interoperable Access to 3D City Models, In: Oosterom, P, Zlatanova, S., Fendel E. M. (editors) Geo-information for Disaster Management, Springer, pages 883-899, 2005.

[KAE04] Kothuri, R. Godfrind A, Beinat E. 'Pro Oracle Spatial', Apress, 2004.

[KSSB99] Kothuri R., Ravada S., Sharma J., and Banerjee J., Indexing medium dimentionality data, in Proc. ACM SIGMOD Int. Conf. On Management of Data, 1999.

[KR01] Kothuri, R., Ravada, S., Efficient processing of large spatial queries using interior approximations, Symposium on Spatio-Temporal Databases, SSTD, 2001.

[L07]. Jung-Rae Hwang, Ji-Hyeon Oh, Ki-Joune Li: Query Transformation Method by Dalaunay Triangulation for Multi-Source Distributed Spatial Database Systems. ACM-GIS 2001: 41-46.

[MP01] Nick Mamoulis, Dimitris Papadias: Multi-way Spatial Joins, ACM TODS, Vol 26, No. 4, pp 424-475, 2001.

[MTP05] Nikos Mamoulis, Yannis Theodoridis, Dimitris Papadias: Spatial Joins: Algorithms, Cost Models and Optimization Techniques. Spatial Databases 2005: 155-184

[MA04] Mohamed F. Mokbel, Ming Lu, Walid G. Aref: Hash-Merge Join: A Non-blocking Join Algorithm for Producing Fast and Early Join Results.

ICDE 2004: 251-263.

[M97] Mortenson, M. Geometric Modelling, second ed. Wiley, New York 523pp., 1997.

[OGC99] Open GIS Consortium, Inc., OpenGIS Simple Features Specification For SQL, Revision 1.1, OpenGIS Project Document 99-049, 5 May 1999.

[OGC06a] Open Geospatial Consortium Inc., OpenGIS Implementation Specification for Geographic information - Simple feature access - Part 1: Common architecture, Version: 1.2.0, Reference number of this document: OGC OGC 06-103r3, 2006.

[OGC06b] Open Geospatial Consortium Inc., OpenGIS Implementation Specification for Geographic information - Simple feature access - Part 2: SQL option, Version: 1.2.0, Reference number of this document: OGC 06-104r3, 2006.

[OGC06c] Open Geospatial Consortium Inc., Candidate OpenGIS CityGML Implementation Specification , Reference number of this document: OGC 06-057r1, 2006.

[POK06] Penninga, F., van Oosterom, P. & Kazar, B. M., A TEN-based DBMS approach for 3D Topographic Data Modelling, in Spatial Data Handling 2006.

[S89] H. Samet, The Design and Analysis of Spatial Data Structures, Addison-Wesley, 1989.

[SE02] Schneider J. P., and Eberly, D.H. Geometric Tools for Computer Graphics, Morgan Kaufman, 2002.

[TS96] Y. Theodoridis and T. K. Sellis, A model for the prediction of r-tree performance, In Proc. of ACM Symp. on Principles of Databases, 1996.

[OT03] P.J.M. van Oosterom, C.W. Quak and T.P.M. Tijssen, Polygons: the unstable foundation of spatial modeling, ISPRS Joint Workshop on 'Spatial, Temporal and Multi-Dimensional Data Modelling and Analysis', Québec, October 2003.

[OT04] P.J.M. van Oosterom, C.W. Quak and T.P.M. Tijssen, About Invalid, Valid and Clean Polygons. In: Peter F. Fisher(Ed.); Developments in Spatial Data Handling, 11th International Symposium on Spatial Data Handling, 2004, pp. 1-16

[YE06] Yohei Kurata, Max J. Egenhofer: The Head-Body-Tail Intersection for Spatial Relations Between Directed Line Segments. GIScience 2006: 269-286 2005.

[YS05] Jin Soung Yoo, Shashi Shekhar: In-Route Nearest Neighbor Queries. GeoInformatica 9(2): 117-137 (2005).

[Z00] Zlatanova, S. On 3D Topological Relationships, DEXA Workshop, 2000.

# Appendix A

**Lemma 1**: Any polygon $P$ with an outer ring $P_o$ and (non-overlapping) interior rings $P_{i1},\ldots,P_{in}$ can always be decomposed into a composite surface $S$ where each polygon has no inner rings with the following characteristics:

- Every edge/vertex in $P$ is an edge in one of the polygons of $S$.
- Area($P$) = Union of Area($Q$) for all $Q$ in $S$.
- No polygon of $S$ has an inner ring.
- Every edge in $S$
  - either belongs to $P$ and is traversed only once in $S$
  - or is an edge inside the outer ring $P_o$ and is traversed twice.

**Proof**: By induction on the number of inner rings. Let n=1. The polygon $P$ has one inner ring. On the inner ring find the two extreme (min, max) points $I_{min}$, $I_{max}$ along a specific axis of the plane of the polygon. Extend the min, max points along the dimension to meet the outer ring $O$ of the polygon at $O_{min}$, $O_{max}$. The polygon P is now $P_1$ and $P_2$: the new edges $<O_{min},I_{min}>$ and $<O_{max},I_{max}>$ are traversed twice: once in $P_1$ and another time in reverse direction in $P_2$. All other edges belong to $P$ and are in either $P_1$ or $P_2$. The new edges do not cross any other edges, as $O_{min}$ and $I_{min}$ are the extreme vertices. Hence $P_1$ and $P_2$ are valid polygons.

**Hypothesis**: Let the lemma be true for n=m-1.

Consider a polygon $P$ with $m$ inner rings. Again, sort the rings along a specific axis of the plane and find the ring that has a vertex with the min value on this axis (i.e. closest to the outer ring). Connect the vertex to either side of the outer ring along this axis. This 'cutting line' cuts some inner rings (at least 1): For each such ring identify the first and the last vertex along the cutting line for the ring. These vertices are the endpoints for that inner ring and are connected to other rings that are cut or to the outer ring boundary. The cutting line cuts the outer ring and some inner rings into two parts. Without loss of generality, let us assume it cuts these rings into top, and bottom halves. To identify these halves precisely for each ring $r$, first determine the extreme (leftmost and rightmost) points $v_L(r)$, $v_R(r)$ on the ring $r$ that are also on the cutting line. The line string $v_L(r)$, $v_R(r)$ in counterclockwise direction determines the top half for ring $r$. The line string $v_R(r)$ to $v_L(r)$ (in counterclockwise direction) determines the bottom half. Figure 19 shows an example. Ring EJHGF is cut at two extreme points E (leftmost) and G (rightmost on ring and cutting line). The ring is split into a linestring EFG (top half) and GHJE (bottom half). The new edges will be AE, EA, GK, KG, MC, CM.

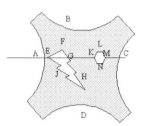

**Fig. 2.19** Example of a cutting line and a Polygon with 1 outer and 2 inner rings.

The linestrings of the top halves along with the new connecting edges between the rings form one polygon. In Figure 19, AE, EFG, GK, KLM, MC, CBA becomes the outer ring of the first polygon $P_1$. Likewise, the bottom half forms along with the new edges in reverse direction form another polygon. In Figure 19, CM, MNK, KG, GHJE, EA, ADC form the outer ring of the second polygon $P_2$.

Note that for a specific ring $s$, the points $v_L(s)$ and $v_R(s)$ may be the same, i.e. the cutting line intersects at a point of the ring $s$. Then only one of the two halves can include the ring, the other half will just include the point. Figure 20 shows an example. The inner ring, KNK, on the right touches the cutting line. The linestring KNK is added to the bottom half and the point K is added to the top half.

If the cutting line cuts at multiple points in a ring $q$, then $v_L(q)$ will be the leftmost and $v_R(q)$ will be the rightmost. Figure 20 shows an example. Even though the cutting line cuts the first ring at E, H, and G, the end points will be E and G. So, the top half will have a linestring EHFG, and the bottom half will have a line string GJE.

The remaining inner rings that are not cut by the cutting line are split between $P_1$ and $P_2$ depending on which outer ring (of $P_1$ or $P_2$) that they are inside. Since there is at least 1 inner ring that is split, $P_1$ and $P_2$ have at most m-1 rings. The inner rings were non-overlapping in $P$, so they will still be

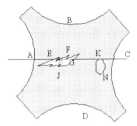

**Fig. 2.20** Example of a Polygon with two inner rings: One ring just touches the cutting line; another inner ring intersects at multiple points

non-overlapping in $P_1$ and $P_2$. Besides the cutting line does not cut the inner rings of $P_1$ and $P_2$: so they still do not touch the exterior rings of $P_1$ or $P_2$.

The only new edges in $P_1$ or $P_2$ are the 'new edges'. Each such new edge is traversed once in one direction in $P_1$. All other edges also existed in $P$ and are traversed only once (in the same direction as in $P$).

Since there is at least one inner ring that is split in this process, the polygons that are formed have at most m-1 inner rings. By induction hypothesis, each of these polygons satisfies Lemma 1 resulting in two composite surfaces one for top half of the cutting line and another for the lower half. A combination of these two composites, is also a composite surface and obeys the above properties thus proving the induction.

For example, in Figure 19, the polygon connecting vertices A, D, B, C is decomposed into a composite surface consisting of two polygons: AEFGKLM-CBA and CMNKGHJEADC. Note that each of them has at most m-1 inner rings, and each of the two polygons is a valid polygon, and the resulting surface combining these two polygons is a valid composite and each new edge is traversed twice and all existing edges are traversed only once, thus proving the lemma.

# Appendix B

| SDO_GTYPE | Name/Type of Element, Value of Etype | Interpretation (has one or more numbers) | SDO_GEOMETRY Representation | Picture |
|---|---|---|---|---|
| 3001 (3005 if N>1) | Point,1 | 1 | SDO_GEOMETRY(3001,NULL,SDO_POINT_TYPE(2,0,2), NULL,NULL) | 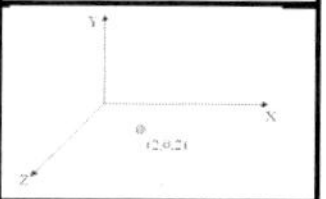 |
| 3002 (linestring) | Line,2 | 1: Connected by Line | SDO_GEOMETRY(3002,NULL, NULL,SDO_ELEM_INFO_ARRAY(1,2,1),SDO_ORDINATE_ARRAY(2,0,2,4,2,4)) | 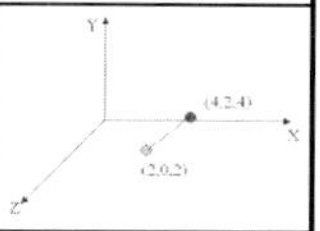 |
| 3003 (surface) | 1 Outer (1003) and [0,n] Interior (2003) | 1: *Planar polygon surface<br>3: Planar rectangle surface<br>4: Planar circle surface | SDO_GEOMETRY(3003,NULL,NULL, SDO_ELEM_INFO_ARRAY(1,1003,1),SDO_ORDINATE_ARRAY(0.5,0.0,0.0,0.5,1.0,0.0,0.0,1.0,1.0,0.0,0.0,1.0,0.5,0.0,0.0)) | 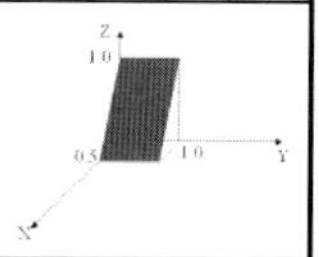 |
| 3003 (surface) | Composite Surface: Simple Outer (1006) and Inner (2006) Planar Polygons | N: number of planar polygon surfaces | SDO_GEOMETRY(3003,0,NULL,SDO_ELEM_INFO_ARRAY(1,1006,2,1,1003,3,7,1003,3),SDO_ORDINATE_ARRAY(2,0,2,4,2,2,2,0,2,4,0,4)) | 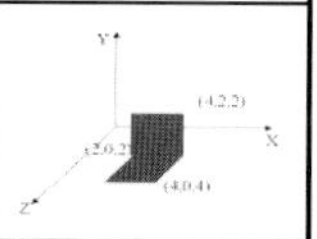 |
| 3008 (solid) | Simple Solid (1007) w/ Inners (2006s) | 1: Explicit format<br>3: *Box | SDO_GEOMETRY(3008,NULL,NULL,SDO_ELEM_INFO_ARRAY(1,1007,3),SDO_ORDINATE_ARRAY(2,0,2,4,2,4)) | 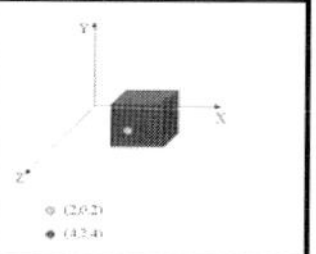 |
| 3008 (solid) | Composite Solid, 1008 | N: the # of solid elements (etype=1007) | SDO_GEOMETRY(3008,NULL,NULL, SDO_ELEM_INFO_ARRAY(1,1007,3,7,1007,3),SDO_ORDINATE_ARRAY(2,0,2,4,2,4,4,-1,3,6,1,5)) | 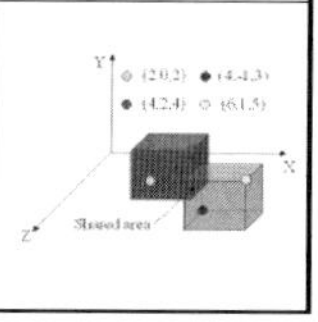 |

**Fig. 2.21** Examples of different types of 3D geometries

# Part II
# Papers

# Chapter 3
# Navigable Space in 3D City Models for Pedestrians

Aidan Slingsby and Jonathan Raper

**Abstract**

This paper explores the state of the art in 3D city modelling and draws attention to the 'missing link' between models of buildings and models of the surrounding terrain. Without such integrated modelling, applications that cross this divide are stalled. In this paper we propose a conceptual approach to this problem and set out a constraint-based solution to three dimensional modelling of buildings and terrains together.

## 3.1 Introduction

3D city models are increasingly considered as important resources for municipal planning and decision-making [2]; examples of 3D City models include Virtual London [2] and Virtual Kyoto [25]. An important aspect of cities is the navigable space within them. In spite of this, we have found no 3D city models which incorporate a model of pedestrian access. Navigable space for pedestrians includes space within buildings and, crucially, the connection between building interiors and exterior space. The majority of 3D city models treat buildings as solid objects which are placed upon a digital terrain model, without any essential integration between them.

In this paper, we argue that there is a need for 3D city models to incorporate topologically-connected navigable spaces, in which space internal to buildings is topologically connected to space outside buildings and in which the terrain is part of this navigable space rather than a simple surface upon which buildings are placed. Published research in this area tends to concern

---

The giCentre, Department of Information Science, City University,
Northampton Square, London EC1V 0HB, UK a.slingsby@city.ac.uk, raper@soi.city.ac.uk

either road vehicles which operate wholly *outside* buildings (transport models) or pedestrians which move *within individual* buildings. Models which operate across multiple storeys of buildings tend to work on a storey-by-storey basis with limited topological links between layers. We describe the target application area and then present a prototype model that addresses some of these requirements.

## 3.2 Brief Review of 3D city modelling approaches

Many 3D city models are implemented in GIS, because this is usually appropriate for the planning application domain and the spatial scale at which this operates. Most geographical information systems (GIS) support a simple but effective modelling strategy in which 2D building footprints are extruded upwards from the terrain to a representative (often LiDAR-derived) building height. Such models can be rapidly produced, offer simple city visualisation opportunities and may be used for a limited set of analyses (Figure 1). This approach of rapid modelling has been successfully used over the Internet through customised web browser plugins and standalone browsers (e.g. Google Earth).

**Fig. 3.1** Extruded block model of part of the Technical University of Delft campus, rendered in Google Earth. *Source: Technical University of Delft.*

In cities, there are often significant landmarks. For visualisation purposes, it is helpful if these buildings are modelled to a higher level of geometrical detail and then inserted amongst the other extruded blocks. Data for such buildings can be hand-modelled or sourced from architectural models as 3D building shells (external surfaces bounding an internal volume). This rather *ad hoc* approach provides a good level of visual realism (especially if photographically texture-mapped), and is supported by many of the software products which support the extruded block model (including ESRI's Arc-

Scene and Google Earth). Models such as these are often used as the basis for graphical applications such as walk- or fly-throughs. However, since the spatial resolution of buildings is essentially the same as their 2D counterparts the range of applications for which the data can be applied is not significantly widened.

Full and detailed 3D models of individual buildings and small groups of buildings are widely used in architecture and construction, but their high spatial resolution, their high geometrical detail and their variable types of semantic definition, often make them unsuitable for use with 3D cities. There has been much research on various aspects of 3D GIS, including different approaches to 3D geometrical modelling, the application of thematic (attribute) data, the creation, maintenance and storage of 3D topology (e.g. [26]) for data validation, algorithms for 3D spatial analysis and potential application areas [7]. However, 3D GIS is not (commercially) fully-realised for a number of reasons including the lack of availability of data and because the individual application area solutions have been developed separately, and no one tool has developed into a general cross sector tool.

Within the last decade, semantically-rich data exchange formats (e.g. IFC) and object-based building modellers (e.g. AutoDesk Revit) have been developed for architecture and construction, designed in part to facilitate the reuse of data for different stages of the design process and for different analysis tasks [6, 10, 17]. Similar approaches have been used for virtual cities; e.g. 'QUASY' [3] and 'Smart Buildings' [5]. CityGML[1]; [8] is an attempt to create a useable and formal standard for the exchange of city models, using this approach. It recognises that many existing 3D city models are rather *ad hoc* creations which neglect semantic and topological modelling aspects. It also recognises the need for a formalised set of levels of detail. CityGML is an XML-based standard which provides a set of object types (through the abstract 'CityObject' class). A building (an instance of an 'AbstractBuilding') comprises building parts, rooms and bounding objects (walls, doors, windows, ground surfaces, ceilings), depending on level-of-detail. The precise geometrical forms of these objects can each be described and classified using codes (based on the German Cadastral Standard: ATKIS). There are also objects which deal with road transport, water bodies and vegetation. Five levels-of-level exist (Figure 2): terrain-only (LoD0), extruded polygons upon a terrain (LoD1), the addition of roof structures and roof textures (LoD2), the addition of external architectural detail such as balconies (LoD3) and the addition of internal rooms (LoD4).

Semantically-rich, object-based modellers underpinned by formal modelling concepts have a number of advantages over the more ad hoc methods described earlier:

---

[1] In this paper, we are using version 0.3.0 of the Candidate OpenGIS standard [8, 11, 12]

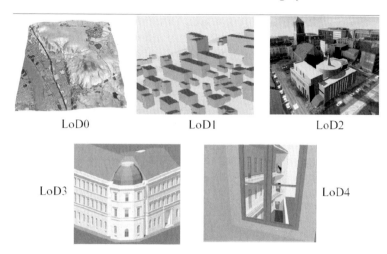

**Fig. 3.2** The five levels of detail defined in CityGML. Source: [12]

1. The formalised levels of detail allow parts of buildings to be modelled at the most appropriate level of detail.
2. Interior spaces in buildings can be modelled where appropriate
3. The 3D city model is exchangeable and reusable
4. The combination of geometrical, semantic and topological information can be used to support a range of analytical tasks.

Models such these can be used to support a wide range of applications [1] such as visualisations, flood-risk modelling, scene generation from different viewpoints, the effect of an explosion or source of noise on individual building parts, analyses of land use and property value, and traffic-related analyses and impacts.

## 3.3 Applications

We have found support from the fire safety sector and the insurance sector for research into how city models can assist in building safety, access, monitoring and planning applications. An important theme in this application area is how spaces navigable to pedestrians are connected, including the connections between buildings and to outside space. Wayfinding, navigation, evacuation and the extent to which various individuals have access to various spaces, are all examples of questions which require an integrated model of navigable space for pedestrians. For this reason, we argue that 3D city models should not neglect navigational aspects of cities, in the same way that they should not neglect semantic aspects. Navigational aspects have a history of being considered in isolation from geometrical and semantic aspects, but all three aspects, considered together are important. There are a diverse range of issues relevant to the emergency evacuation of buildings, involving knowledge of many different aspects of the built environment. This includes [9, 18]:

1. keeping an spatial inventories of the nature and location of damage and hazardous equipment which may contribute to the aggravation of fires;
2. knowing the capacity of escape routes;
3. understanding how fire fumes might spread, how the terrain might affect pedestrian movement and how the variability between individuals may affect their movement and the movement of others.

## 3.4 Navigable space

Navigable space in cities can be considered to be a set of topologically-connected discrete spaces, juxtaposed in three-dimensional space. Access to these spaces is governed by the geometry of these spaces, their semantic details and a microscale description of which pedestrians have access to which spaces and under which circumstances. Such navigable details of space in cities are difficult to obtain, but some of the general-purpose semantically-rich 3D city models may provide opportunities for obtaining this information. Note that in this paper, we are *not* concerned with the *behaviour* of pedestrians in space, just where they are *able* to move – behavioural models can be built in top, in specific application domains. A review and discussion of the modelling of navigable space will follow.

## 3.5 Pedestrian navigation in CityGML

CityGML provides the opportunity to model both space inside buildings and space outside buildings. However, it treats these spaces differently. Interior space is modelled (at the highest level detail) as building parts and the rooms of which they comprise, whereas space outside buildings is modelled as a terrain surface. Buildings contain rooms which are organised hierarchically by building and by building part (e.g. by storey). Kolbe's [12] paper on applying CityGML to various disaster management applications, shows how the connectivity between rooms for pedestrian access can be extracted using the shared openings (doors) between rooms. However, this is both a *by-product* of the geometrical modelling (to reduce the duplication of geometrical description and *optional* [8]. Where such topology is not supplied, it must be derived through the geometrical coincidence of duplicate openings.

Other details which might affect pedestrian access are 'BuildingInstallation' and 'BuildingFurniture' feature types within the 'AbstractBuilding' model and the 'CityFurniture' concept for objects outside. These are classified with ATKIS-based codes classifying their type and function. Those relevant to pedestrian access include stairs, pillars and ramps. Kerbs are another important aspect which; these are part of the transport model.

A fully-populated CityGML model may be able to provide us with some of the information we require to obtain fully-connected pedestrian access networks, though the hierarchical way in which internal spaces are structured in the GML makes a certain amount of restructuring necessary, a task which it is likely to be achievable automatically.

CityGML has been sanctioned by the Open Geospatial Consortium (OGC), and has been evaluated in the OGC Web Services Testbed No. 4[2].

## 3.6 Pedestrian access models

Most published research on pedestrian accessconcentrates either on aggregate measures of accessibly for different user groups (e.g. [4, 19]), or simple network models such as those for transport modelling. Okunuki *et al.* [16] proposed some initial ideas for a pedestrian guidance system – implemented as a web prototype[3] – in which navigable space is represented as a network of single links for corridors, lifts and stairs and gridded meshes of links for open spaces. The prototype was designed to suggest a route for a user taking into account simple preferences (such as the need to avoid stairs). Lee [14] derives a 3D geometrical network by transforming polygons (representing floorspaces on specific storeys) into a connected network, using a modification of a medial axis transform. This 3D geometrical network, which extends over multiple storeys inside buildings and connects to space outside buildings was applied to building evacuation [13].

Neither of these works encodes pedestrian- and time-specific information on pedestrian access at the microscale.

Meijers *et al.* [15] developed a semantic model for describing pedestrian access within buildings. It requires the building to be subdivided into closed and non-overlapping spaces (volumes) called sections, within which pedestrian access is unhindered and whose geometry is described with a set of bounding polygons (a boundary-representation model). Each of the bounding polygons is classified according to its role in restricting or facilitating access; by persistency (presence in time), physical existence (some polygons exist purely to close spaces), access granting (classified as full, semi and limited; those classified as semi may require door keys and those classified as limited may perhaps only allow access in an emergency) and direction of passage (uni- or bi-directional). Using this polygon classification, scheme, each

---

[2] The results are available in an OGC document available from http://portal.opengeospatial.org/files/?artifact_id=21622

[3] http://www.ncgia.ucsb.edu/~nuki/EllisonMenu.html

section is classified into 'end' (with one entrance/exit), 'connector' (with more than one entrance/exit) and 'non-accessible'. From these classified sections, topologically-connected graphs can be derived. This work acknowledges the need for access-granting requirements, but does not describe the details of this can be described.

## 3.7 Relationship of inside and outside space

Traditionally, spaces exterior to buildings and space interior to buildings have been modelled separately, in GIS and CAD-type software respectively. This is due to the different applications domains which primarily use the data, the different scales, and the different semantics. As shown, CityGML which supports the modelling of both inside and outside space models these spaces differently, using the building model for all aspects of inside space, and using the terrain, water, transportation, vegetation, city furniture and land-use models for outside space. However, unlike most of the early 3D city models reviewed in which building blocks are placed on top of a terrain, CityGML allows the 3D geometry of the interface between the building and the terrain to be described, using a 3D polyline (a 'TerrainIntersectionCurve'; Figure 4). Stoter [24] also acknowledges the importance of integrating the terrain surface with the base of buildings. From the point of view of modelling navigable spaces, the way in which the terrain meets the building at access points is of crucial importance.

**Fig. 3.3** CityGML's 'TerrainIntersectionCurve' (shown in black), a 3D polyline representing where the building meets the terrain. Source: [8]

## 3.8 Model design and prototype

Our prototype model for representing navigable space in cities is based on Slingsby's [21] model design, which attempts to combine some of the geometrical, semantic and navigational aspects of cities. Space volumes are implicitly represented by their lower surfaces (ground surfaces), using a 2.5D approach. These surfaces are represented by polygons, tessellated and topologically structured into distinct layers. These layers are topologically-connected to each other where there is pedestrian access (Figure 4).

The three aspects of geometry, semantic and navigational aspects of the model design will be presented (full details of the implementation are in [21]).

## 3.9 Geometrical aspects

We use the 2.5D layered approach (Figure 4) to illustrate the importance that we attach to the topological consistency between layers, in terms of pedestrian access. It is a constraint-based surface model, in which height (e.g. spot heights) and surface morphology constraints (e.g. surface breaklines) primitives are embedded within 2D polygons. These constraints are used to generate a topologically-consistent set of surfaces defined in 3D [23]. As can be seen in figure 4, the topological model must be able to cope with non-2D-manifold joins.

Amongst the point, lines and polygon geometrical primitives, height and surface morphology constraints are embedded, as illustrated in Figure 5. There are two types of point constraint, (absolute and relative heights), two kinds of linear constraint (breaklines and vertical discontinuities called 'offsets') and two areal constraints (ramps and stairs). These constraints all affect the resulting 3D geometry. Examples of all except the ramps can be seen in Figure 5. These constraints are used to generate a 3D geometry which conforms to these constraints and is topologically consistent [23].

These layers and constraints are defined independently of real-world (semantic) meaning. The semantic model allows objects and semantic information about the objects to be defined on top of the geometrical model. Their 3D geometrical forms are parameterised in the object descriptions.

Note that the geometrical model here is only used to represent discrete and connected *navigable spaces*, structured into constraint-controlled 3D surfaces.

## 3.10 Semantic model

The semantic aspect of the model allows feature types (objects) to be defined and attributed a published meaning, taking their geometries from the

3 Navigable Space in 3D City Models

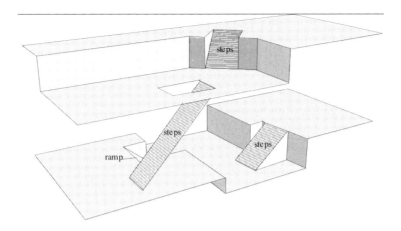

**Fig. 3.4** A small example showing distinct surfaces (layers) which are topologically connected. The surfaces are composed of point, line and polygon primitives (not shown)

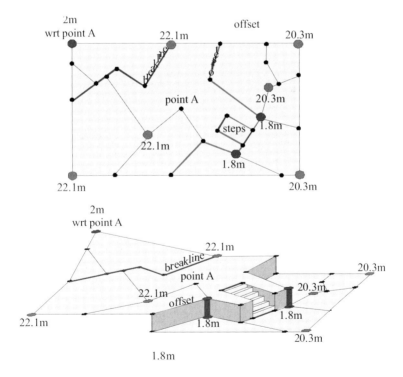

**Fig. 3.5** Height and surface morphological constraints

primitives in the geometrical model and they can have a set of attributes associated. A small set of feature types have been defined, which have particular applicability to pedestrian access. These are 'spaces', 'barriers' (walls and fences), 'portals' (doors and windows) and 'teleports' (lifts). These have attributes which both parameterise their 3D geometries and have access implications (see following section).

## 3.11 Pedestrian navigation model

The pedestrian navigation model [20, 22] follows a similar approach to Meijers *et al.* [15], in that information on persistency, access-granting properties, direction of passage and structural information are described. However, they are attached to objects defined by the *semantic model* of 'barriers' (walls), 'portals' (doors and windows), 'teleports' (lifts) and 'spaces' (specific delineations of space), rather than to the boundaries of building sections of homogeneous access characteristics. An algorithm is then used to delineate space navigable to particular types of pedestrian, at the time and in the context in which access is attempted. These objects have persistency, access-granting, direction of passage and structural properties attached, some of which are time- and pedestrian-dependent. Persistency information is provided through lists of unique or recurring time periods (e.g. some barriers only exist at certain times of day). Access granting information is provided as lists of the times (unique or recurring) at which access is granted and (optionally) a specific or specific type of pedestrian. In this context, a pedestrian has a number of characteristics (e.g. age, gender) and may be in procession of one or more door keys or access cards. Direction of passage information and structural information can be applied to some of these objects. Barriers and openings are classified according to the ease of unauthorised access, e.g. how easily it can be passed with or without damage. Pedestrians have a maximum ease threshold of barriers for which they would be willing to breach. This might be context-dependent, for example if there is an emergency. Using this information, the model attempts to incorporate some of the microscale details of pedestrian access. As stated, a pedestrian has attributes, may have a collection of access cards of door keys has a threshold amount for gaining unauthorised access. Additionally, a pedestrian has a step-height he or she is able to negotiate, which would be zero or very small for a wheelchair user.

## 3.12 Implementation

A prototype implementation of the model design was implemented using ArcGIS for data preparation, editing and visualisation, with the data model

3 Navigable Space in 3D City Models                                                  59

and the 3D generating algorithm implemented in Java. This proof-of-concept
model was used to generate the worked example in the next section. This
example will be extended to full building models of Delft University in the
RGI '3D Topography' project during 2007.

## 3.13 A worked example

Here we present a simple worked example using the model in order to illustrate some of the concepts of this paper. The images shown are annotated actual output from the prototype application. The scenario is a small fictitious example, of a very small area shown in Figure 6. It incorporates a section of road crossed by a bridge, a lower level accessed by a staircase and a ramp from either side of the road and a two storey building whose storeys are connected by a staircase. The main door of the building has a list of pedestrian- and time-dependent restrictions. Figure 7 shows the same area with the walls removed and from a slightly different viewpoint. The entirety of the space of the scenario (Figure 6) is accessible by a pedestrian who can negotiate steps.

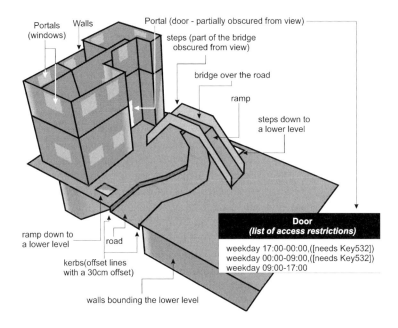

**Fig. 3.6** An annotated image of the entire 3D scenario

The subsequent images (Figures 7, 8 and 9) show the areas of navigable spaces in the context described in the caption.

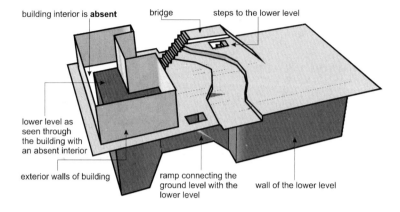

**Fig. 3.7** The space accessible by a pedestrian without a key from outside the building, out of office hours (according to the access rules shown in Figure 6). Since access to the main door is not allowed, the interior of the building is absent, as is the upper storey, because no access has been gained to the internal staircase.

## 3.14 Evaluation and discussion

Although the example scenario is rather simplistic, it serves to illustrate the concepts in the model. It was produced by a prototype model in which navigable spaces by different pedestrians in different contexts can be delineated from a set of topologically-connected spaces with descriptions of objects which affect pedestrian access embedded and are properly attributed.

The semantic model and the attributes used for the navigation model are also simplistic but the approach is intended to allow the full complexity of navigable spaces in the built environment to be encompassed.

As part of the 3D Topgraphy project at Delft, it is our aim to apply this model to the challenge of representing the built environment of the university campus during 2007.

## 3.15 Conclusion

The enormous progress in two dimensional GIS over the last decade currently masks the rather less well-developed situation in three dimensional

3 Navigable Space in 3D City Models                                              61

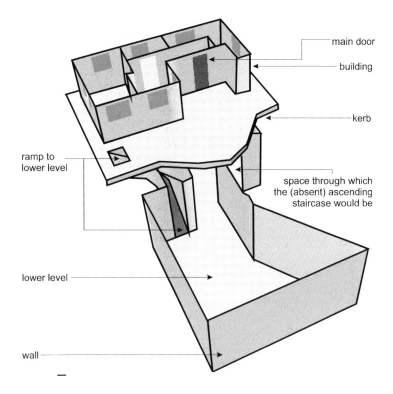

**Fig. 3.8** This shows the space accessible to a pedestrian who starts just outside the building and cannot negotiate steps of any size. Note that all steps are absent; the road, the other side of the road, the bridge and the upper storey of the building are missing because they can only be accessed either by stairs or steps. All elements shown can be accessed without any steps.

GIS. There are a large number of application problems in three dimensions which have been solved in particular application domains. One general problem, for which no acceptable modelling solution appears to have been found, is the connection between buildings and the terrain. Without a solution to this challenge a wide range of applications where interaction has to cross the building-terrain divide are stalled.

# References

[1] Altmaier, A. and T. Kolbe: Applications and Solutions for Interoperable 3D Geo-Visualization. Photogrammetric Week 2003, Stuttgart, Germany (2003)

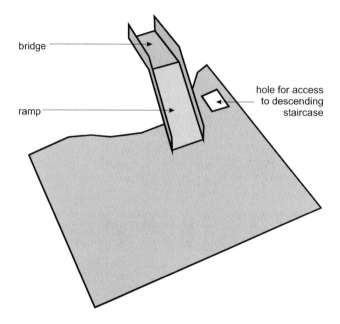

**Fig. 3.9** This figure shows the space accessible to a pedestrian who starts on the opposite side of the road to that in Figure 9 and cannot negotiate steps of any size. Note that most of the bridge is accessible from this side of the road because this side is a ramp, but there is no access over the bridge. Also note that there is no lower level because access to this from this side of the road is by a staircase.

[2] Batty, M.: Model cities. UCL Centre for Advanced Spatial Analysis. Working Paper 113. Online at: http://www.casa.ucl.ac.uk/working\_papers/paper113.pdf (2007)

[3] Benner, J., Geiger, A., Leinemann, K.: Flexible generation of semantic 3D building models. Proceedings of the 1st International Workshop on Next Generation 3D City Models, Bonn, Germany (2005)

[4] Church, R. L. and J. Marston: Measuring Accessibility for People with a Disability. Geographical Analysis 35(1): 83–96 (2003)

[5] Döllner, J., H. Buchholz, Brodersen, F., Glander, T., Jütterschenke, S. and Klimetschek, A.: Smart Buildings – a concept for ad-hoc creation and refinement of 3D building models. Proceedings of the 1st International Workshop on Next Generation 3D City Models, Bonn, Germany (2005)

[6] Eastman, C. M.: Building product models: computer environments, supporting design and construction. Boca Raton, FL, USA, CRC Press (1999)

[7] Ellul, C. and Haklay, M.: Requirements for Topology in 3D GIS.. Transactions in GIS 10(2): 157–175 (2006)

8. Gröger, G., Kolbe, T.H. and Czerwinski, A.: Candidate OpenGIS CityGML Implementation Specification (City Geography Markup Language). Open Geospatial Consortium/Special Interest Group 3D (SIG 3D): 120 (2006)
9. Gwynne, S., E. R. Galea, et al.: A review of the Methodologies used in the Computer Simulation of Evacuation from the Built Environment. Building and Environment 34(6): 741–749 (1999)
10. Khemlani, L.: A 'Federated' Approach to Building Information Modeling. CADENCE AEC Tech News 94 (2003)
11. Kolbe, T., G. Groger and Plümer, L.: Towards Unified 3D-City-Models. ISPRS Commission IV Joint Workshop on Challenges in Geospatial Analysis, Integration and Visualization II, Stuttgart, Germany (2003)
12. Kolbe, T. H., G. Gröger, et al.: CityGML – Interoperable Access to 3D City Models. Proceedings of the Int. Symposium on Geo-information for Disaster Management, Delft, The Netherlands, Springer Verlag (2005)
13. Kwan, M. P. and J. Lee: Emergency response after 9/11: the potential of real-time 3D GIS for quick emergency response in micro-spatial environments.. Computers Environment and Urban Systems 29(2): 93–113 (2005)
14. Lee, J.: A Spatial Access-Oriented Implementation of a 3-D GIS Topological Data Model for Urban Entities. Geoinformatica 8(3): 237–264 (2004)
15. Meijers, M., Zlatanova, S. and Pfeifer, N.: 3D Geo-information Indoors: Structuring for Evacuation. Proceedings of the 1st International Workshop on Next Generation 3D City Models, Bonn, Germany. (2005)
16. Okunuki, K., Church, R. and Marston, J.R.: A Study on a System for Guiding of the Optimal Route with a Hybrid Network and Grid Data Structure. Papers and Proceedings of the Geographic Information Systems Association, Japan (1999)
17. Papamichael, K., Chauvet, H., La Porta, J., Dandridge, R.: Product Modeling for Computer-Aided Decision-Making. Automation in Construction 8 (1999)
18. Pu, S. and S. Zlatanova: Evacuation Route Calculation of Inner Buildings. Geo-information for disaster management. P. van Oosterom, S. Zlatanova and E. M. Fendel. Heidelberg, Germany, Springer Verlag: 1143–1161. (2005)
19. Sakkas, N. and J. Pérez: Elaborating metrics for the accessibility of buildings. Computers, Environment and Urban Systems 30(5) (2006)
20. Slingsby, A. D.: Pedestrian accessibility in the built environment in the context of feature-based digital mapping. Proceedings of Computers in Urban Planning and Urban Management (CUPUM), London, UK. (2005)
21. Slingsby, A. D.: Digital Mapping in Three Dimensional Space: Geometry, Features and Access. Centre for Advanced Spatial Analysis. PhD thesis (unpublished). London, UK, University College London (2006)

[22] Slingsby, A. D. and P. A. Longley: A Conceptual Framework for Describing Microscale Pedestrian Access in the Built Environment. Proceedings of GISRUK, Nottingham, UK (2006)
[23] Slingsby, A.D.: A layer-based data model as a basis for structuring 3D geometrical built-enviornment data with poorly-specified heights, in a GIS context. Proceedings of AGILE07, Aalborg, Denmark, May 2007. (2007)
[24] Stoter, J.: 3D cadastre. PhD Thesis, Delft University of Technology, The Netherlands (2004)
[25] Takase, Y., Yano, K., Nakaya, T., Isoda, Y., Kawasumi, T., Matsuoka, K., Tanaka, S., Kawahara, N., Inoue, M., Tsukamoto, A., Kirimura, T., Kawahara, D., Sho, N., Shimiya, K. and Sone, A.: Visualisation of historical city Kyoto by applying VR and Web3D-GIS technologies. CIPA International Workshop dedicated on e-Documentation and Standardisation in Cultural Heritage, Cyprus (2006)
[26] Zlatanova, S.: 3D GIS for Urban Development. ITC publication 69, ISBN 90-6164-178-0 (2006)

# Chapter 4
# Towards 3D Spatial Data Infrastructures (3D-SDI) based on open standards – experiences, results and future issues

Jens Basanow, Pascal Neis, Steffen Neubauer, Arne Schilling, and Alexander Zipf

## 4.1 Introduction

The creation of Spatial Data Infrastructures (SDI) has been an important and actively studied topic in geoscience research for years. It is also regarded in politics and by decision makers as leveraging technology for reducing thr time and cost of geo services for internal usage as well as for public information services. In Europe, the new INSPIRE (Infrastructure for Spatial Information in Europe) directive 2007/2/EC provides general rules for implementating national spatial data infrastructures for environmental policies. SDIs must rely on open standards specified by the Open Geospatial Consortium (CS-W, WMS, WFS, WCS, WPS, OpenLS, etc.)

Based on the theoretical background of INSPIRE and several discussion drafts of the OGC, we have implemented an SDI for the city of Heidelberg that comprises an array of established OGC services and some new proposed technologies required to extend into the $3^{rd}$ dimension. In this paper, we discuss the components that have been developed for a 3D SDI and some important aspects that must be addressed to make this kind of infrastructure work. For standard services, we could use existing open source solutions; others must be extended or developed from scratch, including new techniques for data preparation and integration. The components have been implemented for several projects with different goals, always with interoperability and reusability in mind.

The central part of the SDI is the OGC Web3D Service (W3DS), which delivers the actual 3D data. The W3DS specification is currently in draft status and is not yet adopted by the OGC. We present our own implementation of this service and some implications when considering different use cases. In our case it was important to use the W3DS not only for producing static

---

University of Applied Sciences Mainz &University of Bonn (Cartography), Germany
<name>@geographie.uni-bonn.de

scenes, but also to request data piecewise in order to stream it to the client; this implements a more dynamic visualization. This is due to the large data quantities, which are not comparable to the 2D bitmaps delivered by a WMS.

A possible extension to the WMS is the support of the Styled Layer Descriptor (SLD) profile for controlling the appearance of maps. It is advisable to separate the geometry or geographic raw data from the visualization rules. Proposals have been made to include further visualization elements directly in CityGML. We suggest using the SLD specification in combination with W3DS services. We describe how SLD can be extended to provide 3D symbolizations - e.g. for 3D points, linestrings, surfaces, and solids.

There must also be an adequate way to describe our 3D data in a catalogue service; we examined different alternatives. We also examined the integration of route services, for which the OGC OpenLS specification can be used, as we will show.

Finally, we discuss future research topics that arise from current trends such as Location Based Services or Service Oriented Architectures (SOA). We need to investigate how these concepts can be applied to mobile 3D navigation services, which have different requirements in terms of visualization and user guidance. In the long term, a higher level concept for defining chains of web services within an SOA could be applied that helps to orchestrate SDI services more flexibly. In particular, the Business Process Execution Language (BPEL) could be used to define scenarios realized through chaining the open GI services that constitute SDIs.

## 4.2 3D Data Management – an overview

Data management is at the heart of an SDI. A powerful database is necessary to manage and administer 3D data efficiently (Zlatanova & Prosperi 2005). Object-relational databases such as PostGIS or OracleSpatial have already been applied successfully to handling geographic information. A lot of work has been done in this respect; in this paper we summarize recent developments in standard-based data sources for 3D visualization services, such as the W3DS. In order to create a 3D data storage layer that can be used as a source for our W3DS implementation, we tested open and commercial database capabilities regarding geometry models, export formats, availability, etc. The following products were assessed:

*iGeo3D* is part of the degree-framework that allows users to manage 3D geodata through the web. It is completely based on OGC standards. The 3D database scheme 'CityFeatureStore' of the data storage module can be used for various database systems (Oracle, PostGreSQL/ PostGis). The degree-WFS offers write- and read access to the data and iGeoSecurity provides access protection. The CityGML format or multiple image formats can be used for exchanging the data.

The *City Model Administration open source toolkit (CAT3D)* was developed within the EU-project VEPS (Virtual Environmental Planning System) by the HfT Stuttgart. It can connect to different data sources and produce several output formats (VRML, KML, Shapefile). The architecture is modular so that additional data sources and formats can be supported by implementing the according modules.

A 3D extension by CPA Geoinformation for commercial *SupportGIS* offers ISO/OGC conform 3D data storage and supports databases such as Oracle, PostgreSQL, MySQL and Informix. Together with SupportGIS-3DViewer and SupportGIS-Web3D a 3D platform is provided with a CityGML central database structure.

Within the VisualMap project (FhG IGD/EML) a database called *ArchiBase* was developed which allows various different 3D geodata to be administered. The modelling tools can come from applications such as 3D Studio Max as well as from GIS. The scheme was realized for Oracle. The data can be managed via a graphical user interface, and exchange formats can be in VRML 2.0 and XML. Principles that evolved from this work can be found within CityServer3D.

*CityServer3D* (Haist & Coors 2005) is a multi-layered application consisting of a spatial database, a server and a client application. The database manages 3D geometries at multiple levels of detail along with the corresponding metadata. The server as the core component provides different interfaces for importing and exporting various geodata formats. The data is structured by a meta-model and stored in a database.

Currently, we use the well known PostGIS extension to PostgreSQL for 2D data and 3D points of the DEM, as well as VRML code snippets, but after evaluating the projects mentioned above, we will extend data management of our server to also support native 3D data types within the database as needed. The exchange of 3D city models through CityGML delivered by a WFS is separate; it is already covered by the above mentioned projects, such as iGeo3D, etc.

## 4.3 Towards a 3D Catalogue Service for 3D Metadata

Within an SDI it is important to record information about available datasets via metadata in order to make it possible to find relevant data. Three metadata standards seemed most relevant for spatial data: ISO 19115 along with its predecessors Dublin Core and CEN-TC287. Nonn et al. (2007) evaluated the suitability of the current metadata standards for 3D spatial data. The authors also investigated which enhancements or supplements might be needed by the most important metadata specification for spatial data, ISO 19115, so that it can be used to describe 3D landscape and city models. We tried to find the highest possible sufficiency for 3D spatial data, city- and landscape models. In particular, we used the present OGC CityGML discussion paper (Gröger et al. 2006) - especially regarding the question of how to allow a semantic description of structures within 3D city models.

As of today there is still no online object catalogue available for CityGML from which attribute values can be derived. If such a catalogue was available online, it would not be necessary to put this kind of information directly into the ISO 19115 standard; instead, the internet catalogue could be referenced. The feature type attributes contain an object type list, also linking the user to the specific parts of the online catalogue.

This work paves the way for future discussions on the needs of 3D-SDI, especially for 3D city models. Although current SDI developments focus on 2D spatial data, we think that in the long run a similar development is necessary for 3D data. Already a range of basic attributes in ISO 19115 apply to 3D data; even so, we found a need to add further specifications to the metadata catalogues. We have made first suggestions for ways to add these missing elements to the ISO 19115. We are aware that these suggestions are a first attempt and need further development. For first results, see Nonn et al. (2007).

## 4.4 Scene-based Visualization with the Web3D Service

Regarding the portrayal of 3D information, a Web3D Service (W3DS) was proposed to the OGC as a discussion draft (OGC 2005). The W3DS delivers 3D scenes of 3D city or landscape models over the web as VRML, X3D, GeoVRML, or similar formats. The parameters are similar to those of the WPVS (Web Perspective View Service), which adds to the well-known WMS interface parameters for camera position, view target, etc. We implemented a server that supports all these parameters, but also provides some noteworthy techniques applied to a W3DS service for the first time in a standard-conforming way. For example, in order to provide techniques that are already state of the art in computer graphics (such as dynamic concepts like continuous LODs for triangle meshes or streaming of geometry parts), we developed a sort of 'pseudo-streaming' using an intelligent client-application and pre-processed DEM-tiles with different resolutions and sizes. This allows faster delivery of scenes compared to typical implementations of the W3DS, which deliver only complete scenes in file documents, covering the entire requested scene. A similar scenario was introduced at Web3D 2002 (Schilling

## 4 Towards 3D Spatial Data Infrastructures

and Zipf 2002). Back then, there were no 3D OGC standards we could use for our scenario. This has changed; we incorporated these new standards into the project. The work presented here is embedded in a larger project that involves a several OGC Web Services (OWS), as well as several clients and the integration of various data sources.

As shown in figure 1, many requests from our Map3D-Client trigger a service chain involving separate OWS necessary to process the request. 2D maps are delivered by a Web Map Service (WMS) for overview maps of the region. 3D information is provided by our Web 3D Service (W3DS) implementation. The Web Feature Service (WFS) standard is or will be the basis for both of these services. The WFS is already integrated for the 2D map data used by the WMS and will also be used to provide the data necessary for creating 3D scenes.

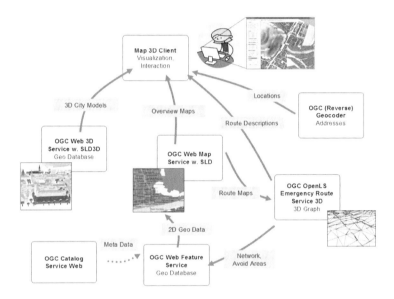

**Fig. 4.1** Components and service chaining in our 3D-SDI (WPS to be added soon for pre-processing of terrain data)

We implemented the OpenLS Route Service Specification (as well as the OpenLS Utility Service (Geocoding and Reverse Geocoding)). The route calculation itself is done on a 2D network graph; however, the resulting route geometry is then replaced by 3D linestrings taken from the 3D network. This 3D network was pre-calculated by mapping the 2D linestrings onto the Digital Elevation Model (DEM) so that the route segments exactly follow the terrain including tunnels and bridges. This so-called Route Service 3D (RS3D) uses exactly the same interface as the already standardized OpenLS Route Service, without needing to extend anything. Due to the more accurate representation of route geometries from the 3D extension, we get a lot more route segments. Practical tests showed that we needed to reduce the geometries further using horizontal and vertical generalization to produce smooth visualizations and animations. A 2D overview map is also produced by our implementation of the OpenLS Presentation Service.

The OGC Catalog Service (CS-W) shown in figure 1 is based on the degree framework and delivers metadata of the actual spatial data. Before adding metadata to this service, we conducted an investigation to determine if the relevant metadata standards such as Dublin Core or ISO 19115 are appropriate for describing 3D spatial data such as 3D city models (Nonn et al. 2007). The purpose of the CS-W is to provide information through search functions such as where to find GI services and spatial data within the spatial data infrastructure or on the Web. We used these for 2D data in former projects, such as OK-GIS or geoXchange (Tschirner et al. 2005).

## 4.5 Streaming and different LODs of DEM using the W3DS

As mentioned earlier, we developed a 'smart' Java3D client that uses pre-processed DEM-tiles served by the W3DS to satisfy state-of-the-art computer graphics with respect to streaming. Further, it uses different LODs when changing the field of view of the viewer. This was done using open standards by OGC, as explained below:

A high-precision (5 meter) DEM, covering an nearly 150 square kilometres, was divided into several groups of smaller, rectangular DEM pieces with different accuracies and point-densities. Each DEM-tile group represents one Level-of-Detail (LOD). This means that those tiles covering wide areas describe the surface more approximately than smaller tiles with a high point density. Each DEM tile is replaced by four smaller tiles in the next higher LOD. This allows the client to retrieve DEM-tiles at different LODs using the W3DS. A dynamic DEM can be processed by requesting the needed tiles depending on the viewer's position, the line of sight and the distance along the line of sight. All changes in the viewer's field of view or position causes a new series of W3DS-GetScene requests delivering new DEM-tiles. These tiles are then added to the scenegraph. Memory is saved by only displaying the tiles in the view and by removing all tiles outside of the view on the fly. An example of the results is available as a video screen capture showing the effect on the DEM when navigating the scene in real time. The videos are available from http://www.gdi-3d.de.

## 4.6 Standard-based Configuration of 3D Visualization through extensions of the Styled Layer Descriptor

In conventional GIS, the raw data is typically separated from the visualisation properties. This provides the possibility of displaying the same data in multiple ways depending on the project use case or user preferences. So far, this separation is not yet established in 3D GIS data, since usually the 3D model is considered as a kind of visualization itself - including all appearance properties. This is the case for all common graphics formats like DXF, 3DS, VRML, and other proprietary CAD formats. In the GIS world we strive to describe only the geometry and the object classes in the raw datasets and to store attribute data and display properties in different files, as is the case with the most popular products.

For 2D web maps, Styled Layer Descriptor (SLD) documents exist, which define rules and symbols controlling the map appearance. The same should be applied to 3D maps, including city models. By using SLD it is also possible to integrate different data sources into a single rendering service like a WMS and to style all data consistently.

We propose an extension to the SLD specification in order to support 3D geometries and appearance properties. As of now, this approach is unique. However, there are considerations on extending CityGML by further visualization elements. If such an extension would also cover pure styling information this would undermine the desired separation of raw data and visualization rules. Therefore, we must be aware of existing OGC specifications and incorporate them into new standards, or simply extend existing ones. In this case, styling information for polygons, lines, and points in SLD is also partly useful in 3D. Therefore an SLD extension seems to be more promising. In the next sections we make some first suggestions for a SLD3D that incorporates standard SLD elements and some new elements only valid in 3D space. The SLD3D was implemented and tested in the 3D-SDI Heidelberg project (Neubauer 2007, Neubauer & Zipf 2007). The SLD files are currently used for configuring the W3DS server; however, in the future the client will be able to specify it as well in order to provide more flexibility of interaction.

Relevant aspects of this extension can be categorized as follows:

- Rotation of elements around all three axes
- Displacements and positions are extended by Z
- SolidSymbolizer for object volume description
- SurfaceSymbolizer for defining surfaces with triangular meshes (tin)
- Integration of external 3D objects into the scene
- Defining material properties
- Billboards for 2D graphics
- 3D legends
- Lines displayed cylindrically (e.g. for routes)

Current WMSs can provide the user with a choice of style options; the W3DS, on the other hand, can only provide style names and not a more detailed scene of what the portrayal will look like. The biggest drawback, however, is that the user has no way of defining his own styling rules. For a human or machine client to define these rules, there must be a styling language that the client and server can both understand. This work focuses on defining such a language, called 3D Symbology Encoding (3D SE). This language can be used to portray the output of Web 3D Services.

3D-Symbology-Encoding includes FeatureTypeStyle and CoverageStyle root elements taken from standard Symbology Encoding. These elements contain all styling, for example, filters and different kinds of symbolizers. As the specification states, Symbolizers are embedded in Rules, which have group conditions for styling features. A Symbolizer describes how a feature will appear on a map or in a 3D scene. The symbolizer also has graphical properties such as color and opacity.

The 3D-SE can be used flexibly by a number of services or applications that style georeferenced information in 3D. It can be seen as a declarative way to define the styling of 3D-geodata independent of service interface specifications, such as W3DS.

## 4.6.1 PolygonSymbolizer

The PolygonSymbolizer describes the standard 2D style of a polygon including *Fill* for the interiors and *Stroke* for the outline, as defined in SLD. Additionally the 3D-SLD extension describes 3D features like BillboardPlacement.

## 4.6.2 LineSymbolizer

A 2D line can be represented in 3D as a pipe feature, with a certain radius and colour. The standard attributes from SLD-specification also can be set (StrokeWidth, StrokeType, etc.).

## 4.6.3 BillboardPlacement

With the BillboardPlacement element, 2D objects (text, images, etc.) can be placed so that they always face the viewer. This is useful for icons, pixel graphics, signs, and other abstract graphics. BillboardPlacement contains 3 sub elements: AnchorPoint, Displacement, and Rotation. The syntax is:

## 4 Towards 3D Spatial Data Infrastructures

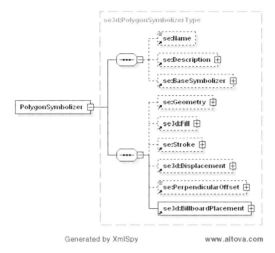

**Fig. 4.2** XML schema for the SLD-3D PolygonSymbolizer

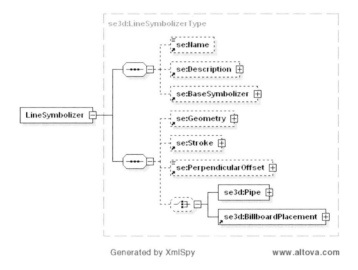

**Fig. 4.3** XML schema of the SLD-3D LineSymbolizer

*AnchorPoint*

The 3D Symbology Encoding Anchor Point element is extended by AnchorPointZ. The coordinates are given as floating-point numbers like AnchorPointX and AnchorPointY. These elements each have values ranging from 0.0 to 1.0. The default point is X=0.5, Y=0.5, Z=0.5 which is at the middle height and length of the graphic/label text. Its syntax is:

*Displacement*

```
<xsd:element name="BillboardPlacement" type="se3d:BillboardPlacementType">
  <xsd:annotation>
</xsd:element>
<xsd:complexType name="BillboardPlacementType">
  <xsd:sequence>
    <xsd:element ref="se3d:AnchorPoint" minOccurs="0"/>
    <xsd:element ref="se3d:Displacement" minOccurs="0"/>
    <xsd:element ref="se:Rotation" minOccurs="0"/>
  </xsd:sequence>
</xsd:complexType>
```

**Fig. 4.4** XSD schema of the SLD-3D BillboardPlacement

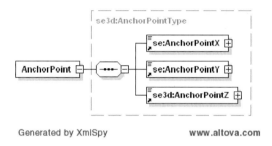

**Fig. 4.5** XSD schema of the SLD-3D AnchorPoint

Displacement is extended by Z like AnchorPoint. The default displacement is X=0, Y=0, Z=0. The schema is visualized in figure 7.

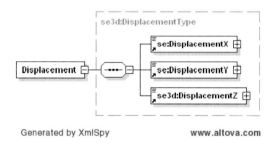

**Fig. 4.6** XSD schema of the SLD-3D Displacement

If Displacement is used in conjunction with Size and/or Rotation, the graphic symbol can be scaled and/or rotated before it is displaced.

## 4.6.4 Material

Due to the more complicated lighting simulation in 3D, it is necessary to replace the simple colour fill element by a Material element describing the physical properties of a surface by simple means. The implementation follows the fixed function pipeline of OpenGL.

```xml
<xsd:element name="Material" type="se3d:MaterialType">
  <xsd:annotation>
</xsd:element>
<xsd:complexType name="MaterialType">
  <xsd:sequence>
    <xsd:element name="DiffuseColor" minOccurs="0">
      <xsd:complexType>
        <xsd:sequence>
          <xsd:element ref="se:SvgParameter"/>
        </xsd:sequence>
      </xsd:complexType>
    </xsd:element>
    <xsd:element name="SpecularColor" minOccurs="0">
      <xsd:complexType>
        <xsd:sequence>
          <xsd:element ref="se:SvgParameter"/>
        </xsd:sequence>
      </xsd:complexType>
    </xsd:element>
    <xsd:element name="EmissiveColor" minOccurs="0">
      <xsd:complexType>
        <xsd:sequence>
          <xsd:element ref="se:SvgParameter"/>
        </xsd:sequence>
      </xsd:complexType>
    </xsd:element>
    <xsd:element name="AmbientIntensity" type="xsd:double" minOccurs="0"/>
    <xsd:element name="Shininess" type="xsd:double" minOccurs="0"/>
    <xsd:element name="Transparency" type="xsd:double" minOccurs="0"/>
  </xsd:sequence>
</xsd:complexType>
```

**Fig. 4.7** XSD schema of the SLD-3D material-description)

The annex shows a simplified basic SLD-3D document containing one NamedLayer and one UserStyle. Several of these can be defined in the document. The examples of extensions given so far give only a first impression of the very large list of extensions to the original SLD schema (Neubauer& Zipf 2007). Further information and the full schema will be made public in late 2007. Currently these new 3D styles are implemented within our W3DS server.

## 4.7 Scene Integration and Server Architecture

The internal architecture of our W3DS implementation is shown in Fig. 9. The service is intended to produce ready-to-use display elements. This means that all integration tasks can be done in advance because the display elements can be processed as far as possible ahead of time. For this reason, we separate functionality into a visualization server, delivering 3D scene graphs in a Web3D format, and a modelling or authoring engine, processing all the raw data into a completely integrated 3D data set of the area. This also prepares tiles that can be quickly streamed to the client.

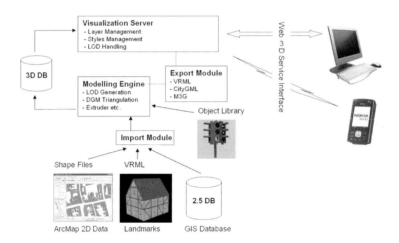

**Fig. 4.8** The Web3D Service is implemented as a visualisation Server

At the moment we use Java3D objects for internal data management in the visualization server. The export module encodes the objects into VRML syntax which are returned to the client as a response to the GetScene request. In a first version we stored all Java3D objects in the server's java heap space. In practice, however, we realized that this is insufficient for larger data sets, so we are now switching to a database implementation that holds all the Java3D objects. In the pre-processing step, we integrate the different data sources like 2D GIS data, terrain data, point objects and VRML landmarks. Fig. 9 describes the integration process of the spatial data, from its original raw state to its on-screen visualization. This data is imported from various sources. In a next step it is converted into 3D objects that comprise a 3D scene and is stored in a database for faster access. This 3D-geodatabase is the data source for the visualization server, which delivers 3D scenes by a W3DS conform web request and offers the possibility of exporting into various formats.

**Fig. 4.9** Thematic 2D layers integrated into the terrain model

We found the integration of thematic 2D area information (forest, streets, water, etc) particularly interesting because these are not applied to the terrain as textures but are instead cut into the TIN triangulation. Fig. 10 shows how the original 2D layers are transformed into 3D geometries. For each layer we create one indexed face set that covers the terrain exactly. The original surface underneath is cut away. The downside of this is that it produces larger geometries. The advantage, however, is that aliasing effects do not occur and that we do not need to transmit additional textures.

## 4.8 Standard-based 3D Route Planning within an 3D SDI

The OpenLS (OpenGIS Location Services) framework consists of five core services (OpenLS 2002). We have implemented three of these services using Java (Neis et al 2006, Neis et al 2007). The Route Service (RS) allows various criteria to be set, such as start and destination, time, distance, travel-type, one-way street information, as well as the possibility to add areas or streets that should be avoided (*AvoidAreas*). Following this standard, we implemented a Route Service 3D (RS3D). The main difference is that RS3D provides GML code that contains Z values for all route geometries and instructions so that the result can be integrated into 3D landscapes without further calculations. Similar ideas have been proposed by Zlatanova and Verbree (2006).

The implementation is based on a complete 3D street network. For the actual route calculations we use the standard Dijkstra-Algorithm. The original 2D network graph and the 3D graph are topologically equal. However, we need to transform the network segments in order to reflect the terrain surface. We do this by mapping the 2D linestrings onto the DEM and adding new line vertices wherever an Edge of the terrain triangulation is crossed, so that it is exactly parallel to the surface. Depending on the terrain accuracy, the amount of data for the network geometry increases, as well as the correctness of the graph. Network segments representing bridges, tunnels, or underpasses still need to be adjusted manually, which is relevant for the visualization and route animation. The RS3D uses the existing OpenLS Location Utility Service for geocoding and OpenLS Presentation Service for generating overview maps. This is a good example of service chaining within an SDI. Our tests show that this approach is faster than calculating the geometries on the fly, which takes a considerable amount of time.

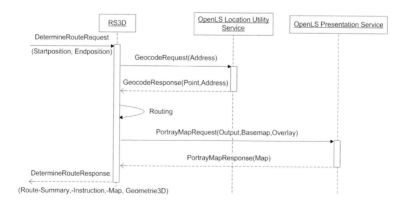

**Fig. 4.10** RS3D UML sequence diagram

## 4.9 Route Presentation within the W3DS Viewer

Route presentations in 3D can be done using either dynamically updated texture maps containing a line string of a certain width, colour, and pattern, or using 3D geometries like tubes along the route. We chose the latter alternative, since texturing is already used for terrain styling. The 3D line string is extruded as a pipe with a certain distance to the ground. Switching to a new route quickly is no problem since the Web3D Services delivering the city model and the route service are independent and the scenegraph part containing the route can be replaced. Waypoints are displayed as 2D labels on top of the screen. By computing key frames for every node and connecting them by a Spline interpolator (KBRotPosScaleSplinePathInterpolator) we created a route animation that moves the viewpoint some distance along the route. Unfortunately, the initial route linestring, which fits the terrain surface perfectly, is not very useful for generating animations. Small features or errors in the TIN, such as little bumps that are not visually dominant, are preserved and lead to jerky movements in the animations. This occurs because the animation moves along a segmented line and the camera view changes at every small segment. Also, the network graph used for the route computation contains sharp corners at intersections. While this is sufficient for 2D maps, there are difficulties when it is applied to 3D. We therefore implemented an additional simplification that filters small features in order to acquire a smooth animation. The results can be seen from the video captures at http://www.heidelberg-3d.de/.

## 4.10 Future Issues in 3D SDIs

### 4.10.1 Orchestrating 3D Web Services

Flexibility and reusability are major goals for complex applications based on OGC Webservices (OWS) that represent a spatial data infrastructure. Through aggregating standardized base services, complex functions and workflows of a certain granularity can be achieved. These new aggregated functionalities can then be used as web services on their own. In order to avoid programming the aggregation of several independent OWS by hand, a higher level solution has been proposed. This alternative is called 'Web Service Orchestration' (WSO) through standardized higher level languages, such as the Business Process Execution Language (BPEL). The promise of WSO is an easy and flexible way to link services in given patterns, especially through configuration. This can be realized through so-called orchestration scripts. Their configuration can then be carried out with graphical tools instead of

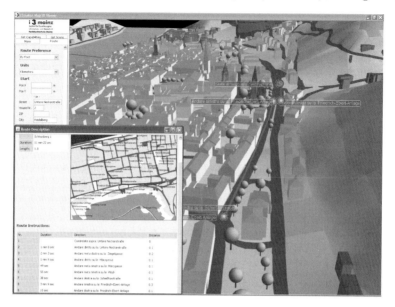

**Fig. 4.11** 3D routing in the W3DS-Viewer

hard-coded programming. The BPEL scripts will be executed in corresponding WSO engines.

First experiences using WSO or BPEL in the context of OGC service have been reported recently (Weiser and Zipf 2007). Discussions within the OGC about service-oriented architectures with OWS started in 2004 (OGC 04-060r1). Although there were some earlier considerations about adjusting the architecture towards compatibility with common web services (OGC 03-014), the OWS2 paper (OGC OWS2) offered the first helpful suggestions in this direction. Similar methods and ideas have been discussed by Kiehle et al. (2006) and Lemmens et al. (2006). Some results from the OWS4 initiative can be found in the recent internal OGC discussion paper 'OWS4 Workflow IPR' (OGC 06-187). The proof-of-concept evaluation presented in Weiser and Zipf (2007) shows that it is possible to create added-value by combining and aggregating OGC Web Services. However it only makes sense to use orchestration where a continuous service chain without human intervention is given. This is why it is necessary to find stable standard chains (small parts of a larger workflow that can act as modular building blocks) as well as to do research on assembling BPEL scripts even more dynamically, even for non-technical users. At the moment, developers must still face many small technical problems when trying to realize WSO for OWS (Weiser et al 2007). User interface concepts are especially needed to ease the highly dynamic orchestration of OWS on the fly. They would also in principle allow the representation of such service chains as presented figure 1 through the use of

WSO-technologies like BPEL. A major obstacle at the moment is the missing SOAP interface for the relevant OWS; this will change in the foreseeable future, as SOAP interfaces must now be added to every new version of an OGC specification.

### *4.10.2 Future 3D Navigation on Mobile Devices*

It makes sense to talk about 3D route planning in a 3D environment. We presented preliminary work on the Mainz Mobile 3D system, a PDA based navigation system that can also send W3DS requests to the W3DS server and display the VRML scene returned (Fischer et al. 2006). Recently, a new project was started with several partners (including Hft Stuttgart and companies like Navigon, Teleatlas, GTA Geoinformatics, Heidelberg Mobil and others), called 'Mobile Navigation with 3D city models' (MoNa3D, http://www.mona3d.de/), which extends mobile navigation systems into 3D (Coors & Zipf 2007). Improved 3D navigation using semantic route descriptions like landmarks will be investigated. 3D visualizations are important for both pedestrian and vehicle navigation systems. 3D city models allow landmarks to be integrated into the route description. Indoor navigation must also be considered – here the choice and visualization of appropriate landmarks is also important (Mohan & Zipf 2005). Most navigation systems today only offer direction and distance information. This is not sufficient to provide the user with ideal orientation information. Studies from cognitive psychology have shown that directions using landmarks are rated higher than direction and distance alone. We aim to gain knowledge about optimizing 3D navigation information to provide users with relevant orientation details. This could lead to safer navigation through reduced stress. In order to achieve sustainable outcomes, 3D city models for navigation support must be available within a functioning 3D geodata infrastructure (3D-GDI), such as http://www.3d-gdi.de/.

## 4.11 Conclusion and Outlook

In this paper, we discussed the first outcomes of a research project that concentrates on implementing the next generation of spatial infrastructures based on open standards currently under discussion in the OGC. We showed how to use these standards to develop a 3D SDI. A demo application using an OpenLS-based 3D route service was introduced.

So far we discussed mostly data management and visualization issues for 3D spatial data. The next step for 3D SDI not yet mentioned in this paper is the standard conform geoprocessing of 3D data. Within the OGC, a draft version of a new specification regarding processing arbitrary spatial data is under development. This so-called Web Processing Service (WPS) (see also Kiehle et al. 2006) still has a way to go, but we have first experiences with this draft gained through implementations of specific processing algorithms (e.g. spatial join and aggregation) (Stollberg 2006, &Stollberg and Zipf 2007) within the project OK-GIS (open disaster management with free GIS, http://www.ok-gis.de/). From these we are confident, that a range of preprocessing steps needed in our scenarios in 3D-SDI Heidelberg can be realized using this new standard.

## 4.12 Acknowledgements

This work was supported by the Klaus-Tschira-Foundation (KTS) GmbH Heidelberg within the project www.heidelberg-3d.de. All spatial data is by courtesy of the Bureau of Surveying Heidelberg and the European Media Laboratory (EML) Heidelberg. We thank all coworkers, in particular A. Weiser, B. Stollberg, U. Nonn and C. Mark for their efforts.

## References

Abdul-Rahman, A.; Zlatanova, S.; Coors, V. (Eds.): Innovations in 3D Geo Information Systems Springer: Lecture Notes in Geoinformation and Cartography , 2006, 760 p.

Chen L, Wassermann B, Emmerich W, Foster H (2006): Web Service Orchestration with BPEL. Dept. of Computer Science, University College London

Coors, V. and Zipf, A, (Eds.)(2004): 3D-Geoinformationssysteme. Grundlagen und Anwendungen. Wichmann - Hüthig Verlag. Heidelberg. 525 pages.

Coors, V. and Zipf, A. (2007 accepted): MoNa 3D — Mobile Navigation using 3D City Models. LBS and Telecartography 2007. Hongkong.

Coors, V. and Bogdahn, J. (2007): City Model Administration Toolkit (CAT3D) http://www.multimedia.fht-stuttgart.de/veps/CAT3D/cat3d.htm

Fischer, M., Basanow, Zipf (2006): http://www2.geoinform.fh-mainz.de/~zipf/MainzMobile3D_Geoinfo3D.pdf Mainz Mobile 3D - A PDA based OGC Web 3D Service Client and corresponding server. International Workshop on 3D Geoinformation 2006 (3DGeoInfo'06). Kuala Lumpur. Malaysia.

Fitzke J, Greve K; Müller M and Poth A (2004): Building SDIs with Free Software - the deegree Project. In: Proceedings of GSDI- 7, Bangalore, India.

Gröger, G., Kolbe, T.H., Czerwinski, A. (2006): OpenGIS City Geography Markup Language (CityGML), Implementation Specification Version 0.3.0, Discussion Paper, OGC Doc. No. 06-057

Haist, J.; Coors, V. (2005): The W3DS-Interface of Cityserver3D. In: Kolbe, Gröger (Ed.); European Spatial Data Research (EuroSDR) u.a.: Next Generation 3D City Models. Workshop Papers : Participant's Edition. 2005, pp. 63-67

iGeo3D: http:www.lat-lon.de/latlon/portal/media-type/html/ language/de/user/anon/page/default.psml/js_pane/sub_produkte _deegree-igeo3d

ISO/TC 211 Geographic information/Geomatics. ISO reference number: 19115 (2002)

Kiehle C; Greve K & Heier C (2006): Standardized Geoprocessing – Taking Spatial Data Infrastructures one step further. Proceedings of the 9th AGILE International Conference on Geographic Information Science. Visegrád, Hungary.

Lemmens R, Granell C, Wytzisk A, de By R, Gould M, van Oosterom P (2006): http://www.agile2006.hu/papers/a051.pdf Semantic and syntactic service descriptions at work in geo-service chaining. Proc. of the 9th AGILE Int. Conference on Geographic Information Science. Visegrád, Hungary

Mohan, S. K., Zipf, A. (2007): Improving the support of indoor evacuation through landmarks on mobile displays. 3rd International Symposium on Geoinformation for Disaster Management. Toronto, Canada.

Müller, M. (ed): OpenGIS Symbology Encoding Implementation Specification version 1.1.0 doc.nr. 05-077r4 http://www.opengeospatial.org/ standards/symbol

Neis, P. (2006): Routenplaner für einen Emergency Route Service auf Basis der OpenLS Spezifikation. Diploma Thesis. University of Applied Sciences FH Mainz.

Neis, P., A. Schilling, A. Zipf (2007): http://www2.geoinform.fh-mainz. de/~zipf/GI4D2007.3DEmergencyRouteService.pdf 3D Emergency Route Service based on OpenLS Specifications. GI4DM 2007. 3rd International Symposium on Geoinformation for Disaster Management. Toronto, Canada.

Neubauer, S., Zipf, A. (2007 accepted): Suggestions for Extending the OGC Styled Layer Descriptor (SLD) Specification into 3D – Towards Visualization Rules for 3D City Models, Urban Data Management Symposium. UDMS 2007. Stuttgart. Germany.

Nonn, U., Dietze, L. and A. Zipf (2007): Metadata for 3D City Models - Analysis of the Applicability of the ISO 19115 Standard and Possibilities for further Amendments. AGILE 2007. International Conference on Geographic Information Science of the Association of Geograpic Information Laboratories for Europe. Aalborg, Denmark.

Nougeras-Iso, J., P.R. Muro-Medrano, F. J. Zarazaga-Soria (2005): Geographic Information Metadata for Spatial Data Infrastructures - Resources, Interoperability and Information Retrieval. Springer. 2005, XXII, 264 p

OGC 2005. Web 3D Service. OGC Discussion Paper, Ref. No. OGC 05-019.

OGC: OWS2 Common Architecture. Hrsg. OGC. RefNum. OGC 04-060r1; Vers. 1.0.0; Status: OGC Discussion Paper.

Open Geospatial Consortium Inc. OWS 1.2 SOAP Experiment Report. Hrsg. OGC. RefNum. OGC 03-014; Vers. 0.8; Status: OGC Discussion Paper.

Open Geospatial Consortium Inc. Towards 3D Spatial Data Infrastrucutres (3D-SDI) based on Open. Hrsg. OGC. RefNum OGC 06-187; Vers. 1.0.0; 2007-09-14 Status: internal OGC Discussion Paper.

OPENLS: OGC Open Location Services Version 1.1
http://www.opengeospatial.org/functional/?page=ols.

OGC: Styled Layer Descriptor (SLD) Implementation Specification V.1.0 doc.nr. 02-070

OGC: Styled Layer Descriptor Profile of the Web Map Service Implementation Specification version 1.1 doc.nr. 05-078

OGC: Web 3D Service. OGC Discussion Paper, Ref. No. OGC 05-019.

Schilling, A., Basanow, J., Zipf, A. (2007): Vector based mapping of polygons on irregular terrain meshes for web 3D map services. 3rd International Conference on Web Information Systems and Technologies (WEBIST). Barcelona, Spain. 2007.

Schilling, A., Zipf, A. (2002): Generation of VRML City Models for Focus Based Tour Animations - Integration, Modeling and Presentation of Heterogeneous Geo-Data Sources. In Web3D Conference, 9-12.03.2003, Saint Malo, France.

Stollberg, B. (2006): Geoprocessing in Spatial Data Infrastructures. Design and Implementation of a Service for Aggregating Spatial Data with the Web Processing Service (WPS). Diploma Thesis. University of Applied Sciences FH Mainz.

Stollberg, B. & Zipf, A. (2007 accepted): OGC Web Processing Service Interface for Web Service Orchestration - Aggregating geo-processing services in a bomb thread scenario. W2GIS 2007: Web&Wireless GIS Conference 2007. Cardiff, UK.

SLD: SLD-Specification (OGC): http://www.opengeospatial.org/docs/02-070.pdf

SupportGIS-3d:
http://www.supportgis.de/Dip2/SupportGIS/3D/SupportGIS-3D.pdf

Weiser, A., Zipf, A. (2007): Web Service Orchestration of OGC Web Services (OWS) for Disaster Management. GI4DM 2007. 3rd International Symposium on Geoinformation for Disaster Management. Toronto, Canada.

Zipf, A., Tschirner, S. (2005): Finding GI-datasets that otherwise would have been lost - GeoXchange - a OGC standards-based SDI for sharing free geodata. 2nd International Workshop on Geographic Information Retrieval (GIR 2005) at the Fourteenth ACM Conference on Information and Knowledge Management (CIKM). November, 2005. Bremen, Germany. Springer Lecutre Notes in Computer Science.

Zlatanova, S. and E. Verbree (2005): The third dimension in LBS: the steps to go. In: Geowissenschaftliche Mitteilungen, Heft Nr. 74, 2005, pp. 185-190.

Zlatanova, S. and D. Prosperi (Eds) (2005): Large-scale 3D Data Integration CRC Press. 256 pages.

# Annex Example Document of SLD-3D Extension

```xml
<?xml version="1.0" encoding="UTF-8"?>
<sld3d:StyledLayerDescriptor xmlns="http://www.opengis.net/sld3d"
    xmlns:sld3d="http://www.opengis.net/sld3d" xmlns:sld="http://www.opengis.net/sld"
    xmlns:se3d="http://www.opengis.net/se3d" xmlns:se="http://www.opengis.net/se"
    xmlns:xsd="http://www.w3.org/2001/XMLSchema"
    xmlns:xsi="http://www.w3.org/2001/XMLSchema-instance"
    xsi:schemaLocation="http://www.opengis.net/sld3d
    ..\3D-SLD\3d-sld\0.1.0\StyledLayerDescriptor.xsd"
    version="0.1.0">
    <se:Name>3D-SLD_1</se:Name>
    <se:Description>
        <se:Title>3D StyledLayerDescriptor 1</se:Title>
    </se:Description>
    <NamedLayer>
        <se:Name>Strassen</se:Name>
        <se:Description>
            <se:Title>Streets and Paths</se:Title>
            <se:Abstract>Polygon Layer</se:Abstract>
        </se:Description>
        <UserStyle>
            <se3d:FeatureTypeStyle>
                <se3d:Rule>
                    <se3d:PolygonSymbolizer>
                        <se3d:Fill>
                            <se3d:Material>
                                <se3d:DiffuseColor>
                                    <se:SvgParameter name="fill">#050505</se:SvgParameter>
                                </se3d:DiffuseColor>
                                <se3d:SpecularColor>
                                    <se:SvgParameter name="fill">#0f0f0f</se:SvgParameter>
                                </se3d:SpecularColor>
                                <se3d:EmissiveColor>
                                    <se:SvgParameter name="fill">#000000</se:SvgParameter>
                                </se3d:EmissiveColor>
                                <se3d:AmbientIntensity>0.3</se3d:AmbientIntensity>
                                <se3d:Shininess>0.0</se3d:Shininess>
                                <se3d:Transparency>0.0</se3d:Transparency>
                            </se3d:Material>
                        </se3d:Fill>
                    </se3d:PolygonSymbolizer>
                </se3d:Rule>
            </se3d:FeatureTypeStyle>
        </UserStyle>
    </NamedLayer>
</sld3d:StyledLayerDescriptor>
```

# Chapter 5
# Re-using laser scanner data in applications for 3D topography

Sander Oude Elberink

**Abstract**

Once 3D information is acquired and used for their initial applications, it is likely that the original source data or its derived products can be re-used. The purpose of this paper is to show the large potential for re-using 3D geo-information. The focus is on the re-use of laser scanner data and its derived products at four major geo-organisations in The Netherlands. Re-using data is not only of interest for end-users but especially for data owners who can better justify the costs for acquisition and maintenance of the data. We analyzed the flexibility of organizations to explore what can be done with the data in their possession. We found that once a 3D data set was acquired with requirements based on initial applications, many 'new' users recognized the added value of 3D data for their own application.

## 5.1 Introduction

Many papers mention the need for 3D building models, and then describe their methods to acquire 3D models [1], [2], [3] or to evaluate their methods. For researchers it is important to explore optimal methods to build, store and analyze 3D data. In many cases, organizations store their original source data, and process it into derived 3D information for their applications using methods and recommendations from research or commercial market. Once 3D information has been acquired and used for their initial applications, it is likely that the original source data or its derived products can be re-used.

---

International Institute for Geo-Information Science and Earth Observation (ITC)
P.O.Box 6
7500 AA, Enschede, The Netherlands
oudeelberink@itc.nl

For example, laser data can be used to derive 3D building models for visualization purposes. For other applications, users can both access the original laser data set or the already derived 3D building models. Little is known about the experiences on the re-usability of 3D geo-information and its consequences for user requirements.

The purpose of this paper is to show the large potential for re-using 3D geo-information, regarding both conventional users, like geo-information departments, as well as new groups of users, like tax departments. The focus is on the re-use of laser scanner data and its derived products. Re-using data is not only of interest for end-users but especially for data owners who can better justify the costs for acquisition and maintenance of the data.

Four cases were compared by information analysis at four major geo-information organizations in The Netherlands. Through interview sessions and a subsequent workshop, we collected and discussed user experiences concerning quality requirements, applications, storage and acquisition of 3D topographical data. Information analysis shows that the re-usability of data strongly depends on the requirements of the source data.

## 5.2 Background

Several years ago, geo-information departments started building up their experiences with laser scanner data, to better and faster acquire DTMs or to support updating topographic maps, as shown in [4]. Laser data and its derived products like 3D city models are relatively new data sources for other departments in many organizations.

Research groups of Delft University of Technology and ITC Enschede work on the efficient modelling and acquisition of 3D topographic models. The focus of the overall research project is on data modelling in TEN structures [5] concerning efficient data storage and analysis, and data acquisition using laser altimetry data [6] focusing on reconstructing objects in 3D. To ensure that in the future their methods will be adapted in practice, an inventory was made on the user requirements of these models and acquisition techniques.

For readability reasons we ignore details of the definitions of whether data is real 3D or 2.5D. If height information is available at a certain location, we speak of 3D geo-information. When necessary, we further specify 3D information as raw source data (like laser scanner data) or derived 3D data (like 3D building models).

## 5.3 User experiences on 3D Topography

Interview sessions were organized between researchers and owners/users of 3D geo-information. We selected four cases, in which the users already gained some experience with the acquisition, storage and analyses of laser scanner data and its derived 3D products. The four organizations are:

- Municipality of Den Bosch;
- Survey department of Rijkswaterstaat (RWS);
- Water board 'Hoogheemraadschap de Stichtsche Rijnlanden' (HDSR);
- Topographic Service of the Dutch Cadastre.

During the interviews we collected information on the necessity for using 3D data instead of the existing 2D data. Limitations of (analyzing) 2D data are important to justify the need for 3D data. The result of this part of the study can be seen in figure 1.

**Fig. 5.1** Initial list of 3D applications

The figure shows an initial list of applications that are based on 2.5D or 3D data. In all of the applications height information is essential to correctly perform the task. Height was determined using laser scanner data at some point during the processing.

### 5.3.1 User requirements

The major purpose of the interviews was to specify user requirements for 3D topography. User requirements should cover topics like specific wishes on data quality, distribution, and analyses. Before we describe the re-usability of the data, we briefly discuss these user requirements first.

#### 5.3.1.1 Municipality of Den Bosch

Den Bosch aims for the production of a large scale 3D geo-database. Their main motive for acquiring 3D data is to perform height-related tasks like volume determination and water management tasks, but also for visualising the "as-is" situation. Visualising models close to reality is an important tool to communicate with their citizens.

Their list with 3D model requirements starts with the modelling of shapes of buildings, followed by the possibility for storing and analyzing multiple objects on top of each other. These requirements are added to the existing requirements for DTM production or determination of height profiles, formerly measured by GPS.

#### 5.3.1.2 Survey Department RWS(Rijkswaterstaat)

The Survey Department is responsible for acquiring and maintaining geo-information of national infrastructures (stored in Digital Topographic Database DTB) and a nation wide height model (AHN).

The Digital Topographic Database (DTB) is a topographic database with a map scale of 1:1.000, containing detailed information about all national infrastructural objects, like highways and national water ways.

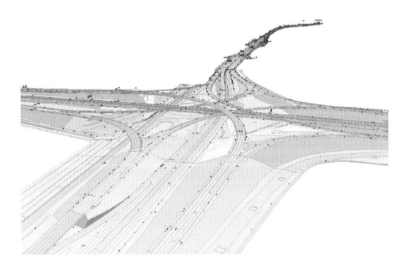

**Fig. 5.2** Points, lines and surfaces in interchange of DTB

Acquisition was done in 3D, by measuring in stereo imagery supplemented with terrestrial measurements at interchanges and tunnels. Points, lines and polygons were classified manually and stored in the database. Quality requirements depend on the idealization precision of the object, e.g. paint strips can be measured with a higher accuracy than a border between two meadow fields. Besides this, user requirements are strongly related to the acquisition method. Demands for terrestrial measurements are higher than photogrammetric demands. In the near future terrestrial and airborne laser scanner data might be introduced as a new data source for fast and automated acquisition of objects. The Actual Height model of the Netherlands (AHN) is a national DTM, initiated by three governmental organizations: Rijkswaterstaat, the provinces and the union of water boards.

User requirements of the AHN changed over time due to the growing number of applications. The most important change is the need for higher point density. In 1996, at the beginning of the project, 1 point per 16 $m^2$ was supposed to be dense enough to fulfil all user requirements. When users started to detect features, or fused the laser data with other detailed datasets, the demand grew for a higher point density laser data set. In 2004, the growing technical possibilities of laser scanners strengthened the idea that the next version of AHN should have at least 1 point per 9 $m^2$. In 2006 it was proposed that if the point density could be increased even more, many new applications could be performed. For example, the state of coastal objects, like dikes, can be monitored by analyzing high point density laser data. Recently, a pilot project began acquiring AHN data with a point density of 10 points per $m^2$. Given the fact that the AHN is a national height model, the pilot project suggests the ambition to cover more parts, if not all, of The Netherlands with a resolution of 10 points per $m^2$ if the pilot is successful.

#### 5.3.1.3 Water board HDSR

For inspection and maintenance of regional dikes, bridges and waterways, the water board needs up-to-date and reliable geo-information. Requirements for a 3D model are that breaklines and objects on top and at the bottom of a dike are measured with high precision. Existing AHN data is not dense enough for detailed mapping purposes. Breaklines are important features for the conditions (shape and strength) of dikes. In the past, parallel profiles were measured with GPS. Water board HDSR decided to acquire a helicopter based laser data set with a point density of more than 10 points per $m^2$, together with high resolution images. Important objects like bridges, dikes, water pipes were measured manually using the laser point cloud for geometric information, and images for detection and thematic information. By using laser data, the water board is able to calculate strength analysis locally instead of globally. This is important for analyzing the behaviour of its dikes. Now that a detailed 3D model of the dike and its neighbouring objects has been captured, analyzing strengths accurately in time and space will be possible when acquiring the next data set.

#### 5.3.1.4 Topographic Service

The Topographic Service of the Dutch Cadastre produces national 2D topographic databases from a scale of 1:10.000 to 1:250.000. Implicit height information has been integrated at specific parts in the 2D topographic maps by:

- Shadowing, visualizing local height differences;
- Symbols, representing a high obstacle like churches, wind mills, etc;
- Building classifications, discriminating between high and low buildings;
- Level code, indicating on which level an object is, when looking from above.

More explicit and absolute height information has been given by:

- Contour lines, representing a virtual line at a specific height;
- Height numbers, representing the local height at a certain location.

Whereas in the past the height information mentioned above was introduced mainly for cartographic purposes, the Topographic Service aims to extend the possibilities and acquire and store objects in 3D. When building up a 3D topographic database the customers of the products of the Topographic Service will be able to perform traffic analysis, volume calculations and 3D visualizations. User requirements can be summarized by the wish to acquire and store rough 3D building models and to add height values to road polygons. A summary of user requirements of all four cases can be found in Figure 3.

**Fig. 5.3** User requirements based on interviews

With these requirements several research activities were set up to specify the optimal way to meet the requirements. For example, after specifying the requirements for a 3D topographic road database for the Topographic Service, a research project was carried out to describe the best way to automatically acquire 3D roads. Results of this research project can be found in [6], where the authors describe a method to successfully combine laser data with 2D topographic map data. During the interviews users mentioned the increasing number of applications, using laser scanner data or its derived products.

## 5.3.2 Re-using data

All four organizations re-used their laser data and its derived products, more than expected. Figure 4 shows the extended list of 3D topography applications. Applications shown in bold and italic represent 'new' applications: they were initiated after the organizations captured their data for their original applications. Total number of applications mentioned in the interviews was 29, whereas the originally planned number was 12.

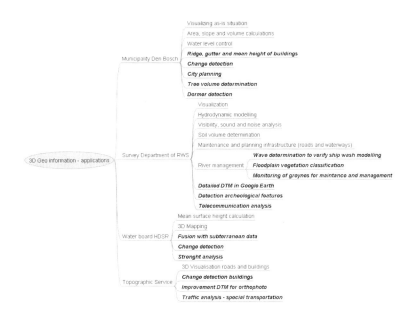

**Fig. 5.4** Extended list of 3D applications

Users mentioned the data-driven character of the new applications. These applications are in the explorative phase, which implies that the users first look at what can be done with the 3D data they have. This can be seen by the fact that the user requirements are characterized by the specifications of the available data. With the maturation of these applications, the requirements will become more application-driven, resulting in a more detailed description of what the specifications of 3D data should be.

The legal and financial consequences of re-using geo-information within the public sector has been explored and described by many others, e.g. [7]. In this section we look in detail at the growing numbers of applications re-using laser data at these four organizations.

### 5.3.2.1 Municipality of Den Bosch

The engineering department re-used parts of laser data classified as 'hard' terrain. They fused it with their existing topographic map and road database to better analyze the drainage of rainwater. The tax department initiated a project to detect dormers more quickly and more accurately, using laser data and imagery. Municipalities are looking for quantitative and fast methods to determine urban tree volumes for various reasons. Therefore, research has been done to detect individual trees and calculate urban tree crown volume in the city of Den Bosch, using their existing laser data [8]. Intermediate results can be the extraction of laser data on trees, as shown in Figure 5.

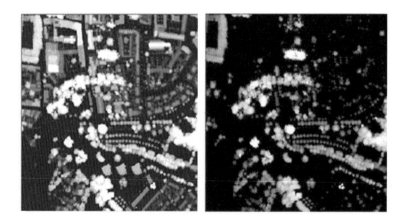

**Fig. 5.5** Gridded laser data (left), laser data classified as trees (right), [source: Lim, 2007]

### 5.3.2.2 Survey Department of RWS

In [9] it has been shown that RWS performs various river management applications with one high point density dataset.

The authors of [9] conclude that time-consuming terrestrial measurements, visual inspections and mapping from imagery, can be replaced by laser altimetry. In case of extreme low-water levels laser altimetry enables RWS to acquire detailed morphologic information about the groyne fields which usually cannot be measured. In combination with multi beam echo-sounder data, acquired at high-water level, behaviour of the river bed and groyne fields can be analysed simultaneously. A combined DEM has been shown in figure 6.

AHN has intensively been used by archaeologists. Large scale morphological structures, possibly indicating historical objects or activities, which

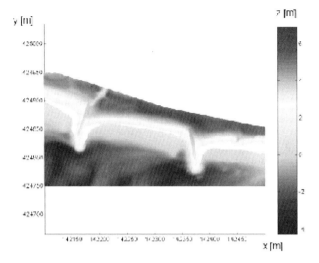

**Fig. 5.6** DEM of laser altimetry data in combination with multi-beam echo-sounder data at the river Waal [source: Brügelmann and Bollweg, 2004]

cannot be seen from the ground, may clearly be visible in the DTM. Besides this, slopes can indicate where to look, using the knowledge that historical objects tend to slide to lower parts in the terrain.

### 5.3.2.3 Water board 'HDSR'

Information about topography can be combined with subterranean information, to better analyze the strength of a dike. The use of laser scanner data is essential to correctly fuse topographic features with the (also 3D) subterranean information. Change detection is a hot topic in the maintenance of dikes. Already existing data sets are as important as future laser data sets when looking at differences between them.

### 5.3.2.4 Topographic Service

The Topographic Service seeks methods for fast and reliable change detection. In [4] it has been shown that laser data can be used to automatically detect changes in the 2D map. The authors explain that changes between laser data and map data should be handled with care. Changes can be caused by misinterpretations of aerial photographs, or by differences in generalization of the map.

## 5.4 Factors increasing re-usability

In the interviews users mentioned various factors that had a positive influence on the re-use of laser data.

### 5.4.0.5 Availability and distribution

GIS based intranet applications make it possible to show geo-data to the organization. Google Earth already has shown the success of simple visualizing, navigating and zooming of 2D geo data. When visualizing the as-is situation in 3D, it generates an alternative perspective for new user groups, including tax departments and citizens who want to walk through their streets in the model. Eye opening is the first and most important step in using new kinds of data for existing or even new applications.

### 5.4.0.6 Data fusion

Combining data sources not only delivers information on the similarities and differences between the two datasets, it also can use complementary aspects to create new or better products. Examples can be found in the fusion of map and laser data [6], where map data delivers thematic and topologic information and laser data adds geometric information.

### 5.4.0.7 Generalization and filtering

Although several authors use both terms Generalization and Filtering as being the same activity, we distinguish between generalizing 3D data, focusing on the representation of the output (reducing derived 3D data), and filtering laser scanner data, focusing on data reduction of the input (reducing raw data). In [10,11] the authors describe the need for 3D generalization. The need has been explained by the fact that generalized 3D models are easier to render. More important in our context, is their motivation that generalization allows organizations to use 3D geo data multiple times at multiple scales, thus reducing the costs of acquiring 3D data. For water boards a special kind of generalization is important, because objects close to dikes have to be represented in more detail than objects located farther away.

Although high point density laser data is useful for reliable classification of buildings, vegetation and other objects, and for the extraction of breaklines in the terrain, it is clear that for large parts in the terrain the point density is too high to allow efficient processing of a DTM. Filter algorithms help the user to reduce laser data at an early stage of the process, making the huge datasets much more flexible for their application.

## 5.5 Discussion

The reason for the increasing numbers of users of 3D data instead of 2D data is that 3D better represents the as-is situation. From this situation many users perform their activities. For example, city planners can add features to the as-is situation, civil engineers are able to calculate volumes and strengths at given situations, etc. Whereas 3D information started at the geo information departments to create a faster way to detect 2D objects automatically and to add value to the existing 2D information, it is for many other departments the first contact to geo information. It has to be noted that for a number of applications represention in 2D is still the most convenient way to reach their goal. Examples can be found in route descriptions and assessing parcel information.

Although airborne laser data is a good method to quickly acquire detailed information, it cannot replace all terrestrial measurements for purposes like measuring and monitoring point objects.

User requirements of 3D objects and databases are still under development. One of the reasons is that the number of applications and users is still growing. On the other hand, the technical possibilities of airborne imagery and laser altimetry are increasing in terms of geometric and radiometric resolution. With the growing offer of detailed information, the user requirements get more specific and the demand for more detailed information grows. Scientific projects in data acquisition, data fusion and storage are essential for users to show the re-usability of their data.

## 5.6 Conclusions

In our study we analyzed user requirements on 3D geo information in four major organizations. The user requirements were based on originally expected applications. We recognized the flexibility of organizations to explore the limits of their existing data. Therefore, the most important insight was the large potential for re-using existing 3D geo information. Once a 3D data set had been acquired, many 'new' users recognized the added value of 3D data for their application.

With a growing of number of users, the number of requirements also grows. A good example is the desired point density of the national height model AHN, which has increased from 1 point per 16 m$^2$ in 1996 to 10 points per m$^2$ in 2006.

Even information analyses can be re-used for different purposes. The actual purpose of analyzing the interview information was to specify user requirements, whereas the re-used version was to show the advantages of re-using geo-information.

# Acknowledgement

This research is partially funded by the Dutch BSIK research program Space for Geo-Information, project 3D Topography. The authors would like to thank Bram Verbruggen of the municipality of Den Bosch for providing additional information on the re-use of data, and the other persons who cooperated in the interviews and at the workshop.

# References

1. Henricsson, O. and Baltsavias, E. (1997). 3-d building reconstruction with aruba: A qualitative and quantitative evaluation. In: Gruen, Baltsavias and Henricsson (Editors), Automatic Extraction of Man-Made Objects from Aerial and Space Images (II). Birkhauser, Ascona, pp. 65-76.
2. Haala, N., C. Brenner and Anders, K.-H. (1998). 3D Urban GIS From Laser Altimeter and 2D Map Data, ISPRS Commission IV – GIS Between Visions and Applications.Ohio, USA.
3. Maas, H.-G., 2001. The suitability of Airborne Laser Scanner Data for Automatic 3D Object Reconstruction, Third International Workshop on Automatic Extraction of Man-Made Objects from Aerial and Space Images, Ascona, Switzerland.
4. Vosselman, G., Kessels, P., Gorte, B.G.H. (2005).The Utilisation of Airborne Laser Scanning for Three-Dimensional Mapping International Journal of Applied Earth Observation and Geoinformation 6 (3-4): 177-186.
5. Penninga, F., van Oosterom P., Kazar B.M. (2006). A Tetrahedronized Irregular Network based DBMS approach for 3D Topographic Data Modeling, the 12th International Symposium on Spatial Data Handling (SDH 2006), Vienna, Austria.
6. Oude Elberink, S. and Vosselman, G. (2006). Adding the Third Dimension to a Topographic Database Using Airborne Laser Scanner Data, ISRPS Vol 36, Part 3, "Commission III symposium", Bonn, Germany.
7. Loenen, B. van (2006), Developing geographic information infrastructures; the role of information policies. Dissertation. Delft University of Technology. Delft: DUP Science.
8. Lim, C. (2007) Estimation of urban tree crown volume based on object - oriented approach and LIDAR data. Master thesis. ITC Enschede, The Netherlands.
9. Brügelmann, R. and Bollweg, A. E.: Laser Altimetry for River Management. International Archives of Photogrammetry, Remote Sensing and Spatial Information. Vol XXXV (part B2), p 234-239, Istanbul, Turkey.
10. Meng, L. and Forberg, A. (2006): 3D building generalization. Chapter 11, 211-232. In: Mackaness, W., Ruas, A. and Sarjakoski, T. (Eds): Chal-

lenges in the Portrayal of Geographic Information: Issues of Generalisation and Multi Scale Representation. VTEX, Vilnius.
11 Thiemann, F. (2002). Generalization of 3D building data, ISPRS Vol 34, Part 4, 'GeoSpatial Theory, Processing and Applications', Ottawa, Canada.

# Chapter 6
# Using Raster DTM for Dike Modelling

Tobias Krüger and Gotthard Meinel

**Abstract**

Digital Terrain Models are necessary for the simulation of flood events. Therefore they have to be available for creating flood risk maps. River embankments for flood protection have been in use for centuries. Although they are artificial structures that actually do not belong to the natural elements of the land surface they are usually implicitly embedded in digital terrain data. Being elongated and elevated objects, they appear – depending on the used colour ramp for visualisation – as bright stripes on the surrounding background.

For purposes of flood protection it might be useful to gain data about crest levels, especially if these information are not available from other sources. High resolution Digital Terrain Models (DTM) can be used as highly reliable sources for deriving dike heights. Using laser scanner technique a general height accuracy of about 10–15 cm can be achieved for elevation models. Thus, by analysing DTM data relevant geometrical information on dikes can be directly derived.

## 6.1 Introduction

The last decades have shown a high frequency in the appearance of severe flood events in Central Europe. This tendency is continuing after the turn of the millennium, and the problem will probably become more serious in the future due to global warming.

Flood protection has therefore seen a change of paradigm within in the recent years and decades. In former times it was common to count only on technical protection strategies as building dikes, reservoirs, or flood polders.

---

Leibniz Institute of Ecological and Regional Development (IOER), Dresden
t.krueger@ioer.de, g.meinel@ioer.de

Recent flood events have shown the limited capacity of these measures. Today a more integrative view of flood protection is being adopted. The strategic focus here, which is of vital importance, lies on risk assessment and risk management. The current state of treating flood risks is given in [11].

**Table 6.1** Selected European flood events since 1978. Source: [11]

| Time | Event description |
| --- | --- |
| 1978 | Flood event in Switzerland changes Swiss flood protection strategies towards integrated approaches |
| 1993, December | Rhine flood event, later declared as "Hundred Year Flood" |
| 1995, January | Rhine flood event, overtopping the December event of 1993, overall damage 1993/95: > 5.5 billion EUR |
| 1997, Summer | Oder Basin flood in Germany, Poland, Bohemia, more than 100 casualties, damage about 5.5 million EUR |
| 1999 | "Whitsun Flood" in Southern Germany, five casualties, damage 335 million EUR |
| 2002, August | "Hundred Year Flood" in Central Europe, esp. in the Elbe and Danube basins, in Germany 21 casualties, overall damage of about 11.8 billion EUR |
| 2003 | Winter flood on the Elbe River |
| 2005, August | Flood in Switzerland, the most expensive damage event of the last hundred years, overall damage about 2.6 billion CHF [16] |
| 2006, March/April | Springtime flood along the Elbe River, partly with higher gauges compared to 2002 (esp. in the lower Elbe due to less dike breaches in the middle river stretch); Danube flood in Romania |
| 2007, August | Flood event caused by heavy rain in Germany (Rhine), Switzerland, Austria |

The list of recent flood events shown in Table 6.1 demonstrates the necessity to deal with flood risk management, especially in densly settled areas like Central Europe. Flood risk management can be seen as the effort to optimise the relation of hazard reduction – as erecting protection buildings – and vulnerability mitigation.[1] The latter can be achieved by the interaction of several components, e. g. to adopt resistant and resilient building structures. It is also of high importance to establish an efficient disaster management system which provides communication tools capable of working under hard pressure.

Another way of reducing vulnerability is to withdraw from natural floodplain areas. This improves the ecological capability and complies the natural conditions of a seasonally flooded river regime (see [4]). In [3] it is claimed to provide rivers with retention areas which have been successively reduced

---

[1] For detailed information on flood protection terms see [9].

6 Dike modelling using raster DTM                                    103

to a fraction of their original sizes. The Elbe River has preserved much of its natural conditions and shows a relatively high ecological potential compared to other Central European riparian landscapes. Nevertheless, more than 80 % of the Elbe floodplains have been cut from the river during the last centuries (see [5]). A coarse map showing the differences between the former flooding area and the recent floodplain is shown in [15].

The research project VERIS-Elbe

The research project VERIS-Elbe [8] examines the changing flood risk along the German Elbe River due to land use change, climate change, and other factors using the scenario technique prospecting into the next one hundred years.

The potential flood area of the Elbe River in Germany covers about 5 000 square kilometres. Within the project it is intended to determine flood risks under varying conditions, which includes to remove dikes and rebuild them on other places.

## 6.2 Digital elevation data

### *6.2.1 Dikes as terrain model objects*

Depending on the objective of the model one speaks of Digital Terrain Models or Digital Surface Models. The latter depict the surface including elevated objects while the former contain information only on the very earth's surface. Therefore it is useful and necessary not only to talk about Digital Elevation Models but exactly to determine what kind of elevation is meant.

The fertile floodplain soils are favourable for agrarian use and require protection. Therefore the beginning of dike formation dates back for centuries. Whereas flood dams were already erected by Roman soldiers the planned installation of dikes in Central Europe began in the early Middle Ages (see [14]).

The derivation of Digital Terrain Models includes the clearing up raw the data from elevated items like buildings, bridges, or trees. Contrary to this, dikes usually remain as land surface elements in the terrain model datasets. Depending on the visualisation colour scheme they appear as bright bands. Dikes therefore turn out to be a kind of hybrid objects which are man-made on the one hand, but on the other hand are considered as belonging to the earth's surface.

If one needs information about geometrical properties of dikes such as length, width, and height it is necessary to collect external data. Length and width can quite easily be obtained by using measurement tools as provided by standard GIS[2] software. Height information must be provided by terrestrial survey data or can be extracted directly from the DTM.

## 6.2.2 Available Digital Terrain Models

The research project VERIS-Elbe examines the flood risk on nearly the full length of the German Elbe River. The investigation area ranges from the German-Czech border to the gauging station Neu Darchau which is situated in Lower Saxony and is to be considered as the last gauge not influenced by the tides (see [1, p. 55]).

### 6.2.2.1 High resolution DTM

One of the project partners is the German *Bundesanstalt für Gewässerkunde*[3] *(BfG)* which is providing a high resolution Digital Terrain Model for the Elbe including the hydrologically relevant earth's surface along the river channel. That means that all flood protection dikes are included in the model. The model's acronym is DGM-W[4] and it is divided into three sections called South, Middle, and North. The spatial resolution is 2 m in section South – covering the Saxon part of the Elbe River – and 2 m in section Middle – covering the Elbe in Saxony-Anhalt as well as the area of the Havel River. Section North data have not been processed so far but will be at 2 m resolution as well once available. All these datasets were derived from laser scanner data. The river bed information origins from sonar measurements. The height accuracy is indicated as 0.15 m.

Another high resolution DTM is available for the Saxon Elbe section. It has been provided by the Saxon *Landestalsperrenverwaltung*[5] *(LTV)* and has a resolution of 2 m. It also covers the immediate neighbourhood of the river and has a height accuracy better than 0.10 m. This model has no specific acronym, but it is referred to as *HWSK data*[6] by the LTV.

The *Landesbetrieb für Hochwasserschutz und Wasserwirtschaft Sachsen-Anhalt*[7] *(LHW)* has provided a high resolution DTM for a projected flood polder site near Lutherstadt Wittenberg. It has a spatial resolution of 1.0 m and a height accuracy of 0.15 m.

---

[2] Geographic information system

[3] Federal Institut of Hydrology, Koblenz, http://www.bafg.de/

[4] Digitales Geländemodell – Wasserlauf, engl.: DTM Watercourse

[5] State reservoir authority, Pirna, http://www.talsperren-sachsen.de/

[6] HWSK: Hochwasserschutzkonzeption, engl.: Flood protection conception

[7] State Agency for Flood Protection and Water Management Saxony-Anhalt, Magdeburg

6 Dike modelling using raster DTM

**Table 6.2** Available high resolution DTM datasets

| DTM dataset | Spatial resolution [m] | Height accuracy [m] |
|---|---|---|
| HWSK data | 2 | 0.10 |
| DGM-W Middle | 2 | 0.15 |
| DGM-W South | 2.5 | 0.15 |
| Polder DTM | 1 | 0.15 |

#### 6.2.2.2 Medium resolution DTM

Unfortunately, the whole inundation area of the Elbe River cannot be covered with a high resolution DTM. Thus for the remaining regions a DTM provided by the German *Bundesamt für Kartographie und Geodäsie*[8] is being used. The DGM-D[9] is part of the ATKIS[10] dataset and has a spatial resolution of 25 m. The height accuracy varies within a quite large range. As stated in the dataset's manual [2] the accuracy is determined as ranging from 1 m to 8 m – depending on the quality of the underlying data. This quite high inexactness of the data is caused by the very different sources which have been used to compile the DTM that serves the whole country. In Germany the survey authorities are under the responsibility of the Federal States. The federal survey agencies are supplying data which is used by the BKG to compile datasets covering Germany as a whole. The data originate from very diverse sources and show different spatial resolution and accuracy. Some parts of the data are collected by laser scanning, stereographic interpretation of aerial imagery, or even might originate from digitising contour lines from large-scale topographic maps.

## 6.3 Dike extraction

### *6.3.1 Object recognition*

Because dikes can be perceived as elevated objects, dike extraction leads to object recognition methods which are common in raster image processing.

Identification of dikes can generally be done by two different approaches. The first possibility uses pure image processing. These methods base on the analysis of elevation differences in the model. Fulfilling certain criteria causes the identification of pixels as belonging to an elevated object or not. The

---

[8] Federal Agency for Cartography and Geodesy, Frankfurt/M. and Leipzig, http://www.bkg.bund.de/

[9] DGM Deutschland, engl: DTM Germany

[10] Amtliches Topographisch-Kartographisches Informationssystem, engl.: Authoritative topographic cartographic Information system

second method uses pre-information. If vectors depicting the lineage of dikes are available the raster model can selectively be investigated. Using vector information it is no longer necessary to examine the whole terrain model for identifying dikes. In this paper only the first approach mentioned is discussed. In all cases an interpolation of the base heights of the detected dike bodies has to follow. The final step to establish the *Digital Dike Model (DDM)* is to calculate the actual crest levels. This leads to a raster based model which can be used as the basis for ongoing analysis.

Object recognition in DTM are based on the finding of sudden level leaps. If a given difference threshold $\Theta$ is exceeded the pixel is considered as an elevated object (see [10]). The further editing will appear as follows: The elevated flagged pixels are being erased from the terrain model and form a mask of non-ground points *(Non-Ground Model)*. Afterwards the remaining holes in the Digital Terrain Model must be filled with approximated ground height values, which have to be interpolated from the surrounding edge pixels. The actual crest level values can be obtained by subtracting the interpolated surface from the original elevation data.

## 6.3.2 Adapted Filter method

The principle of detecting elevated objects is to examine the surface level differences within a certain neighbourhood. If the difference between a pixel and its neighbouring minimum exceeds a defined threshold it is being marked as belonging to an elevated object.

It is useful to apply combinations of several filter sizes and threshold settings. The result of the filtering is being cleaned and will be used for building objects which are classified by shape parameters. As a result the actual crest level can be directly derived.

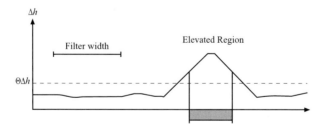

**Fig. 6.1** Detection of elevated objects in DTM (inspired by [10]).

The principle of the filter method is illustrated in Fig. 6.1. Each pixel in the given DTM scene will be compared to its neighbourhood minimum whose

# 6 Dike modelling using raster DTM

extent is indicated by *filter width*. If the difference exceeds the threshold $\Theta_{\Delta h}$ the pixel will be flagged as elevated.

The filter method described in [10] was already applied by the author [7]. For the use with one of the above-mentioned high resolution DTM it had to be adapted and realised in a programming language available at the IOER[11]. The programme allows the user to adjust any options concerning the appliance of the filter to adequately fit the current conditions of the investigated DTM scene.

To detect elevated objects of different dimensions it is useful to combine the use of several filter sizes in combination with different threshold values. It can easily be seen that bigger filter widths combined with higher $\Theta$-values will detect large elevated objects that have a relatively wide extent while a small filter size with lower thresholds would yield smaller objects of little height. In order to detect most of the elevated objects a combination of two option settings should be applied.

Height thresholds and filter sizes

The following facts have been used to preset the thresholds for discriminating elevated from non-elevated pixels:

**Table 6.3** Flood alert levels in relation to dike height. Source: [13]

| Alarm level | Event/characteristics for declaring |
|---|---|
| 1 | Bankfull riverbed, little overflowings occurring here and there. |
| 2 | Beginning overflow, water level reaches dike base. |
| 3 | Water level reaches half crest level, beginning dike defence measures if necessary. |
| 4 | Dike-overtopping threat, endangered dike stability. |

Considering the parameters in Table 6.3 the dike's crest levels can be estimated as the difference of the water levels that belong to Alarm levels 4 and 1. Table 6.4 indicates the Alarm levels of selected water level gauging station along the Elbe River.

The differences between the values of alarm levels 4 and 1 in Table 6.4 suggest that dikes rise at least 2 m above the surrounding surface. Therefore the threshold $\Theta_{\Delta h}$ should not be bigger than 2 m.

In [7] dike width values were detected ranging from 12 m to 25 m that can be considered as indicatory values. The filter has to ensure that at least one ground point is inside the search window while passing over the DTM raster. Assuming a maximum dike width of 25 m the filter size then has to be 13 m.

---

[11] Leibniz Institute of Ecological and Regional Development, Dresden

**Table 6.4** Flood alarm levels of selected Elbe gauges. Source: [12] and [6]

| Gauging station (River km from Czech border) | Alarm level [cm] | | | |
| --- | --- | --- | --- | --- |
| | 1 | 2 | 3 | 4 |
| Schöna (2) | 400 | 500 | 600 | 750 |
| Dresden (55.6) | 350 | 500 | 600 | 700 |
| Torgau (154.1) | 580 | 660 | 720 | 800 |
| Wittenberge (453.9) | 450 | 550 | 630 | 670 |

For dikes do not elevate abruptly out of the surface, a smaller filter size with smaller threshold is to be applied in order to detect the lower parts of the dike slope. The application of the second filter can be reduced to the regions neighbouring to the pixels that have been detected by the larger filter window. Therefore these regions will be buffered and used as mask for the second filtering.

Object Selection

Some regions in the DTM might be detected which are not dikes or embankments. These include single pixels or small pixel groups which do not belong to any dike body. Therefore the Non-Ground Model have to be classificated if its objects can be dikes or not. That's why the recognised objects are described by form parameters:

*Direct Parameters*

are basic geometrical attributes which are calculated directly from the raster data:

- *Area*: The Area $A$ is calculated by cumulating the count of pixels that form one object. This number is depending on the spatial resolution of the Digital Terrain Model used for dike detection.
- *Perimeter*: The perimeter $P$ is formed by the outline of the surrounding pixels of one object and is therefore a multiple of the pixel width.

*Indirect Form Parameters*

are calculated from Area and Perimeter and describe the object shape independent from the actual object size:

- *Form Factor*: The Form Factor $F$ is defined by the ratio of the squared Perimeter $P$ and Area $A$: $F = P^2/A \geq 4\pi$. It describes the figure's deviation from the circle of which the Form Factor is $F_0 = (2\pi r)^2/2\pi r^2 = 4\pi$.

- *Contour Index*: The Contour Index $C$ is given by the ratio $C = P/P_C \geq 1$, where $P$ is the object perimeter and $P_C$ is the circumference of an equal-area circle. For circular objects the Contour Index $C_0 = (2\pi r)/(2\pi r) = 1$.

The parameters $F$ and $C$ are related and can be converted by $F = 4\pi C^2$. Hence it is possible to describe the found objects by just one of the indirect parameters.

Once the objects have been build and their form parameters have been calculated they are classified as possible dikes or non-dikes by thresholding the two parameters Area and Form Factor.

1. One object must consist of a minimum number of pixels, that it has to cover a minimum area to be considered as possible dikes.
2. The Form Factor has to be greater than a certain threshold value which is typical for elongated features like dikes.

### 6.3.3 Filter appliance

The application of the two filters will lead to two *Non Ground Models* $\text{NGM}_1$ and $\text{NGM}_2$. Afterwards both models can be merged to produce the *Combined Non-Ground Model* $\text{NGM}_{comb}$.

Fig. 6.2 shows the consecution of the filtering. The analysed scene has dimensions of 2 000 m on both axes.

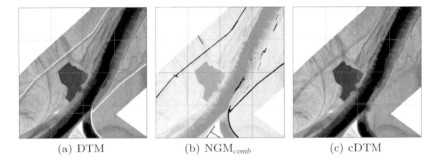

(a) DTM      (b) $\text{NGM}_{comb}$      (c) cDTM

**Fig. 6.2** Demonstration of DTM filtering

The different stages of filtering are shown in the sub-pictures (a) to (c) of Figure 6.2:

(a) Original DTM. Dikes on both sides of the river are well exposed.
(b) Detected object areas for ground point interpolation after cleaning filter result.
(c) DTM after removing dikes (cDTM).

## 6.3.4 Ground level interpolation

Removing the detected objects includes the calculation of the underlying base heights in order to fill the masked pixels in the DTM. This can be accomplished quite easily by using the triangulation technique. The pixels bordering the masked areas function as mass points for the triangulation. The masked areas are of longish shape. So the distance which has to be filled by interpolation is not too far and the result of the triangulation can be considered as reliable. Figure 6.3 shows the principle of triangulating.

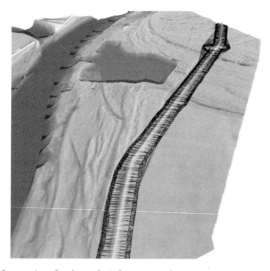

**Fig. 6.3** TIN creation for base height interpolation

The resulting model is a Digital Terrain Model cleaned from dikes. For differentiation from the original version it should be abbreviated with cDTM (see Fig. 6.4).

Once all objects (dikes) have been erased and replaced by their ground heights the final *Above Ground Model* (AGM) can be calculated by subtracting cDTM from DTM: $AGM = DTM - cDTM$.

This model can be considered as the desired DDM (see 6.3.1). At this state it forms a raster where each pixel value represents the above ground level. To make clear it is of raster format the abbreviation should be extended to *DDM-R*. The single dikes are surrounded by no-data regions[12].

The focal maxima within the distinct objects give information on the real object height. Dike lineage can be derived by thinning and vectorising the data. The stored features of the polylines will hold attributes which indicate

---

[12] Depending on the further analysis it might be useful to use zero values to fill non-elevated areas instead of no-data.

6 Dike modelling using raster DTM

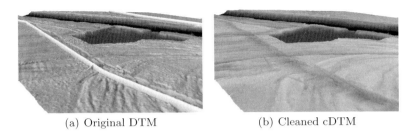

(a) Original DTM   (b) Cleaned cDTM

**Fig. 6.4** DTM scene before and after detecting and removing dikes

important geometrical information as object heights and widths. This dataset is referred to as *DDM-V*. Because it consists of vector data it can easily be edited and manipulated with common GIS methods. Changes can be reconverted into raster format for further use in flood simulations.

## 6.4 Outlook

The technology described in this paper offers a quite effective method to extract dikes from high resolution DTM data. This is useful especially if no geometrical data concerning the dikes is available. Another aspect is to verify given information on crest levels, e.g. on medium and large scale[13] topographic maps. The procedures have been programmed in *Arc Macro Language (AML)* scripts. A user friendly version for ArcGIS is intended which includes dike detection and dike removal as well as the establishment of new dikes. Therefore the user will have to digitise the new dike lineage in vector format. The vectors' attributes include information on crest levels and dike widths and/or slope ratios from which a new dike can be modelled and merged with the underlying DTM.

This will provide a useful toolbox to estimate the effect of dike building measures or dike breaches on the flood risk of a certain area. The main problem will remain the limited availability of high resolution DTM which is the most important pre-condition of object recognition.

---

[13] Medium scale: $\geq$1:10 000–>1:100 000, large scale: >1:10 000

# References

[1] Boehlich MJ (2003) Tidedynamik der Elbe. J Mitteilungsblatt der Bundesanstalt für Wasserbau, 86:55–60. http://www.baw.de/vip/publikationen/Mitteilungsblaetter/mitteilungsblatt86/boehlich.pdf (2007/03/22)

[2] Bundesamt für Kartographie und Geodäsie (2006) Digitales Geländemodell für Deutschland DGM-D. http://www.geodatenzentrum.de/docpdf/dgm-d.pdf (2006/09/12)

[3] Frerichs S and Hatzfeld F and Hinzen A and Kurz S and Lau P and Simon A (2003) Sichern und Wiederherstellen von Hochwasserrückhalteflächen. Umweltbundesamt, Berlin

[4] Jährling KH (1994) Bereiche möglicher Deichrückverlegungen in der Elbaue im Bereich der Mittelelbe – Vorschläge aus ökologischer Sicht als Beitrag zu einer interdisziplinären Diskussion. In: Guhr H and Prange A and Punčochář P and Wilken RD and Büttner B (eds) Die Elbe im Spannungsfeld zwischen Ökologie und Ökonomie: Internationale Fachtagung in Cuxhaven vom 8. bis 12. November 1994 / 6. Magdeburger Gewässerschutzseminar. Teubner, Stuttgart & Leipzig

[5] Jährling KH (1998) Deichrückverlegungen: Eine Strategie zur Renaturierung und Erhaltung wertvoller Flußlandschaften? Staatliches Amt für Umweltschutz, Magdeburg

[6] Koll C (2002) Das Elbehochwasser im Sommer 2002: Bericht des Landesumweltamtes Brandenburg im November 2002, vol. 73 of Fachbeiträge des Landesumweltamtes. Landesumweltamt Brandenburg (LUA), Potsdam

[7] Krüger T and Buchroithner M and Lehmann, F (2005) GIS-gestützte Kartierung hochwasserschutzrelevanter topographischer Informationen mit HRSC-Daten. J Photogrammetrie-Fernerkundung-Geoinformation 2/2005:129–133

[8] Leibniz Institute of Ecological and Regional Development (2006) Change and management of risks of extreme flood events in large river basins – the example of the Elbe River. http://www.veris-elbe.ioer.de/

[9] Loat R and Meier E (2003) Dictionary of Flood Protection. Haupt, Bern

[10] Mayer S (2003) Automatisierte Objekterkennung zur Interpretation hochauflösender Bilddaten. Thesis, Humboldt University of Berlin

[11] Merz B (2006) Hochwasserrisiken. Grenzen und Möglichkeiten der Risikoabschätzung. Schweizerbart, Stuttgart

[12] Sächsisches Landesamt für Umwelt und Geologie (2002) Hydrologisches Handbuch – Teil 1 – Pegelverzeichnis Stand Januar 2002. http://www.umwelt.sachsen.de/de/wu/umwelt/lfug/lfug-internet/veroeffentlichungen/verzeichnis/Wasser/PVZ_Internet.pdf(2007/03/07)

[13] Sächsisches Landesamt für Umwelt und Geologie (2007) Bedeutung der Alarmstufen. http://www.umwelt.sachsen.de/de/wu/umwelt/lfug/lfug-internet/wasser_9562.html (2007/03/26)
[14] Schmidt M (2000) Hochwasser und Hochwasserschutz in Deutschland vor 1850. Eine Auswertung alter Quellen und Karten. Oldenbourg, Munich
[15] Voigt M (2005) Hochwassermanagement und Räumliche Planung. In: Jüpner R (ed) Hochwassermanagement. Magdeburger Wasserwirtschaftliche Hefte, vol. 1. Shaker, Aachen
[16] Willi HP and Eberli J (2006) Differenzierter Hochwasserschutz an der Engelberger Aa. J tec21 36:4–7. http://www.tec21.ch/pdf/tec21_3620063740.pdf (20070/03/07)

# Chapter 7
# Development of a Web Geological Feature Server (WGFS) for sharing and querying of 3D objects

Jacynthe Pouliot, Thierry Badard, Etienne Desgagné, Karine Bédard, and Vincent Thomas

**Abstract**

In order to adequately fulfil specific requirements related to spatial database integration with 3D modeling tools, this paper presents the development of a generic and open system architecture called Web Geological Feature Server (WGFS). WGFS provides direct access through Web services to 3D geological models. WGFS is based on a three-tier architecture: a client (Gocad), an application server (Apache Tomcat and Deegree) and a DBMS (MySQL). This architecture takes advantage of standard-compliant spatial applications such as WFS and GML standards stemming from OGC and spatial schema from ISO TC/211-Geographic Information. Before introducing the architecture and motivations of some geoinformatics choices, we will remind some important issues that have to be taken into account when such development is planned.

## 7.1 Introduction

By interpreting field observations and integrating available geophysical and geochemical data to determine the 3D configuration of rocks, their temporal relationships and causal processes, geoscientists are now frequently using GeoModels[1](Bédard 2006; Fallara et al. 2006; Jessell 2001; Kessler et al. 2005; Mallet 2002;). GeoModels, thanks to various modeling software, can be geometrically designed and investigated visually or by performing quantitative

---

Centre for Research in Geomatics and Geomatics Department,
Laval University,
Quebec, Canada
jacynthe.pouliot@scg.ulaval.ca

[1] Geological Models

analyses (Lee 2004; Pouliot et al. 2007; Zlatanova et al. 2004). A great variety of 3D geological modeling tools exists such as EarthVision, Gocad, GEMS and Vulcan²³ model construction, they present important limitations. For example, and even if they store geospatial data, they generally do not manage coordinate projection systems, which could hinder potential integration of various geospatial data sources. They are not very accessible as they require advanced knowledge in modeling and in computer science which can only be acquired after several hours of training. They are also not easily extensible and interoperable since they are mainly closed systems with their own development language, are not fully compliant with standards and are proposing few import and export capabilities (file formats, database connectivity ... ). Moreover, the modeling tools provide few functions to support advanced selection and query of spatial objects. Today with the great emphasis on database's implementation (and more specifically geospatial databases), several governmental agencies look at such system that could improve efficiency of data management, exchange and analysis. In addition, online services are now seen as a must and promoting Web based development will contribute to this end.

Geospatial information system (GIS) can also be evaluated for 3D geospatial modeling and management. GIS and related spatial database management system (DBMS) provide several capacities for spatial and non spatial data management, transformation, and analysis while controlling the organization, the maintenance, the storage and the retrieval of data in a database. However, no GIS currently propose the concept of volumetric (solid) objects. They are limited to the presentation of 2.5D models, like digital elevation model (DEM), where we can drape raster data onto the model and extrude some regular objects such as buildings. This considerably limits the capabilities of such system to manage 3D (volume) Geomodels.

From this report and in order to develop an appropriate 3D GIS for exploration assessment and efficient management of mineral resources, a research team involved in a GEOIDE⁴ project propose the extension of a 3D modeling tool by enhancing its functionalities to perform spatial data selection and query over the Web. It consists of accessing 3D GeoModels organized in a coupled environment integrating a 3D modeling tool, a 3D spatial DBMS and a server. It tries to take advantage of Web-based and standard-compliant spatial applications (such as WFS and GML standards stemming from OGC⁵

---

[2] Respectively: http://www.dgi.com/earthvision,http://www.gocad.org/, http://www.gemcomsoftware.com/products/GEMS, http://www.vulcan3d.com/.
  However, when we consider these modeling tools, we note that, even if they are powerful for 3D

[3] 3D (objects or model) will refer to objects placed in a 3D universe (x,y,z), no matter which geometric dimension the object has. To mark the difference with the geometric dimension of the object, we will refer to punctual, linear, surface or volumetric object.

[4] GEOIDE (GEOmatics for Informed Decision), http://www.geoide.ulaval.ca/

[5] Open Geospatial Consortium, http://www.opengeospatial.org/

and spatial schema from ISO TC/211-Geographic Information). Theses concerns also drive some software selection towards open source solutions. The article will present these experiments of designing a Web Geological Feature Server (WGFS). WGFS is based on a three-tier software architecture: a client (Gocad), an application server (Apache Tomcat and Deegree) and a DBMS (MySQL). We will finally discuss the first and future experiments in the context of coupling a 3D geological tetrahedral model and SIGEOM [6] database.

## 7.2 Review of 3D GIS development and standards

Since the attempts made by Raper in 1989, several authors have been addressed the development of 3D GIS. More recently, we can first report the work of Apel and Frank (Apel 2006; Frank et al. 2003). They proposed a 3D GIS framework based on existing 3D Geomodeling theory and software, an integrated data model and a XML database server. There works were specifically designed for geological observation data and geomodel construction. This development was of particular interest for us because closed needs were addressed mainly related to 3D Geomodels and Web based information system. Personal communications in 2006 with Apel and Frank demonstrated us that even if promising concepts and ideas were at that time proposed, some constraints were not solved and the proposed system is no more available. For example, XML and XQuery were not specifically designed for storing and querying geospatial data and the application server (3DXApps) was only supporting surfaces objects, no solids, which is a crucial constraint.

Van Oosterom, Arens and Stoter (Arens et al. 2005; van Oosterom et al. 2002) from Delft University of Technology, the Netherlands are also quite involved in the development of 3D GIS components. Among others, they proposed, according to OGC 'Simple Feature Specification for SQL', some concepts helping the design and the implementation of a real 3D primitive in a spatial DBMS. As they stated, this implementation is a first experiment where a spatial DBMS (Oracle Spatial 9i Spatial) supports a 3D primitive (Stoter and van Oosterom, 2002). Of particular interest, they cover aspects such as validation of 3D functions, 3D indexing (and benchmarking) and 3D visualization. There works and because it was not a priority, do not link spatial DBMS with 3D Geomodeling tools and are not based on Web protocols.

Zlatanova has presented in 2000 the design of a 3D topological model specifically designed for Web-oriented query and visualisation (Zlatanova, 2000 and Zaltanova et al. 2004). Her works was particularly of interest regarding the optimisation of data transfer over the Web and strategies to assemble such coupled system. She developed a GUI interface, used CGI scripts

---

[6] Système d'information géominière du Ministère Ressources naturelles et Faune du Québec, http://sigeom.mrnfp.gouv.qc.ca/

(on the Apache server side), VRML and HTML languages and MySQL as a DBMS. Her thesis even if it was in 2000 and specially oriented for municipal applications was a valuable source of information for 3D GIS design and applications.

A specific group of researchers from the Earth Sciences Sector of Natural Resources Canada and Mira Geoscience are also working to extend Gocad software (Sprague et al. 2006). A query framework with application to mineral exploration is proposed supporting for example proximity query, property query (only numeric attribute), feature query (dome, depression, curvature), and intersection query. This framework is interesting but actually restricted to Gocad environment.

Breunig presented GeoToolKit (Balovnev et al., 1997; Breunig, 1999 and 2001), an open 3D database kernel system implemented by using the object-oriented database management system ObjectStore. This system provides specific algorithms, spatial access methods and visualization for 3D geosciences modeling. GeoToolKit contains C++ class library extensible and offers the representation and the manipulation of simple and complex 3D spatial objects. Even if this architecture is open and extensible and can be assembled into a ready-to-use application by geoscientists with little programming experience, it still represents a new environment to learn and this could limit its capacity to penetrate this community of specialists. In Shumilov and Breunig (2000), they tried to connect Gocad with GeoToolKit/ObjectStore ODBMS and a CORBA/ODBMS adapter. GeoToolKit was not specially designed for web based applications, but this coupled environment was particular interesting as it offers great advantage of using a common database which gives the opportunity to geoscientists to perform a cooperative work under one 3D GeoModel.

If we now have a close look on 3D modeling from the standardization point of view, we can resume it as follow. The International Organization for Standardization (ISO), particularly ISO/TC 211 – Geographic information/Geomatics, proposes a spatial schema (ISO 19107 2002) that describes geometric objects in universes having up to three dimensions. Geometric primitives used to build objects are the points, curves, surfaces and solids. The solids are considered as the basis of the 3D geometry of the spatial schema and they are defined by a boundary which is an envelope composed of surfaces.

On its side, the Open Geospatial Consortium (OGC) works on the edition of specifications with the objective to make frequently used spatial data and services accessible and interoperable. The 'Simple Feature Specification for SQL' (OGC SFS 1999) defines an SQL schema which supports the storage, querying and updating of simple geospatial features. It enables the definition of punctual, linear and surface geometry but no volumetric or solid object are described. Another specification stemming from OGC is the 'Geography Markup Language (GML) Encoding Specification' (OGC GML 2004) that allows the modeling and the storage of geospatial information through a standard XML (Extensible Markup Language) encoding. GML 3.x now supports solids as geometric objects but the previous version (2.x and older) only supported points, curves and surfaces. GML 3.x is thus closely related to the ISO spatial schema as most of the objects present in the schema have a corresponding object in GML. However, it does not mean that an application using GML 3.x will offer the management of volumetric object. In fact, GML 3.x defines a way to encode information that allows 3D data but it is always possible to encode 2D data with GML 3.x; presently, very few applications produce complete volumetric exchanges using GML 3.x. If GML is a generic language, we could find for geoscience applications, GeoSciML, a GML application language for geoscience in which a suite of feature types based on geological criteria (units, structures, fossils) or artefacts of geological investigations are proposed. GeoSciML helps the representation of geoscience information associated with geologic maps and observation, actually including no volume.

The OGC also proposes a wide variety of protocols and standards for service exchange between applications, based on a client/server approach. These Web services facilitate interoperability between diverse tools working on different platforms. The 'Web Feature Service' (OGC WFS 2004) proposes a protocol to query/insert/delete/update distributed geospatial databases through a standardized and neutral interface. Technologies which are used to effectively store the data are hidden by such interface: responses and requests exchanged between the client and the service are encoded in XML and geospatial data are delivered to the client over the Internet in GML 3. All these notions, spatial schema, SFS, GML and WFS, are important as they represent a standard way to manage data and operators in geospatial tools. The standardization organisms are only starting to revise their standards and specifications in order to better integrate 3D notions. GML 3.x is a good example of such an evolution, there is still work to be done to bring the other standards to a complete tridimensional stage.

At the time we made an inventory of GIS/CAD/DBMS capabilities to perform volumetric data manipulations (Pouliot et al. 2006), we notified some elements. Although GIS and spatial DBMS commonly allow the data storage of X, Y, Z coordinates, the majority only enables the management of features with a point, a curve or a surface geometry or a possibly heterogeneous collection of the previous types such as proposed by the SFS for SQL (OGC SFS 1999). Last summer, Oracle announced Oracle 11g in which they will support three dimensional data types (a solid); geometry consists of multiple surfaces and completely enclosed in a three-dimensional space. Very few GIS and spatial DBMS software propose models of the raster type (i.e. voxels), except perhaps MapInfo's Engage 3D. The 3D spatial analyses are generally limited to visibility, slope (gradient/orientation) and convexity (profile/plane) computations. Finally no GIS or DBMS allow 3D topological analyses.

Computer-aided design software (CAD) offer a wide range of geometric models (wireframe, surface, volume) but from a volumetric point of view, they are not actually linkable to a database. Solid objects are usually represented as a collection of surfaces forming closed volumes such as B-Rep. Other modeling's software in the CAD category have been developed for specific applications. For example, Gocad software is a 3D modeling tool specifically designed for earth sciences. 3D representation in Gocad is based on some 3D structures (B-Rep, voxel and tetrahedral models).

## 7.3 Web Geological Feature Server (WGFS)

From this review and having in mind the specific requirements of the governmental agency[7] partner in the GEOIDE project, we proposed a Web Geological Feature Server (WGFS). WGFS is a generic and open coupled environment integrating an existing 3D modeling tool, a 3D spatial DBMS and a server while take advantage of Web-based and standard-compliant spatial applications.

At present, 3D Geomodels of our partner are built in Gocad software with the help of information such as attribute and spatial data coming from an external geospatial database (GeoDB). They are quite familiar and pleased with the performing 3D modeling tool and do not want to change their modeling procedure. Nevertheless, once the 3D Geomodel is built, they have no mean, mainly because the spatial DBMS in which the GeoDB is stored does not manage volumetric objects, to return back this new 3D information in the GeoDB and share this valuable knowledge with a larger community of users. The first issue in this context is then to extend the GeoDB to be able to manage volumetric object. Then, and because we wanted to take full advantage of the 3D modeling tool and not having to develop every 3D functions for data handling, the system will be able to interactively link a performing 3D modeling software with a GeoDB through functions of:

- being informed of available 3D Geomodel stored in a GeoDB;
- data selection and querying the GeoDB over the Web;
- importing selected descriptive (qualitative and quantitative attributes) and geometric data (point, line, surface and solid) into the 3D modeling software;
- modification and creation of new spatial features (actual functions of the 3D modeling tool), while having access to descriptive data;
- returning back the new 3D feature (attribute and spatial data) into the GeoDB.

WGFS was developed with the objective of linking 3D modeling software to a DBMS because this kind of system ensures a reliable, fast and multi-user

---

[7] Québec Ministère Ressources naturelles et Faune (MRNF).

access to data. In order to provide interoperable capabilities to the system, geospatial data access and delivery is performed through an interoperable Web Protocol. The interoperable approach we adopted relies on ISO and OGC standards for the definition of geometric objects and spatial operators (ISO 19107 – Spatial schema, OCG GML) as well as for the online services used (OGC WFS). WGFS allows thus to provide access, via the Web, to geospatial data and operators located on a server for multiple users who do not have the same modeling tools.

In accordance with our partner's priorities, first tests and experiments should be performed with the 3D modeling software Gocad, a complex 3D GeoModel of the Porcupine-Destor region (north-western Québec, Canada) containing surfaces and tetrahedral solids built in Gocad© by Fallara et al. (2004) and partial information (mainly attributes of lithology and stratigraphic classes) extracted from the geospatial database SIGEOM.

## 7.4 System architecture and implementation

WGFS is based on a three-tier software architecture composed of a client (Gocad), an application server (WFS compliant software) and a database management system (MySQL). This type of architecture allows many clients to access and share the same data because data are centralized in a common 'database'. Each user can access data with the help of a client-application linked to the server via a network, either local or Internet. The application-server enables the connection between the client and the DBMS and it hosts the components which hold the system's applicative logic where data processing is performed. In addition, such a three-tier software architecture facilitates the deployment and maintenance of the system: the application does not require to be installed on all client's computers, only the server has to be updated when an upgrade or a change in the application is performed. The architecture of the WGFS is presented in figure 1.

As, at present, DBMS do not support volumetric geometries, two alternatives were examined to store 3D objects. We could have stored our geological models in a 3D topological data structure adapted to geological context such as the GeoTEN designed by Lachance (2005) or TEN (Tetrahedral Network Structure) by Pilouk (1996). For simplicity and efficiency reasons it has been decided to directly store geological models as GML documents in a relational table of the DBMS. Storing GML documents in a native XML DBMS could also have been possible but at present they are not as fast as relational DBMS. The well-known and widely used DBMS MySQL has been chosen for performance, deployment rapidity and simplicity reasons. The geometry of the objects is stored as GML objects (text) and their attributes are stored as fields in the relational table. To accelerate spatial queries, a pseudo spatial index was implemented by storing the 'bounding box' (spatial envelope) for

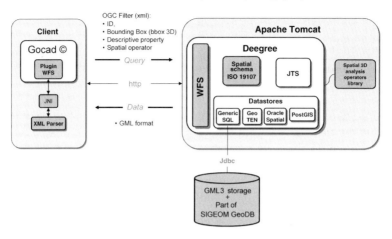

**Fig. 7.1** Architecture of the Web Geological Feature Server (WGFS) prototype

each GML object. It allows a faster retrieval of the objects in the table. Figure 2 illustrates the structure of the relational table which stores the GML documents. Even though this storing strategy is not optimal for data management, it enables the rapid and easy implementation of the 3D models in the database.

| id | geometry | minx | miny | minz | maxx | maxy | maxz |
|---|---|---|---|---|---|---|---|
| 1237 | <gml:Solid | 64509 | 45674 | -600 | 86768 | 556456 | 210 |
| 1248 | <gml:Solid | 56743 | 45767 | -600 | 45676 | 456654 | 121 |

**Fig. 7.2** Example of the storage of the bounding box coordinates of objects in the relational table of the DBMS

GML version 3 allows 3D objects to be represented with composite geometries. In GML 3.x, objects are thus composed of a collection of homogeneous primitives such as compositeCurves, compositeSurfaces and compositeSolids. It is then possible to construct a tetrahedral model as a collection of compositeSolids where each solid is a tetrahedron composed of surfaces. Figure 3 illustrates the composition relationships which occur in a 3D model: it is composed of one to many compositeSolids that are themselves composed of one to many tetrahedrons. This model is compliant with Gocad data model where a solid (referred as TSolid) can be modelled as a specialised tetrahedron.

7 Development of a Web Geological Feature Server 123

**Fig. 7.3** Composition of a 3D model in GML 3.x

The application server is the core of this three-tier system; it is where geospatial processing is performed. This part of the architecture is based on an open source WFS implementation, Deegree from the University of Bonn in Germany. Deegree is a Java Framework offering the main building blocks for Spatial Data Infrastructures. Deegree has been chosen because its entire architecture is developed using standards of the Open Geospatial Consortium (OGC) and ISO/TC 211 (ISO Technical Committee 211 - Geographic Information/Geomatics). Especially, it allows the handling of geospatial objects in compliance with the ISO 19107 – Spatial schema standard (ISO 19107 2002). Nevertheless, some additions have to be performed to the source code of Deegree (which is possible because it is an open source solution) in order to enable the support of 3D objects. Indeed, Java classes allowing the modeling of 3D objects were not available (empty classes) when we have implanted the WGFS system. For instance, some GM_Solid subclasses have been coded in order to enable the generation of truly 3D GML 3.x messages. Remember, it is not because a system can produce GML 3 compliant messages that all these messages necessarily contain 3D geospatial objects! GML 3 is just an encoding. Our modified version of the Deegree WFS has been deployed through the well-known and widely used Servlet and JSP container, Tomcat from the Apache Foundation.

In addition, we have designed and developed a specific interface (a DataStore) in order to enable the connection with the DBMS. Deegree already offers different DataStore to connect to commonly used geospatial DBMS (Oracle, PostGIS, . . . ), but as mentioned previously, none of these DBMS allows the storage of 3D objects. The DataStore we developed enables the consistent processing of queries dealing with 3D objects and sent to the WGFS system by the client (selection, insert, updates). In order to be able to process all kind of selection queries, spatial operators (union, intersection, etc.) should have been extended for 2D to 3D. At present, no API (Application Programming Interface) fully offers such 3D processing capabilities. For experiments purposes, we have only extended the basic BBox (Bounding Box) operator. It allows us to provide responses to selection queries as 'give me all objects contained in the following 3D bounding box'.

Finally, the client part of the architecture is composed of the client modeling software and a WGFS plug-in. Because our partner was using Gocad© modeling software and thus it is a valuable 3D geological modeling tool, our system architecture was built with Gocad© as a client. In order to allow Gocad users to build and send queries and then receive results from the server, a plug-in and a data converter has been designed. The plug-in provides a standard interface for those operations. Each operation is performed in a transparent way for the user. The client is linked to the application server over a network and the query system between both tiers is implemented following the WFS specification (OGC WFS 2004). When a query is sent to the server from the client-end, the data returned as the result of the query are in GML format. The XML Parser receives the GML document and then translates it to Gocad object format and vice-versa. The Java Native Interface (JNI) is used to enable the data transfer between the XML parser (in our case in Java) and the Plugin in C++. These GML documents contain spatial as well as descriptive data. However, it is not usually possible to query descriptive information in CAD systems. We have developed a tool in order to be able to query those descriptive data in Gocad. It is thus possible to directly access the attributes of each object once they are imported in Gocad.

The architecture has the advantage that other 3D modeling software packages could easily be linked to the application server with only the development of a new plug-in for the translation of GML standard documents into software formats.

## 7.5 WGFS in action

As quoted previously, the prototype allows multiple clients to access and share data located on a server. Diverse operations are performed in order to provide the client with geological data. First, the user sends a query through the Gocad plug-in to know what data are available on the server. A WFS *getCapability* query is thus sent to the server which then relies on its DataStore to get the information. This information is subsequently sent to the client who receives a list of all the available data in XML format. From there, the user can select the data he/she wants to get into his/her system, again with the help of the Gocad plug-in. The client selects a dataset and he can optionally specify constraints for the importation such as a *bounding box*(tridimensional envelope) or selection by attributes characterizing the data. The figure 4 presents the WGFS query dialog to get available 3D GeoModels and perform a selection query based either on a portion of the GeoModels (BBOX) or on values of descriptive attributes stored in the GeoDB.

When the user sends the query, the plug-in transmit a *getFeature* WFS query to the application server. The server then sends an SQL query to the database containing the data to retrieve the information asked by the client. Data are consequently extracted from the database and sent to the client in GML documents. The figure 5 illustrates an example of SIGEOM GeoDB content.

From there, the plug-in translates the GML exchange into the Gocad 3D data structure. The next figure illustrates a group of tetrahedral displayed in

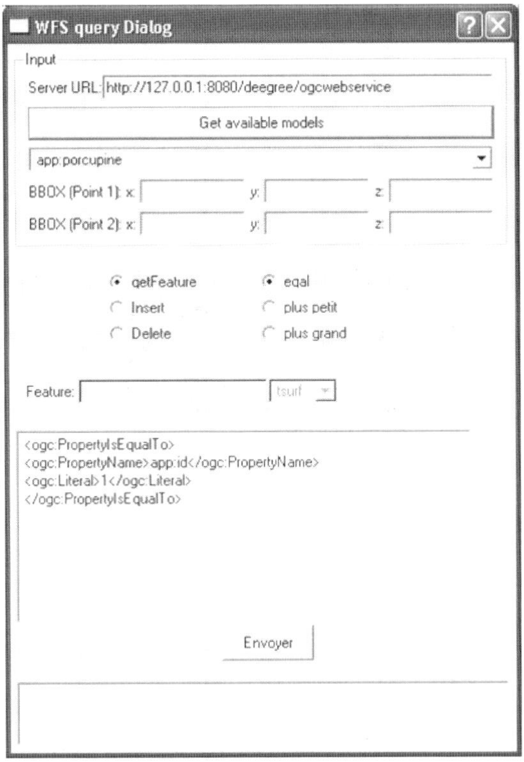

7 Development of a Web Geological Feature Server 125

**Fig. 7.4** WGFS plug-in interface for Gocad. A getFeature query is built to get the complete 'porcupine' odel in Gocad

| ID | geometry | lithology | litho desc | stratigraphy | stra desc |
|---|---|---|---|---|---|
| 3 | <gml:Solid xmlns:gml="http:/I3A | GABBRO | [arch]3 | | Archéen |
| 5 | <gml:Solid xmlns:gml="http:/V3B | BASALTE | [arch]dg1 | | Archéen, Formation de Deguisier |
| 2 | <gml:Solid xmlns:gml="http:/I3A | GABBRO | [prot] | | Protérozoïque |
| 1 | <gml:Solid xmlns:gml="http:/V2J | Andésite | [arch]dg4 | | Archéen, Formation de Deguisier |
| 4 | <gml:Solid xmlns:gml="http:/V2J | Andésite | [arch]dg4 | | Archéen, Formation de Deguisier |

**Fig. 7.5** Example of SIGEOM GeoDB descriptive data

the Gocad interface with descriptive data made available by the Interrogation (I) button specifically added in the Gocad menu.

Up to now, the WFS *getFeature* and the transactions *insert, delete* and *update* operations have been implemented. Some tests and experiments were made with a complex 3D GeoModel (11.4km x 4km x 1km). We used a small portion (2.5km x 1.5km x 900m) of the complete GeoModel to build 267 tetrahedral solids (see figure 7).

A database containing the solids and their attributes (lithology and stratigraphy) derived from the SIGEOM database was also built. This allows a user

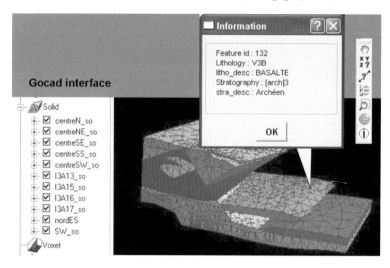

**Fig. 7.6** Example of 3D Geomodel retrieved in the WGFS and displayed in the Gocad environment with available identification data and descriptive attributes coming from the SIGEOM GeoDB (see Information window)

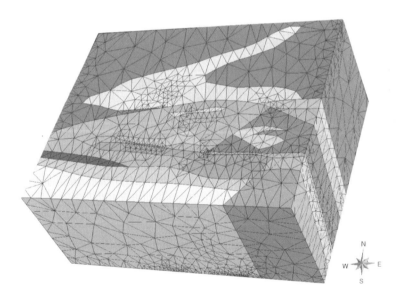

**Fig. 7.7** Portion of the 3D GeoModel of the Porcupine-Destor model (Fallara et al. 2004) used to test the WGFS

to build a query with filter expression based on descriptive data such as lithology = 'V3B' AND stratigraphy = '[arch]3'.

In order to validate our solution, we implemented a prototype version of the WGFS in the office of the Québec MRNF. Preliminary results demonstrated good response from geologists and spatial data managers. It helped us to demonstrate the importance of using such a geospatial database approach, thus improving the efficiency of data management, exchange and analysis. Next steps will concern testing the performance of such system.

## 7.6 Conclusion

As we specified in the literature review, proposal for 3D GIS development is not new. Several authors agree on the necessity of having a connection between 3D Geomodels (or 3D modeling tools) and actual spatial database systems. Nevertheless, we did not have access to interconnected 3D geomodeling tools and DBMS, neither on commercial solutions and R&D activities.

As far as we know, the innovative part of our work is to have implemented a system that fully integrates a 3D modeling tool, a spatial DBMS (with 3D primitive), and this especially over the Web (Web services). OGC and ISO/TC211 standards and open solution choices insure portability and easy extensibility. At present, a plug-in has to be installed in Gocad. Other 3D modeling software packages could easily be linked to the application server with only the development of a new plug-in for the translation of GML standard documents to software formats.

Until now, simple 3D primitive (polyhedron), few 3D objects and a pseudo spatial index storing the 'bounding box' of each GML object were chosen to experiment the feasibility of this open and generic architecture. Next steps have to estimate the efficacy and limitations of this architecture. In addition, the integration of some conceptual aspects of our development with the GeoSciML application language would improve the exchange and sharing of geoscience data.

## Acknowledgements

We want to thank GEOIDE Network (http://www.geoide.ulaval.ca/) and NSERC (http://www.nserc-crsng.gc.ca/) which funded the research project that leaded to the ideas presented in this paper. We also want to thank our partners, R. Marquis, F. Fallara and G. Perron respectively from Québec MRNF, Université du Québec en Abitibi Témiscaminque and Mira Geoscience, and finally, Tobias Frank at that time at Technische Universität Freiberg, for helpful comments for Gocad developments.

# References

Apel M (2006) From 3d geomodelling systems towards 3d geoscience information systems: Data model, query functionality and data management. Computers & Geosciences 32: 222-229

Arens C, Stoter J, van Oosterom P (2005) Modelling 3D objects in a Geo-DBMS using a 3D primitive. Computers & Geosciences 31: 165-177

Balovnev, O., Breunig, M., Cremers, A.B., 1997. From GeoStore to GeoToolKit: the second step. Proceedings 5$^{th}$ International Symposium on Spatial Databases, Berlin, Lecture Notes in Computer Science 1262. Springer, Berlin : 223-237.

Bédard K (2006) 3D geological model construction in the standardization era (in French). Masters dissertation, Université Laval, Canada

Breunig M (1999) An approach to the integration of spatial data and systems for a 3D geoinformation system. Computers & Geosciences 25: 39-48

Breunig, M., 2001. On the way to component-based 3D/4D geoinformation systems. Lecture Notes in Earth Sciences 94, Springer, Berlin, Heidelberg

Fallara F, Legault M, Cheng LZ, Rabeau O, Goutier J (2004) 3D model of a segment of the Porcupine-Destor Fault, metallogenic synthesis of Duparquet (phase 2/2) (in French). Ministère des Ressources naturelles et de la Faune du Québec. 3D 2004-01. CD

Fallara F, Legault M, Rabeau O (2006) 3-D integrated geological modeling in the Abitibi Subprovince (Québec, Canada): Techniques and applications. Exploration and Mining Geology 15: 27-41

Frank T, Apel A, Schaebec H (2003) Web Integration of gOcad Using a 3d-Xml Application Server. 23$^{rd}$ Gocad Meeting, Nancy, June 10-11

Gong J, Cheng P, Wang Y (2004) Three-dimensional modeling and application in geological exploration engineering. Computers & Geosciences, 30: 391-404

Jessell M (2001) Three-dimensional geological modeling of potential-field data. Computers & Geosciences 27: 455-465

Kessler H, Lelliott M, Bridge D, Ford F, Sobisch HG, Mathers S, Simon Price S, Merritt J, Royse K (2005) 3D geoscience models and their delivery to customers. Annual Meeting of the Geological Society of America, 3D geologic mapping for groundwater applications. Salt Lake City, Oct 15

Lachance B (2005) 3D topological data structure development for the analysis of geological models (in French). Masters Dissertation, Université Laval, Canada

Larrivée S, Bédard Y, Pouliot J (2005) How to enrich the semantics of geospatial databases by properly expressing 3D objects in a conceptual model. The First International Workshop on Semantic-based Geographical Information Systems (SeBGIS 05), Ayia Napa, Cyprus, November

Mallet JL (2002) Geomodeling. Oxford University Press, New York

OGC SFS (1999) Simple Features Specification For SQL, Revision 1.1, OpenGIS Project Document 99-049, Release Date: May 5

OGC GML (2004) Geographic information – Geography Markup Language (GML). Open Geospatial Consortium, ISO/TC 211/WG 4/PT 19136

OGC WFS (2004) Web Feature Service Implementation Specification. Open Geospatial Consortium, OGC 04-094

Pouliot J, Desgagné E, Badard T, Bédard K (2006) SIG 3D : Où en sommes-nous et quelles sont les avenues de développement ? (in French) Géomatique 2006, Montréal, Oct 25-26

Pouliot, J, Bédard K, Kirkwood D, Lachance B (2007 accepted). Reasoning about geological space: Coupling 3D Geomodels and topological queries as an aid to spatial data selection. Computers and Geosciences.

Raper, J. (Ed.), 1989. Three Dimensional Applications in Geographical Information Systems. Taylor and Francis, London.

Pilouk, M., 1996. Integrated modelling for 3D GIS. Ph.D. Thesis, ITC, The Netherlands.

Shi W, Yang B, Li Q (2003) An object-oriented data model for complex objects in three-dimensional geographical information systems. International Journal of Geographical Information Science 17: 411-430

Shumilov S. and Breunig M (2000) Integration of 3D Geoscientific Visualisation Tools with help of a Geo-Database Kernel, 6th EC-GI & GIS Workshop 'The Spatial Information Society - Shaping the Future Lyon', France, 28-30.

Sprague, K., de Kemp, E., Wong, W., McGaughey, J., Perron, G., Barrie, T., 2006. Spatial targeting using queries in a 3-D GIS environment with application to mineral exploration. Computers & Geosciences 32 (3), 396-418.

Stoter, JE, van Oosterom, PJM (2002) Incorporating 3D geo-objects into a 2D geo-DBMS. In: Proceedings FIG ACSM/ASPRS, Washington DC, USA

van Oosterom P, Stoter J, Quak W, Zlatanova S (2002) The balance between geometry and topology. In: Richardson DE, van Oosterom P (eds) Advances in Spatial Data Handling. Springer, Berlin, pp 209-224

Zlatanova S (2000) 3D GIS for urban development. Ph.D. thesis, Graz University of Technology, Austria

Zlatanova S, Rahman AA, Shi W (2004) Topological models and frameworks for 3D spatial objects. Computers & Geosciences 30: 419-428

# Chapter 8
# Using 3D-Laser-Scanners and Image-Recognition for Volume-Based Single-Tree-Delineation and -Parameterization for 3D-GIS-Applications

Jürgen Rossmann and Arno Bücken

**Abstract**

Today, 2D-GIS applications are standard tools in administration and management. As with all state-of-the-art systems certain environmental parameters were found to be difficult to perceive and convey, such that a novel 3D-presentation of the environment was developed to be more intuitive. The cost-effective graphics performance of current computers enables the move towards 3D-GIS and in turn, more detailed views of the environment. While current forestry-2D-GISs provide information on areas or collection of trees, the next generation of forestry-3D-GISs will store and make information available at level of an individual tree. In this paper we introduce an approach to bring vision technology into the forest. We present a volumetric algorithm based on the well-known watershed algorithm and use it to detect trees in laser-scanner point-clouds and four-channel aerial views. Based on this data, maps and virtual environments, 'virtual forests' are generated which can be used in a 3D-GIS for forest management, disaster management, forest machine navigation and other purposes.

## 8.1 Introduction

In recent years, laser-range-finders and image-recognition algorithms have been continuously enhanced responding to the requirements of modern applications in the fields of automation, robotics and last but not least surveyor technology. Currently, laser-scanners are fast and powerful enough to generate airborne-scans of large areas. LIDAR-sensors can provide up to 150.000 measurements per second at helicopter flight level [1] or 100.000 measure-

Institute of Man-Machine-Interaction, RWTH Aachen
rossmann;buecken@mmi.rwth-aachen.de

ment points per second at airplane flight level. [2] These measurements are then turned into 3D maps of an environment at a resolution of several points per square meter. Combining the laser-scanned 3D maps with high-resolution photos, it is possible to generate true-ortho-photos. [3] In a true-ortho-photo, parallactic distortions have been eliminated; each pixel shows the color information virtually as a photo taken from an infinite distance.

In this paper we will show how to adapt these technologies to the area of GIS in forestry management. We will present an approach to map generation for huge regions on the level of an individual tree using a combination of LIDAR-data and true-ortho-photos. The goal of this development is the creation of a so called 'virtual forest', a virtual world representing a real forest that can be used to administrate biological resources, to navigate through the corresponding real forest and — in the near future — to control autonomous forest machines. Our current test forest already includes about 120.000 trees; more will be added with future flight campaigns.

## 8.2 Data Acquisition

All examples in this paper are based on data recorded in spring 2004 and summer 2005 using a Toposys Falcon II scanner [4]. The data shows the area known as 'Glindfeld', which is located close to Winterberg, Germany. The size of the recorded area was about 82km$^2$. The resolution of the LIDAR data is one point per square-meter, the true-ortho-photos are specified to have a resolution of four pixels per square-meter. In addition, a high density data set of a small part of the area was generated during the 2005 flight session. That data has a LIDAR-resolution of four points per square-meter.

Due to the radius of the laser-point at ground level, approximately 80cm, multiple echoes may occur (Fig. 1). For each laser-pulse the first echo and the last echo of a total of eight measured echoes were recorded. The first echo gives information about the canopy surface (DSM – Digital Surface Model) whereas the last echo is used to generate a topological model of the ground (DTM – Digital Terrain Model).

The point-cloud recorded during the flight was converted into a grid-based representation with a cell-size of 1m x 1m using a Toposys internal software-package while the DTM was filtered to eliminate plants, buildings and other disturbances in a partly automated process. Due to hte impenetrability of dense canopy and of buildings, the resulting DTM shows gaps. A Filled Digital Terrain Model (FDTM) was generated by interpolating the gaps. In a last step a Differential Model (DM, also known as normalized digital surface model nDSM or canopy height model CHM) was calculated as the difference between the DSM and FDTM. The DM gives the height of the crown surface of a tree above the ground. This is the most important LIDAR data set for single tree delineation.

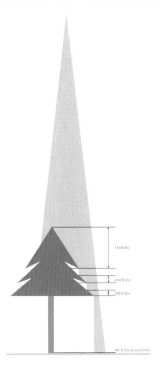

**Fig. 8.1** Multiple Echoes in LIDAR Recording

The image data recorded by the Falcon II 4 channel line scanner was normalized and converted to true-ortho-photos in RGB (red, green and blue channel) and CIR (near-infrared, red and green channel)(Fig. 2).

## 8.3 LIDAR Processing

The LIDAR DM can be seen in two ways: As a three-dimensional model of the canopy or as a two-dimensional height-map (Fig.2). The latter view leads to the use of a watershed algorithm for single tree delineation. [5] With a standard watershed-algorithm the z-axis of the three dimensional data is only used to generate gradients and calculate affiliations, resulting in a set of areas, each annotated with its size. Thus, the size of the region would be the only criterion to decide whether a region represents a tree or only a branch of a tree.

We decided to use the full 3D-information and therefore, to look at the volume the peaks pointing out of the canopy. One way to get an approximation of all peak-volumes is to modify the watershed-algorithm to work on

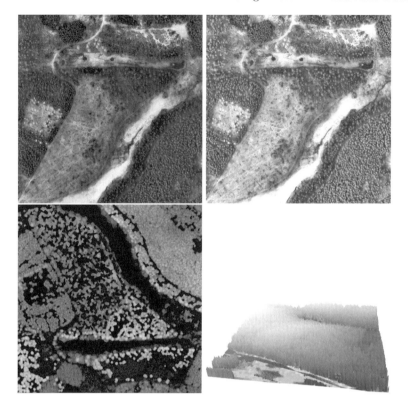

**Fig. 8.2** The Source Data: Aerial Photos in RGB and CIR, Differential Model in Height Map-Representation and in 3D-Visualisation

three dimensional data. For the illustration of the following algorithm we will use a sectional drawing through a three dimensional DM. Fig. 3a shows some trees and the sectional drawing above them. To make it easier to imagine rainfall and water-flow we turned the sectional drawing upside down in the subsequent images with the most significant points – the maximum heights in the original data that may represent tree-tops – as local minima of the graph. Fig. 3b illustrates the idea of a standard watershed algorithm. Water is poured over the area uniformly. The water-flow is simulated and the amount of arriving water is measured at all local minima.

To get volumetric information we figuratively fill the DM with water. Then, in each cycle, we puncture the point with the most water-pressure acting on it and measure the amount of water streaming out of this opening. (Fig. 3c) The result is a value which is always higher than or equal to the real volume of the peak. Interestingly, the result is far from the real volume for the most extreme points (areas that are most likely treetops) but very close to the real volume for the critical peaks that are hard to decide. For each opening

# 8 Volume-Based Single-Tree-Delineation and -Parameterization

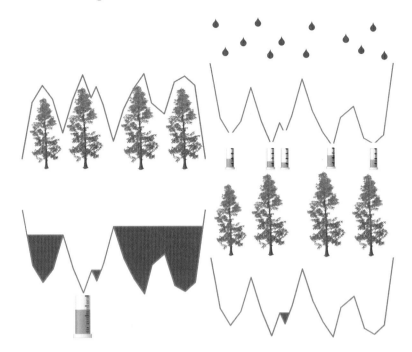

**Fig. 8.3** Single Tree Detection in LIDAR data. a) Trees and LIDAR Information, b) Watershed-Algorithm, c) Volumetric Algorithm – First Cycle, d) Last Cycle

with a volume greater than a user-specified threshold, a tree is generated in the map using the x and y position of the puncture - which is always the highest point within its peak –, the z value taken from the FDTM. The tree is annotated with its height, which can be determined from the DM. Fig. 3d shows a situation where only one peak is left. The remaining volume is below the threshold so no tree will be generated at this position.

In contrast to the well known 2D approach, the volumetric approach adds another dimension to the data used for calculation and makes it easier to decide whether a peak is a tree or just a tree branch. This is especially valuable for the available data because the z-axis of the grid-based DM-model has a resolution of 1cm compared to the 1m-resolution of the x- and y-axis.

Using the volumetric algorithm, the percentage of correctly detected tree-tops increased significantly. We compared the two available resolutions of LIDAR data and, not surprisingly, the higher resolution proofs produced better detection results.

The data-points measured with a Falcon II scanner were inhomogeneously distributed over the surface. The average rate was specified to be one measurement per square-meter, but the deviation was rather high. There were some grid cells measured with as much as six laser points whereas other cells were not hit at all by the laser beam. Unfortunately, there were connected areas of up to 25 square meters occurred that had not been hit. This led to the question how the results may look on a homogeneous distributed data set. We decided to use the true-ortho-photos for this.

**Fig. 8.4** RGB Image and Image after Color Tone based Brightness Reduction

## 8.4 RGB and CIR Processing

A key idea of the presented approach is that **the same volumetric approach** used for LIDAR data for processing is also applied to the true-ortho photos. Substituting DM height values with brightness information derived from the aerial photos allows the algorithm to focus on bright spots in the picture. These may be tree-tops or white marks on a street. There are two ways of eliminating unwanted bright spots: Classification of the detection results or filtation of the picture before the detection starts. We chose to filter the picture because we did not want to lose the ability to overestimate the most likely positions, which may happen if artificial marks are brighter than tree-tops in the selected input-channel. We applied a color tone based brightness reduction filter. (Fig. 4) The color tone was calculated using all four available channels (red, green, blue and near-infrared). The brightness and color of a pixel were adjusted based on the distance of its original color from a specified area of colors within the color-space that represents trees.

In the resulting image we chose the green-channel of the RGB image – which is characteristic for trees – as input for the detection-algorithm. Following the optical impression, this data looks similar to the LIDAR-data. We associated height-levels with brightness-levels in the green-channel and used the described volumetric algorithm on the camera data. The true-orthophotos are geo-referenced so the position of a pixel can be associated with

8 Volume-Based Single-Tree-Delineation and -Parameterization   137

a geo-coordinate giving the x- and y-coordinate of the selected brightness-maximum in the algorithm. The z-coordinate and height of the tree are generated the same way as during calculation with LIDAR data: they can be read out of the FDTM respectively the DM. These tree candidates are filtered by using the LIDAR-information. Thus, if the height of the tree is outside a user-selectable range, the tree is neglected.

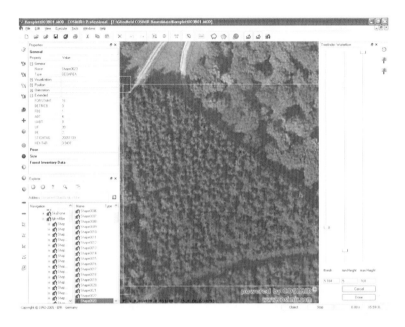

**Fig. 8.5** The Threshold for the Volumetric Algorithm can be set interactively using a Slider

## 8.5 Results

The described algorithms were implemented in a 3D-GIS. Because the threshold needed for the volumetric algorithm depends on the density and age of the forestry unit, it can be set interactively. (fig. 5)

The result of changing the threshold is displayed in real-time to help the user find the correct value for each forest unit. In general, older units require higher thresholds because smaller peaks will most likely represent branches whereas a peak with the same volume in a younger unit will most likely be a tree-top.

Fig. 6 shows the results for the algorithm with LIDAR data. The left image shows the detected trees as points on a grayscale image that displays

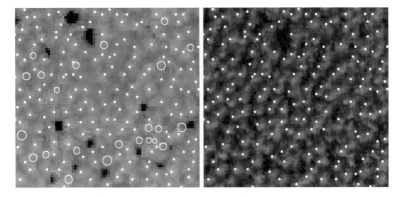

**Fig. 8.6** Results of the volumetric Algorithm on LIDAR data displayed on a grayscale representation of the LIDAR data and on an rgb true orthophoto

the values of the DM. The detected points seem to describe the trees in the real world. However, upon comparison to the rgb aerial photo of the same area it turns out that a large number of trees are not detectable in the LIDAR data – even for a human operator. Some examples are marked with white circles in the left image.

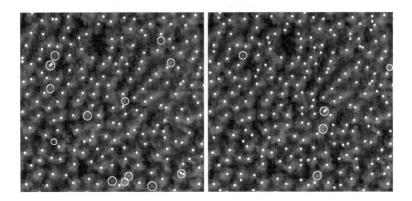

**Fig. 8.7** Results of the watershed Algorithm (left) and the volumetric Algorithm (right) on aerial photos

A comparison between the watershed algorithm and the volumetric algorithm presented in this paper is shown in fig. 7. The results of both algorithms seem to match the scenery. The white circles in the pictures mark positions where either a tree was detected twice or a human operator would expect additional trees.

For one unit we had ground truth data measured with an absolute precision of 15cm. In this unit we compared the results of the algorithm with the real situation. The volumetric algorithm detected 95 percent of the trees correctly (Fig. 8) while the watershed algorithm only delivered 84 percent. With LIDAR data the detection rate was very poor, about 60 percent. The visual impressions on aerial photos of units where we do not have ground truth data support these values.

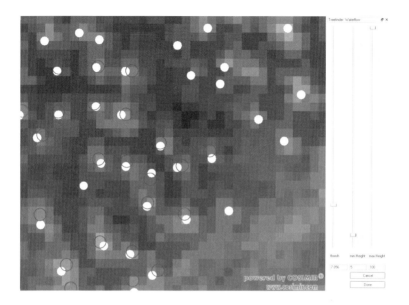

**Fig. 8.8** Results of the Volumetric Algorithm compared to Ground Truth Data. Solid Marks indicate recognized Trees, white Circles mark known positions of a Set of Sample Trees

Although the detection rate gets better using the visual data, LIDAR-data will typically deliver better tree-positions. RGB and CIR shots depend on the actual light situation. At noon with the sun in a very high position the tree-tops are the brightest point of a tree, giving accurate results. However, if an image is taken in the late afternoon, the sun only lights one side of a tree, moving the center of the brightest spot away from the tree-top. Additionally, within the test-area of $82km^2$ we found several places where lightning conditions made it hard to recognize trees – even for a human operator.

## 8.6 Applications

### 8.6.1 Forestry GIS

The single tree-delineation algorithm was developed for a 3d forestry GIS. This system can support forestry experts in planning the economical usage of a forestry area, from seeding to harvesting. Ownership and size of a tree can be monitored, felling activities can be planned and ideal trees for a certain commission could be searched within the 3D-GIS.

For these activities not only tree-positions, but also properties of the individual tree are important. The most important key-parameter is the (stem-) diameter at breast height (DBH). This parameter is usually measured on-site, i.e. on the ground, using a pincer, so there are a lot of publications showing statistical relations between the DBH and a number of other parameters of a tree [6][7]. In particular, the timber volume, which is critical for the value of an individual tree, depends on the DBH and the height of the tree.

According to Hyyppä [8] the DBH can be calculated using the height of the tree and the diameter of its canopy:

$DBH = \alpha L + \beta h + \gamma$

In this equation $\alpha$, $\beta$, and $\gamma$ are parameters depending on the local situation of the tree, L is the crown-diameter and h represents the height.

Using the known position of each tree in a forestry unit the crown diameters are calculated by a gradient descent algorithm starting at the treetop of each tree. A point reachable by a gradient descent from multiple treetops is added to the most likely tree or – if the point is located at a height minimum between two trees – it is added in parts to all adjacent trees. This calculation has been implemented using a stepwise calculation on each of the discrete height levels and presorted height points.

Knowing the parameters of the individual trees the value of the biological repository can be estimated more precisely. This may help in selling standing wood and getting better prices.

The virtual forest can be used to plan harvesting actions. Simulations can show how to harvest most economically. Depending on the slope of the ground and the distribution of the interesting trees a lumberjack or a harvester could be the better choice.

When looking for trees with special attributes, a virtual forest – a tree-oriented geo-database – can show where to find appropriate trees or groups with a significant number of such trees.

Tools like SILVA [10] can simulate forestry growth. The virtual forest is an excellent foundation for growth prediction because the position and the important growth parameters of every tree in the forest are known exactly.

## 8.6.2 Visual GPS

Using particle filters it is possible to find groups of trees in the virtual forest, even given a significant uncertainty between the real stem positions and the derived positions. In a first example, we found that it was possible to detect a group of 34 trees in a forest of about 15.000 trees, even if the sample trees showed an average position error of 6m. This may lead to using a virtual forest for special purpose navigation systems. The virtual forest database can be used as a map. While GPS systems often deliver poor results in dense forests, a visual GPS using additional information out of the virtual forest and a laser or visual sensor for local tree detection could provide improved position information.

## 8.6.3 Accessibility-Check

Using the 3D-GIS including the filled digital terrain model and positions of individual trees it becomes possible to check whether a point in the forest is reachable by huge forest machines. The slope lateral to the path can be checked as well as the horizontal clearance towards the individual trees. Combining both information a route can be declared accessible or not. The 3D-GIS supports visual feedback for the user in order to understand the results.

**Fig. 8.9** Applications of the Virtual Forest in a Virtual Reality Environment – Forest Machine and Disaster Simulation

## 8.6.4 Forest Machine Simulation

Today, forest machines are complex vehicles featuring multiple driver controllable degrees of freedom. For example, a harvester is equipped with a manipulator with up to 10m grip-range, the harvesting head has multiple knifes, wheels and a chain-saw and the chassis may have several hydraulic cylinders controlling the level of each wheel. Due to the number of simultaneous controlled joints, the handling of a modern harvester becomes complex and difficult. It has been shown that driver training on forest machine simulators makes it easier to handle the real machine. Damage and breakdown of the machine were reduced dramatically as well as damage to the trees, increasing the value of the harvested wood.

The data stored in the forestry 3D-GIS is used to generate virtual environments for out forest machine simulators, enabling a very realistic training scenario. The slope of the environment is again taken from the FDTM and the individual trees are generated with their parameters as species, height and DBH and placed at the appropriate positions in the simulation.

## 8.6.5 Disaster Management

The virtual forest gives a very realistic impression of the environment. Accessibility and a precise idea of the local situation are vital facts especially for use in a rescue procedure as part of disaster management. The data used in the 3D-GIS could be exported for use in a virtual reality system where all other features of our simulation system may be used – for example several types of vehicles and a fire-simulation displaying actual situations in order to discuss several possibilities of intervention.

During the rescue mission the virtual GPS could provide better localization information to the rescue team, helping them to find the scene faster and more efficiently. In combining the virtual GPS with the accessibility check an efficient forest navigation system can be implemented.

## 8.7 Conclusions

The volumetric algorithm presented in this paper shows a significantly better result compared to the well-known 2D-watershed-algorithm. It reaches detection rates up to 95 percent on homogeneous data with a resolution of 0.5m per measurement point. The results scale with the resolution and the homogeneity of the data, so working on camera data delivering a homogeneous distribution with a high resolution delivered the best results.

# 8 Volume-Based Single-Tree-Delineation and -Parameterization

The single-tree delineation shown in this paper was implemented in our 3D-GIS in order to generate the data needed for a virtual forest. Several graphical representations (view from above, 3D using symbols instead of trees and high-definition rendering of scenes) were added to the system (Fig. 10).We have already populated a number of forestry units with a total of about 120.000 trees. During this work we discovered several points which will still require continuing effort:

The results were best using high-resolution data with a homogeneous distribution of the individual measuring points. Promising candidates to deliver appropriate data may be stereoscopic cameras like the DLR HRSC [9] or LIDAR scanners with a swinging or rotating mirror, which are capable of delivering a higher resolution. During the next steps of the project we will import this data in the 3D-GIS and evaluate its quality for the volumetric algorithm.

We currently use a recent official forestry unit classification. The approach works well on units with a homogeneous age and species structure of the trees while the usage becomes difficult if there are several species or age classes of trees in the same unit. In order to increase comfort and usability we are integrating algorithms to classify and segment the area into appropriate units, mostly featuring a homogeneous tree age and species structure. For the remaining units we need to look at colors and shapes of the tree-tops to group them into younger and older trees and divide them by species.

**Fig. 8.10** A view into the Virtual Forest – a computer graphics frontend to the database

The Hyyppä formula we currently use to calculate individual tree properties is based on local constants $\alpha$, $\beta$ and $\gamma$ which must be determined for each forest section. It turned out that the variance of the three parameters is rather high. A promising approach for important forestry units is combining the aerial view with ground measured data. Matching the positions of airborne detected trees with the ground detected trees will result in complete sets of parameters (DBH, height and crown diameter) for these trees. The complete sets can be used as samples for calculating the Hyyppä-parameters for the unit by using regression formulas. We are going to use ground-based laser-scanners to gather the requested calibration data in the next phase of the project.

# References

1. Leica ALS50-II Technical Specification, Leica Geosystems, Heerbrugg Switzerland, www.leica-geosystems.com
2. Riegl LMS-Q560 Technical Specification, Riegl Laser Measurement Systems, Horn, Austria, www.riegl.com
3. Katrin Schnadt, Rolf Katzenbeisser: Unique Airborne Fiber Scanner Technique for Application-Oriented Lidar Products, International Archives of Photogrammetry, Remote Sensing and Spatial Information Sciences, Vol. XXXVI - 8/W2
4. Toposys Falcon II Technical Specification, Toposys GmbH, Biberach, Germany, www.toposys.de
5. O. Diedershagen, B. Koch, H. Weinacker, C. Schütt. Combining LIDAR- and GIS-Data for the Extraction of Forest Inventory Parameters, 2003
6. Kramer, H. und Akça, A. Leitfaden zur Waldmesslehre. 3. erw. Aufl. Frankfurt a.M., Sauerländer. 266 S., 1995
7. Landesanstalt für Ökologie, Landschaftsentwicklung und Forstplanung Nordrhein-Westfalen. Hilfstafeln für die Forsteinrichtung. 3. Auflage, 1989
8. J. Hyyppä, M. Inkinen. Detecting and estimating attributes for single trees using laser scanner. The Photogrammetric Journal of Finland, Vol. 16, No. 2, s. 27-42, 1999
9. F. Scholten, S. Sujew, F. Wewel, J. Flohrer, R. Jaumann, et al.: The High Resolution Stereo Camera (HRSC) - Digital 3D-Image Acquisition, Photogrammetric Processing and Data Evaluation. "Sensors and Mapping from Space 1999" ISPRS Joint Workshop, Hannover, 27.-30.9.1999, International Society of Photogrammetry and Remote Sensing (ISPRS), International Society of Photogrammetry and Remote Sensing Proc. of Joint Workshop "Sensors and Mapping from Space 1999" (CD-ROM), (1999)
10. H.-J. Klemmt, P. Biber, H. Pretzsch: Mit SILVA in die Zukunft des Waldes blicken, LWF aktuell, Bayrische Landesanstalt für Wald und Forstwirtschaft, No. 46, 2004

11. J. Holmgren. Estimation of Forest Variables using Airborne Laser Scanning, 2003
12. J. Holmgren, A. Persson, U. Södermann. Identification of tree species of individual trees by combining very high resolution laser data with multispectral images. Workshop on 3D Remote Sensing in Forestry, Vienna 2006
13. E. Naesset. Determination of mean tree height of forest stands using airborne laser scanner data, 1996

# Chapter 9
# Automatic building modeling from terrestrial laser scanning

Shi Pu

### Abstract

We present an automatic approach to create building models from terrestrial laser points. Our method starts by extracting important building features (wall, window, roof, door, extrusion) from segmented terrestrial point cloud. Then visible building geometries are recovered by direct fitting polygons to extracted feature segments. For the occluded building parts, geometric assumptions are made from visible parts and knowledge about buildings. Finally solid building models can be obtained by combining directly fitted polygons and assumptions for occluded parts. This approach achieves high automation, level of detail, and accuracy.

## 9.1 Introduction

The topic of building reconstruction has been a research of interest, due to the increasing demand for accurate three-dimensional building models from various fields, such as urban planning, construction, environment safety, navigation, and virtual city tourism. Manual creation of a building model is a slow procedure. For example, it is necessary to measure the length of all the wall edges to make a wall face. This operation can be rather time-consuming when the target building contains too many edges, and/or there are many buildings to be modeled. Manual building modeling is also an inaccurate procedure, because visual measurement of geometric properties (distance/size/area) may depend on human operator.

---

International Institute for Geo-information Science and Earth Observation
P.O. Box 6, 7500AA, Enschede, the Netherlands
spu@itc.nl

After several years' work, a lot of algorithms and systems have been proposed towards the topic of automatic building reconstruction. However, a versatile solution has not been found yet, with only partial solutions and limited success in constrained environments being the state of art [9]. This is mainly because digital imagery is the only data source used for the reconstruction for a long time, and it is still hard to recover 3D building structures from 2D image. Recent studies ([2] [6]) show that laser scanning data can be a valuable data source for the automatic building reconstructing. Comparing to digital imagery, airborne and terrestrial laser scanning give explicit 3D information, which enables the rapid and accurate capture of the geometry of complex buildings. In particular, terrestrial laser scanning is able to provide very dense point clouds of building facades, which gives enough raw data from which high detailed 3D building models can be obtained automatically.

We propose a bottom-up building reconstruction process based on terrestrial laser scanning as the following steps:

1. Feature extraction, where important building features (walls, windows, doors, etc.) are recognized and extracted;
2. Modeling, where recognized features are fitted to geometric shapes (polygon, cylinder, etc), and then combined to complete geometric models;
3. Refine, where models are verified, and improved with information from other data sources if necessary;
4. Texturing, where geometry models are textured with selections from digital image or pre-defined textures.

So far the research has been done up to the modeling stage. This paper presents our approach to automatic creating building models from terrestrial laser scanning. Section 2 explains how features are extracted from terrestrial laser point cloud. Section 3 elaborates modeling of building geometry by fitting feature segments and estimating occluded faces. Some conclusions and recommendations for future work are given in the last section.

## 9.2 Feature extraction

An algorithm of building feature extraction from terrestrial laser scanning has been demonstrated in [8]. We improved this method by refining the feature constraints and integrating a second extraction method, to make the extraction more accurate. Our approach starts with segmentation, where laser points are grouped roughly according to the plane they belong to; then each segment is checked through some pre-defined feature constraints, to determine which building feature (wall, roof, door, extrusion) it is; finally window feature are extracted particularly from wall holes.

## 9.2.1 Segmentation

A couple of segmentation algorithms based on laser point cloud are available ([4,5,7,9,10]). We adopted the planar surface-growing algorithm by [10] because it is more suitable for segmenting planar surfaces. A description about this algorithm is given here because of its strong relevance to our feature extraction method.

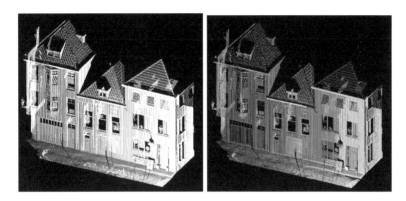

**Fig. 9.1** Left: terrestrial laser scanned building facade; right: segmentation result

The planar surface-growing algorithm starts by choosing seed surfaces. A seed surface consists of a group of nearby points that fit well to a plane. The algorithm selects an arbitrary unclassified point and tests if a minimum number of nearby points can be fitted to a plane. If this is the case, these points constitute the seed surface; otherwise, another arbitrary point is tested. Then seed surfaces tries to grow to their nearby points. Only the point within certain distance to the seed surface, and with perpendicular distance to the seed plane below some threshold, can be added to the seed surface to make it grow.

Figure 1 left gives a terrestrial laser scanned building façade. The data acquisition of this data set is done in one scan, by a Leica terrestrial laser scanning machine which was placed a couple of meters in front of the building. The point density is around 500 points per square meter. Figure 1 right gives the segmentation result.

## 9.2.2 2.2 Feature constraints

Humans understand building features by analyzing their characteristics such as size, position, direction and topology. These characteristics can be 'taught' to machines, so that they can also 'understand' the laser point cloud of buildings, and extract features automatically.

We list different building features' constraints in Table 1, which describes the most significant characteristics to distinguish one feature from another. Note that although ground is not a real building feature, it is still recognized because ground provides important clue for recognizing wall feature and door feature.

|  | Size | Position | Direction | Topology |
|---|---|---|---|---|
| **Ground** | Segment(s) with large area | Lowest | | |
| **Wall** | Segment(s) with larger area | | Vertical | May intersect grouond |
| **Roof** | Segment(s) with large area | Above wall | Not vertical | Intersects a wall |
| **Door** | Area with certain area | On the wall | Vertical | Intersects the ground |
| **Extrusion** | | A little outside the wall/roof | | |

Table 9.1 Constraints for building features (The unit for position is meter, for area is square meter, and for direction is degree)

The feature constraints listed in Table 1 are pretty robust. First, all the constraint values are independent of data sets, because these constraints are semantic based. The size/length/direction of any feature is the same no matter where is the data acquisition position, and no matter how dense the point cloud is. Second, some constraints are relatively determined, which make constraint values dynamic. For example, 'the WALL feature segments have the larger area'. This is always true because walls are always bigger than door and extrusion. And in terrestrial laser scanned point clouds, the wall part is also usually bigger than the roof part, as the data acquisition position is on the ground.

## 9.2.3 Feature recognition

Five building features: wall, roof, door, extrusion and window, are recognized and extracted in our approach, because we believe they are the most important elements on building surface. Each segment is checked through the feature constraints defined in Table 1, to determine which of the four features (wall, roof, door, extrusion) it is. Windows are specially extracted from the holes on wall segments, after filtering out the holes caused by extrusions and doors.

9 Automatic building modeling from terrestrial laser scanning

Sometimes a feature might be over-segmented, which means, a same feature is segmented into multiple segments, due to discontinuous laser points or wrong segmentation parameters. Over-segmentation can be corrected automatically by some simple geometry checking: if two segments are on the same plane, and they are attached to each other, then the algorithm treats them as over-segmented results, and merge them to one segment. For example, the wall in Figure 1 is segmented into two parts, due to over segmentation. They will be treated as one part (merge) in the modeling stage.

Figure 2 shows the feature recognition results of the building façade in Figure 1.

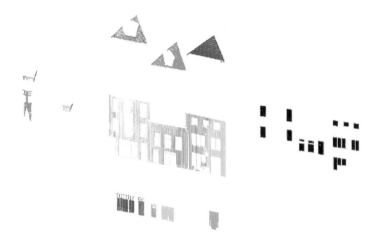

**Fig. 9.2** Feature recognition results (extrusion, roof, wall, door, window)

## 9.3 Geometry modeling

A straightforward thinking of the geometry modeling would be: first fitting the extracted feature segments to some simple geometric shapes such as polygon, and then combine them to the final building model. However, the actual procedure of geometry modeling is not simple as fitting plus grouping. This is because:

1. Sometimes feature segments contain incomplete geometry information. Due to scanning or segmentation error, an extracted feature segment may only contain points from part of the whole feature. For example, the roof segment in Figure 3 left has the lower part missing, because an eave

blocked laser beam when scanning. Direct geometry fitting will result in gaps between patches (the area between bottom green edge and yellow edge in Figure 3 right).
2. Terrestrial laser points are only available for the building facade, which means, there is no direct geometry information available for left, right, up and back sides of wall, roof and dormer window. Assumptions have to be made to obtain a solid building model instead of only geometries of facade patches.

**Fig. 9.3** Left: laser points of a roof, with the part near eave missing. Right: directly fitted shape (yellow) and actual shape (green)

Knowledge about buildings is helpful to solve these problems, too. For example, we know that roof must intersect a wall, so the low contour of a roof should be parallel and on the same height with (a part of) wall contour. We know that a dormer window is extrusion on a slope roof, so there must be connection parts between the dormer window and its projection on the slope roof. These missing parts can be well estimated from existing laser points and knowledge about buildings. The actual steps in building modeling stage are: geometry fitting, geometry estimation, and combining.

## *9.3.1 Geometry fitting*

Based on the hypothesis that most building surfaces are planar, we try to fit planar shapes such as polygon to all feature segments. Although there are curved surfaces on some buildings, we are unable to deal with them yet. Further research is needed to fit curved surfaces in addition to planar surfaces, so that more accurate building models can be obtained.

### 9.3.1.1 Wall

As the most important feature for buildings, a wall provides the outline for a building model. Our scan-line algorithm first extracts the upper contour points from wall segments; then line segments are fit out of the contour points, to combine the building outline. The steps are given as follows:

a. Select all the points where wall segment and ground segment intersect. These points fit a 2D contour line of the wall on the ground plane.
b. Pick sample points from the 2D contour line by scanning this line from left to right, with 10 centimeter as the step. This step length is determined after a series of experiments. Shorter step will result in more details in the fitted wall outline, and longer step may result in missing of details. As long as the density of point cloud, particularly in the wall part, is higher than 100 points per square meter, we will be able to select point every 10 centimeters. The density capabilities of mainstream terrestrial laser scanners are much higher than this minimal requirement.
c. For each sample point, select all the surrounding points from the wall segment, which are within 5 centimeter from this sample point. Find the highest point among these surrounding points.
d. Repeat step c for all sample points, determine all the highest surrounding points, and group them to form a 3D upper contour of the wall (Figure 4 left).
e. Scan the points in 3D upper contour from left to right. Keep all the extreme points, which are far away from their previous points, or causing sharp angle change in this 3D upper contour line.
f. Connect all the extreme points to make the upper outline for the wall. Then project the most left point and most right point to the ground plane, to make the left and right outline of the wall. These two outlines are vertical, based on the hypothesis that walls are vertical.
g. Give the height of ground plane to 2D contour line (in step a) to make it 3D, which is also the lower outline of the wall.
h. Finally the whole building outline (Figure 4 right) is combined from upper outline, left outline, right outline, and lower outline.

### 9.3.1.2 Roof and extrusion

In our approach, each roof/extrusion segment is simply fit to a convex polygon, based on the hypothesis that most roofs/extrusions have the geometry of convex polygon. Concave polygon fitting is still to be researched in the future work. The Quick Hull algorithm is used to fit the convex polygons in fitted planes. A detailed explanation of this algorithm can be found in [1]. Figure 5 shows the fitting results of roof and extrusion.

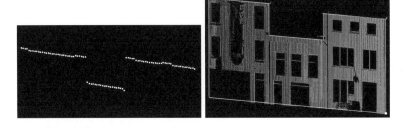

**Fig. 9.4** Left: upper contour points of a building façade. Right: fitted building outline

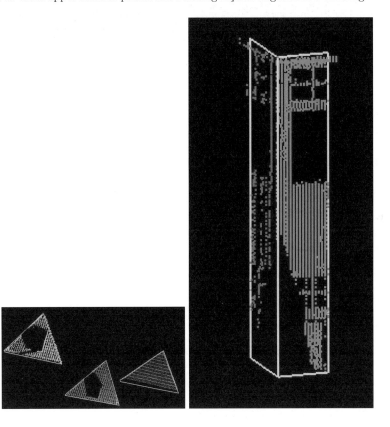

**Fig. 9.5** Fitted roof and extrusion polygons

9 Automatic building modeling from terrestrial laser scanning 155

**9.3.1.3 Door & window**

Each door segment and window segment is simply fit with a minimum bounding box, based on the hypothesis that most doors and windows are rectangular and vertical. Figure 6 shows the fitted door and window rectangles.

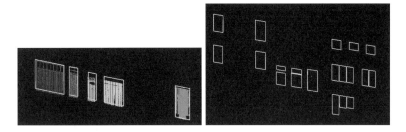

**Fig. 9.6** Fitted door and window rectangles

## 9.4 Occlusion assumption

As mentioned earlier, terrestrial laser scanning can only retrieve points from building façades. The occluded area should be filled based on existing laser points and knowledge about buildings, so that solid building model can be obtained. In particular, the following parts should be filled:

- **Left, right, top, bottom and back sides of a building.**
  We first construct a line that is perpendicular with wall plane. Then we 'push' the existing wall outline back along this perpendicular line for a certain distance. This results in an assumed building backside, which is parallel with the façade outline. The left, right, top and bottom sides can be generated by connecting corresponding vertices on wall outline and estimated backside outline. The offset distance can be either a fixed value, or derived from 2D ground plan.
- bf Parts on roofs blocked by eaves.
  Based on the hypothesis that the blocked area is also on the same plane with the whole roof, we extend the directly fitted polygon till it intersects the wall outline. Or in another words, we first construct a horizontal plane which has the same height with nearby wall outline, then replace the bottom edge of directly fitted polygon with intersection edge between roof patch and this horizontal plane.
- **All the missing parts between extrusion and its supporting wall/roof.**
  So far we just project the directly fitted polygons to its supporting

wall/roof plane, then connect all vertices between origin polygon and projection polygon, to generate the estimated polygons. We are aware this only works for simple extrusion types. For example, direction projection will lead to wrong geometry for the stair extrusion shown in Figure 7. Further research about extrusion structure knowledge is needed for more accurate estimation.

**Fig. 9.7** An incorrect extrusion

## 9.5 Modeling result

Figure 8 shows the final model of the building façade in Figure 1. This model contains both directly fitted geometries and estimated geometries. The whole procedure takes 3.5 minutes, including segmentation, feature recognition, direct geometry fitting, and geometry estimation. No manual interaction is needed throughout the process. The raw laser point cloud contains 381057 points, and the final model contains around eighty points, which are just the vertices of the polygons. Some faces share vertices and these points are not stored more than once.

Comparing with reality, we are able to model the wall facade, roof facade, and most extrusions, windows and doors accurately. This is mainly because we have sufficient terrestrial laser information for reconstructing reliable geometries.

9 Automatic building modeling from terrestrial laser scanning        157

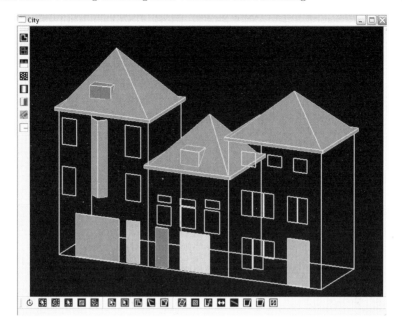

**Fig. 9.8** Final geometry model of the building façade in Figure 1

Wrong shapes appear in the following parts:

- The left, right and back sides of the wall. Due to no terrestrial laser points available for these areas, assumptions have to be made to combine a solid model, as mentioned is section 3.2. These assumptions can be inaccurate.
- There is a curved surface patch under an extrusion on the wall. This patch is not modeled because we haven't fit curved surfaces yet.

Figure 9 shows another buildings model example. The wall outline, roof and some windows are accurately modeled.

Two main modeling errors are:

- The extrusions' shapes are incorrect. There are two extrusions in this building facade: a dormer window on the roof and a balcony on the wall. The dormer window has its own roof on top of it, which is missing in the final model because this 'dormer-roof' feature is not supported yet. The balcony results in a bit extruded shape, because so far we model an extrusion's geometry by projecting the convex hull of extrusion façade to its supporting wall. This naturally leads to wrong result for extrusions with concave facade, such as this balcony.
- The three holes are recognized and modeled as windows. This is be-cause in one hand, we haven't deal with holes on walls yet; and in the other hand, we define the window feature as 'big holes on wall segments, after filtering out the holes caused by extrusions and doors'.

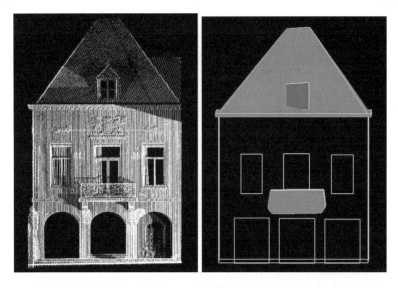

**Fig. 9.9** Another building example (left: raw terrestrial laser data; right: geometry model)

Figure 10 shows another building example which can not be correctly modeled yet. Besides the wrong shape of dormers, the awnings are also missing in the final model. This is because the shapes of these awnings are curved, and curved geometry is not yet supported in our method.

## 9.6 Summary and recommendations

In this paper we presented an automatic building modeling technique based on terrestrial laser scanning. The raw laser points are segmented first, so that all the points belong to the same plane are grouped together. Then some important building features (wall, door, window, roof and extrusion) are extracted out of the segments, by checking each segment with some knowledge-based feature constraints. In the modeling stage, boundary polygons are derived from the extracted feature segments first; then geometries of occluded area are estimated from existing boundary and human knowledge about buildings.

Two recommendations are given for future research:

First, our modeling method is a bottom-up process, where building models are combined from patches. For buildings with complex shapes, the feature recognition may fail as a result of lacking context information among features. Concepts of top-down approach, or grammar based building modeling, should

9 Automatic building modeling from terrestrial laser scanning 159

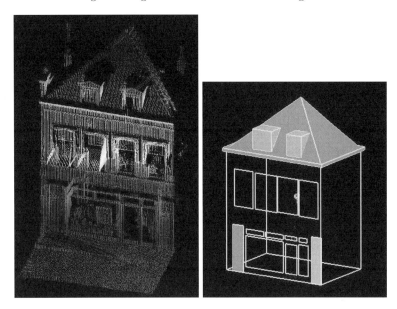

**Fig. 9.10** Another building example with some awnings (left: raw terrestrial laser data; right: geometry model)

be helpful to increase the feature recognition ability, and in turn improve the modeling.

Second, fusing of other data sources can be helpful throughout our modeling process. For example, ground plan gives explicit 2D contour of buildings, which can be used to verify the fitted building outline; it is relatively easy to extract edge information from digital imagery, which can be combined together with laser segments, to obtain more confident wall outline and roof boundary.

# References

1. Barber, C. B., Dobkin, D. P., and Huhdanpaa, H. T., 1996, The Quickhull Algorithm for Convex Hulls, *ACM Trans. Mathematical Software* 22, 469-483
2. Brenner, C., 2000, Towards Fully Automatic Generation of City Models, XIX ISPRS Congress IAPRS, Amsterdam, the Netherlands, pp. 85-92
3. Brenner, C., 2003, Building Reconstruction from Laser Scanning and Images, Proc. ITC Workshop on Data Quality in Earth Observation Techniques, Enschede, the Netherlands

4. B Gorte, N Pfeifer, 2004, Structuring Laser-scanned Trees Using 3D Mathematical Morphology, International Archives of Photogrammetry and Remote Sensing, Vol. XXXV, Istanbul, Turkey.
5. H. Woo, E. Kang, Semyung Wang and Kwan H. Lee, 2002, A New Segmentation Method for Point Cloud Data, International Journal of Machine Tools and Manufacture, January, vol. 42, no. 2, pp. 167-178 (12)
6. Maas, H.-G., 2001, The Suitability of Airborne Laser Scanner Data for Automatic 3D Object Reconstruction, Third International Workshop on Automatic Extraction of Man-Made Objects from Aerial and Space Images, Ascona, Switzerland.
7. Rabbani, T., Heuvel, F.A. van den, Vosselman, G., 2006, Segmentation of Point Clouds Using Smoothness Constraints, International Archives of Photogrammetry, Remote Sensing and Spatial Information Sciences, vol. 36, part 5, Dresden, Germany, September 25-27, pp. 248-253
8. Pu, S., Vosselman, G., 2006, Automatic extraction of building features from terrestrial laser scanning, International Archives of Photogrammetry, Remote Sensing and Spatial Information Sciences, vol. 36, part 5, Dresden, Germany, September 25-27, 5 p.
9. Suveg, I., Vosselman, G., 2004, Reconstruction of 3D Building Models from Aerial Images and Maps, ISPRS Journal of Photogrammetry and Remote Sensing 58 (3-4), p202-224.
10. Vosselman, G., B. Gorte, G. Sithole and Rabbani, T., 2004, Recognizing Structure in Laser Scanner Point Clouds, International Conference NATSCAN, Laser-Scanners for Fores and Landscape, Processing Methods and Applications, ISPRS working group VIII/2, Freiburg im Breisgau, Germany.

# Chapter 10
# 3D City Modelling from LIDAR Data

Rebecca (O.C.) Tse, Christopher Gold, and Dave Kidner

**Abstract**

Airborne Laser Surveying (ALS) or LIDAR (Light Detection and Ranging) becomes more and more popular because it provides a rapid 3D data collection over a massive area. The captured 3D data contains terrain models, forestry, 3D buildings and so on. Current research combines other data resources on extracting building information or uses pre-defined building models to fit the roof structures. However we want to find an alternative solution to reconstruct the 3D buildings without any additional data sources and pre-defined roof styles. Therefore our challenge is to use the captured data only and covert them into CAD-type models containing walls, roof planes and terrain which can be rapidly displayed from any 3D viewpoint.

## 10.1 Introduction

We have successfully addressed this problem by developing a several-stage process. Our starting point is a set of raw LIDAR data, as this is becoming readily available for many areas. This is then triangulated in the x-y plane using standard Delaunay techniques to produce a TIN. The LIDAR values will then show buildings as regions of high elevation compared with the ground. Our initial objective is to extrude these buildings from the landscape in such a manner that they have well defined wall and roof planes. We want to know how to extract building outlines from the triangulation when it is not available from the national mapping department. We do this by superimposing a coarse Voronoi cell structure on the data, and identifying wall segments within each.

---

University of Glamorgan, Pontypridd, Wales, UK rtse@glam.ac.uk, cm-gold@glam.ac.uk, dbkinder@glam.ac.uk

There are two ways to examine the triangulated interior (roof) data. The first method is to find out the folding axis of the roof, but it may not be suitable for a complex roof. The second is to identify planar segments and connect them to form the final surface model of the building embedded in the terrain. This is done using Euler Operators and Quad-Edges with preserved topological connectivity. This was successfully developed by [10, 11]. In this paper, we will focus on discussing how to extract building blocks from raw LIDAR data and how to find out the roof shape of those building blocks.

## 10.2 What is LIDAR?

ALS (or so called LIDAR) is a new independent technology which is highly automated to produce digital terrain models (DTM) and digital surface models (DSM) [1]. It is a laser-based technology to emit and capture the returned signal from the topographical surface.

A laser scanning system, a global positioning system (GPS) and an inertial measuring unit (IMU) are the three main units in an ALS system. The laser scanning system is mounted on an aircraft, a helicopter or satellites and emits pulses toward the earth's surface and measures the distance reflected from the earth's surface and other objects on the surface back to the aircraft. IMU and GPS are important for determining the absolute position and orientation of the LIDAR sensors. The Inertial navigation system is used to correct the errors from the pitch, roll and yaw of the plane. GPS monitors the altitude and the flight path of the aircraft which observes the three dimensions data. A high accuracy GPS is installed in the plane and a ground control based station is established. Figure 10.1 shows an aircraft scanning over a piece of land.

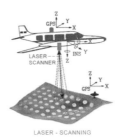

**Fig. 10.1** Airborne laser scanning

## 10.3 Building Construction from LIDAR Data Alone

LIDAR provides an efficient way to capture 3D data; however it is not easy to extract building information from the data. Much research focuses on extracting building outlines, and they may combine different data sources, for example photogrammetric data or existing landline data [6, 7, 8, 13] . It does not work if there is no other data available. Our approach is to use LIDAR only to reconstruct the 3D buildings and remodel the roof structure without using any pre-defined models.

### 10.3.0.4 Building Blocks Identification

Our method is to identify building blocks from the terrain rather than searching for the building footprints directly. It separates the high-elevation data (building block) from low-elevation data (terrain surface). We make use of the duality and connectivity properties of Delaunay triangulation, and its dual Voronoi diagram.

A Delaunay triangulation is created using the original high density LIDAR data (Figure 10.2). Then we sample it to a lower resolution triangulation (Figure 10.3. In Figure 10.3 each big Voronoi cell contains many data points (about 50 - 100), some with only the ground points (low elevation) and some with only the building points (high elevation). Voronoi cells with low and high points are extracted for further modification because building segments can be found in those cells. We are using some made-up data with an L-shaped building to illustrate the method.

**Fig. 10.2** Raw LIDAR data points

The extracted cells contain low and high points which will be split into two. The direction of the splitting line is found by calculating the eigenvalues and eigenvectors of the 3 x 3 variance-covariance matrix of the coordinates of the points within each cell. The result of three eigenvectors "explain" the

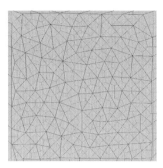

**Fig. 10.3** Lower resolution LIDAR data points

overall variance, the left-over and the residue. For example, a wrinkled piece of paper might have the first eigenvector (the highest eigenvalue) oriented along the length of the paper, the second (middle eigenvalue) along its width, and the third (the smallest eigenvalue) "looking" along the wrinkles. Thus the eigenvector of the smallest eigenvalue indicates the orientation of a wall segment, if present, and looks along it. Figure 10.4 shows the eigenvector with the smallest eigenvalue which shows the orientation of the splitting line between the low and high points.

**Fig. 10.4** The eigenvector with the smallest eigenvalue

With the orientation of the splitting line, the next step is to find the best location to put the line and split the cell. This is achieved iteratively, by testing various positions of the line parallel to the smallest eigenvector in order to find the greatest difference between the low and the high points. Figure 10.5 shows a thick line which separates the high points from the low points. In order to minimize the effect of sloping roofs or terrain, only those elevations close to the line are used. If this maximum difference is not sufficiently large then no wall segment was detected. Therefore walls have a specified minimum height and this height difference is achieved within a very few "pixels".

10 3D City Modelling from LIDAR Data

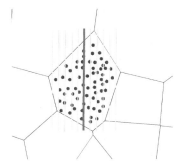

**Fig. 10.5** The thick red line separates the low and high data points

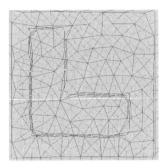

**Fig. 10.6** Building segment in each Voronoi cell

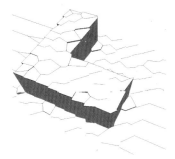

**Fig. 10.7** Vertical Building Walls formed by split Voronoi Cells

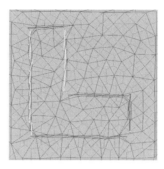

**Fig. 10.8** The split Voronoi edges in six groups

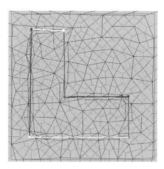

**Fig. 10.9** Corners of the building

We locate the splitting line and add a generator on each side of this line, at the mid-point to split the cell. Figures 10.6 and 10.7 show the 2D and 3D view of the split Voronoi cells. A set of "high" Voronoi cells are split and surrounded by "low" ones. In Figure 10.6 building boundaries are then determined by walking around the cells and connecting the Voronoi boundary segments (the splitting line) or the immediate Voronoi edges to form a closed region. If a closed high region is found, it is considered to be a building.

The Voronoi boundary segments are used to estimate the building outline. We use the Voronoi boundary segments which are created with the eigenvector technique (but not all the Voronoi edges to form the closed high region. The Voronoi segments are clustered according to their orientation (Figure 10.8). A best fit line is found to represent each group of the clustered Voronoi segments. Figure 10.9 forms the building outline by intersecting the best fit lines.

## 10.4 Roof Modelling

Many systems use pre-defined building models for reconstruction [2, 5], but we would like to reconstruct the buildings without any pre-defined models. Two methods are used to remodel the roof structure. The first is simple but works only with simple gabled roofs. The second is more complicated but can solve the problem of the complex roof structure.

### *10.4.1 Simple Roof*

No other assumptions are made about the form of the roof in our approach, except that the roof is made up of planar segments. This method may be extended to detect other basic shapes if required [14]. When the building boundary is determined, the interior points are extracted to model the roof structure. The extracted points are used to create a triangulation and each of the interior triangles has an associated vector normal (Figure 10.10. The vector normals are used to calculate and find the "smallest" eigenvector (described in the section 10.3). We project the vector normals on a right-hand coordinate system according to the "smallest" eigenvector (Figure 10.11).

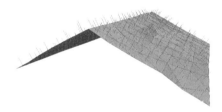

**Fig. 10.10** Each of the interior triangles has an associated vector normal

In Figure 10.12 all the vector normals are plotted on a unit semicircle. The plotted vector normals are close to each other if they have the same orientation. They can be clustered into different groups (Figure 10.13). This works well even if the data is fairly noisy because the scatter of the vector normals is fairly large (Figure 10.14). If there are two or more parallel planes on the roof, these may be separated at this stage by constructing the Delaunay triangulation in x-y space for the data points of the cluster, extracting the Minimum Spanning Tree (MST), and separating the two or more parallel roof portions. The general technique is described in the next section.

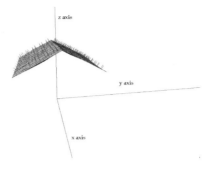

**Fig. 10.11** Projected right-hand coordinate system according the smallest eigenvector

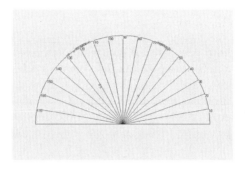

**Fig. 10.12** All vector normals are plotted on a semicircle

**Fig. 10.13** Clustered vector normals with its associated triangle

10 3D City Modelling from LIDAR Data

**Fig. 10.14** A clustered simple roof created by noisy data

## 10.4.2 Complex Roof

The above method only works for roofs with a simple axis. If the roof has many differently oriented segments, the vector normals have to be projected onto the unit hemisphere. [4] used different projections to find out the roof segments.

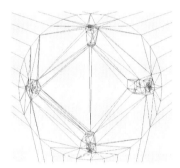

**Fig. 10.15** 2D view of vector normals on the unit hemisphere

A cross-hipped L-shape roof building is used to illustrate the clustering methods. The extracted data points (inside the building boundary) are used to create a Delaunay triangulation. The vector normals of the interior triangles are project onto the unit hemisphere. Figures 10.15 and 10.16 show the 2D and 3D views of projected vector normals on the unit hemisphere. Then the vector normals are clustered by their orientation and geographical location. The result is several sets of clustered triangles (vector normals) which share a single roof plane with a common description of the plane. The planar descriptions are used to form roof planes and intersect the building walls which produce the building model.

**Fig. 10.16** 3D view of vector normals on the unit hemisphere

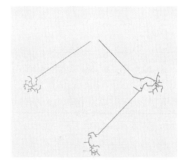

**Fig. 10.17** Orientation clustering of vector normals in 2D view

**Fig. 10.18** The darker triangles face toward the same direction

# 10 3D City Modelling from LIDAR Data

Orientation Clustering is the first method used to separate the vector normals. The location of the vector normals on the unit hemisphere represent their orientation on the roof (direction). They are clustered into groups using the Minimum Spanning Tree (MST). If the vector normals are close enough, they will be assigned to the same group. Figure 10.17 shows four groups of vector normals which means the triangles face four different directions. However the roof may contain more than four roof planes which means some roof planes face the same way. Figure 10.18 shows that triangles with the same orientation are clustered. Further clustering is needed to separate the same direction roof planes.

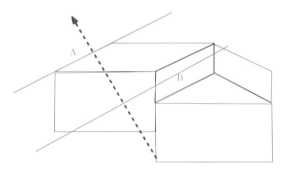

**Fig. 10.19** Triangles on roof A and B projected onto the averaged vector normal (thick dash line)

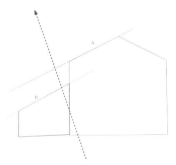

**Fig. 10.20** Triangles on roofs are projected on its averaged vector normal (thick dash line)

In the second clustering method, we average the vector normals (the thick dashed line in Figure 10.19) and project the centre point of the triangles (solid thin lines in Figure 10.19) onto its averaged vector normal. They are in the same group if the projected centre points are close to each other. The same method is used to separate the building extension due to the height difference between the main and the extension buildings. In Figure 10.20 triangles on Roof A (main building) and B (extension building) are projected and clustered into two groups because of their locations.

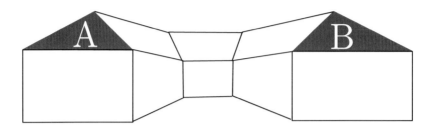

**Fig. 10.21** A building with complicated hipped roofs

Figure 10.21 shows roof planes A and B in the same group after two clustering; however they are two different roof planes. The geographical clustering method is used to separate roof planes A and B. The centre points of the triangles are extracted to create a Delaunay triangulation. We then cluster the centre points using the MST. Roofs A and B will be separated into two groups.

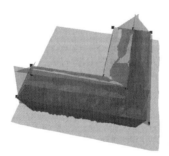

**Fig. 10.22** Square points are the intersection points

Finally each cluster of triangles represents a roof plane. Once we have got all the roof planes, we need to consider the intersection of the roof planes. The relationships between roof planes may be represented as a dual triangulation. We use the dual triangulation to intersect every adjacent three to four roof planes. Figure 10.22 shows the intersection points (square points) between the roof planes and the vertical walls of the building.

## 10.5 Tools for Building Reconstruction

Our approach is to reconstruct a 3D city model with preserved topological connectivity. We have successfully developed a set of tools for building reconstruction [12, 9]. Few steps are used for the building reconstruction:

- Delete all the data points which are inside the building boundary.
- Insert the intersection points (include the building outline and the roof structure points) using the constrained Delaunay triangulation.
- Use Euler Operators to extrude the building to its height and remodel the roof structure.

The result of an extruded L-shape building is shown in Figure 10.23.

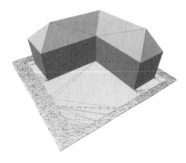

**Fig. 10.23** An extruded L-shape building

## 10.6 Conclusion

We have outlined a procedure for the direct extraction of building exteriors from LIDAR data without any additional data sources and pre-defined building models. More complicated buildings have been built. With the help of the research of [3] may be able to model the roof like the Wales Millennium Centre which is an arch shape roof.

## References

[1] F. Ackermann. Airborne laser scanning - present status and future expectations. *ISPRS Journal of Photogrammetry & Remote Sensing*, 54(1):64–67, 1999.

[2] Claus Brenner. Interactive modelling tools for 3D building reconstruction. In D. Fritsch and R. Spiller, editors, *Photogrammetric Week '99'*, pages 23–34, Wchmann Verlag, Heidelberg, 1999.

[3] H. A. K. Charlesworth, C. W. Langenberg, and J. Ramsden. Determining axes, axial places and sections of macroscopic folds using computer-based methods. *Candian Journal Earth Science*, 13:54–65, 1975.

[4] Alexandra D. Hofmann, Hans-Gerd Maas, and Andre Streilein. Derivation of roof types by cluster analysis in parameter spaces of airborne laserscanner point clouds. In *IAPRS International Archives of Photogrammetry and Remote Sensing and Spatial Information Sciences*, volume 34, Part 3/ W13, pages 112–117, Dresden, Germany, 2003.

[5] F. Rottensteiner and C. Briese. Automatic generation of building models from LIDAR data and the integration of aerial images. In H.-G. Maas, G. Vosselman, and A. Streilein, editors, *Proceedings of the ISPRS working group III/3 workshop '3-D reconstruction from airborne laserscanner and InSAR data'*, volume 34 Session IV, Dresden, Germany, 2003. Institute of Photogrammetry and Remote Sensing Dresden University of Technology.

[6] Gunho Sohn and Ian Dowman. Building extraction using lidar DEMS and IKONOS images. In H.-G. Maas, G. Vosselman, and A. Streilein, editors, *Proceedings of the ISPRS working group III/3 workshop '3-D reconstruction from airborne laserscanner and InSAR data'*, volume 34 Session IV, Dresden, Germany, 2003. Institute of Photogrammetry and Remote Sensing Dresden University of Technology.

[7] Gunho Sohn and Ian J. Dowman. Extraction of buildings from high resolution satellite data and LIDAR. In *ISPRS 20th Congress WGIII/4 Automated Object Extraction*, Istanbul, Turkey, 2004.

[8] I. Suveg and G. Vosselman. Reconstruction of 3D building models from aerial images and maps. *ISPRS Journal of Photogrammetry & Remote Sensing*, 58(3–4):202–224, 2004.

[9] R. O.C. Tse. *Semi-Automated Construction of fully three-dimensional terrain models*. PhD thesis, The Hong Kong Polytechnic University, Hong Kong, 2003.

[10] R.O.C. Tse and C.M. Gold. Terrain, dinosaurs and cadastres - options for three-dimension modelling. In C. Lemmen and P. van Oosterom, editors, *Proceedings: International Workshop on "3D Cadastres"*, pages 243–257, Delft, The Netherlands, 2001.

[11] R.O.C. Tse and C.M. Gold. Tin meets CAD - extending the TIN concept in GIS. In P.M.A. Sloot, C.J.K. Tan, J. Dongarra, and A.G. Hoekstra, editors, *Computational Science - ICCS 2002, International Conference, Proceedings of Part III. Lecture Notes in Computer Science*, volume 2331, pages 135–143, Amsterdam, the Netherlands, 2002. Springer-Verlag.

[12] R.O.C. Tse and C.M. Gold. Tin meets CAD - extending the TIN concept in GIS. *Future Generation Computer Systems (Geocomputation)*, 20(7):1171–1184, 2004.
[13] George Vosselman and Sander Dijkman. 3D building model reconstruction from point clouds and ground plans. In *International Archives of the Photogrammetry, Remote Sensing and Spatial Information Sciences*, volume 34, part 3/W4, pages 37–43, Annapolis, MA, USA, 2001.
[14] George Vosselman, B.G.H. Gorte, G. Sithole, and T. Rabbani. Recognising structure in laser scanner point clouds. In *International Archives of Photogrammetry, Remote Sensing and Spatial Information Sciences*, volume 46, part 8/W2, pages 33–38, Freiburg, Germany, 2004.

# Chapter 11
# First implementation results and open issues on the Poincaré-TEN data structure

Friso Penninga and Peter van Oosterom

**Abstract**

Modeling 3D geo-information has often been based on either simple extensions of 2D geo-information modeling principles without considering the additional 3D aspects related to correctness of representations or on 3D CAD based solutions applied to geo-information. Our approach is based from the scratch on modeling 3D geo-information based on the mathematically well-defined Poincaré-TEN data structure. The feasibility of this approach still has to be verified in practice. In this paper, the first experiences of loading a reasonable sized data set, comprised of about 1,800 buildings represented by nearly 170,000 tetrahedrons (including the 'air' and 'earth'), are discussed. Though the Poincaré-TEN data structure is feasible, the experience gained during the implementation raises new research topics: physical storage in one (tetrahedron only) or two tables (tetrahedron and node), effective clustering and indexing improvements, more compact representations without losing too much performance, etc.

## 11.1 Introduction

### 11.1.1 Motivation

This paper presents the first implementation results of the Poincaré-TEN data structure, as presented earlier in [1]. This structure is developed within a research project 3D Topography and a prototype is being developed within

---

Delft University of Technology, OTB, section GIS Technology,
Jaffalaan 9, 2628 BX the Netherlands
F.Penninga@tudelft.nl, oosterom@tudelft.nl

Oracle Spatial. The theoretical strengths of this concept (a compact topological DBMS approach based on a solid mathematical foundation) were demonstrated in previous papers [1, 2, 3]. Despite these strengths, the applicability of the new approach depends heavily on whether the approach is feasible in terms of storage requirements and performance. Therefore, implementing and testing these new ideas is essential. The first implementation results will provide insight to the number of TEN elements and provide some preliminary ideas on storage requirements (as future optimization steps will affect these requirements). At the same time implementing the approach raises new design questions and these open problems will be presented.

### 11.1.2 Related research

Research in the field of 3D GIS has been performed over the last two decades. Zlatanova et al. [4] gave an overview of the most relevant developments during this period. Related to the topics discussed in this paper, Carlson [5] can be seen as the starting point as he introduced a simplicial complex-based approach of 3D subsurface structures. However, this approach was limited to the use of 0-, 1- and 2-simplexes in 3D space. Extending this into higher dimensions (as indicated by Frank and Kuhn [6]) is mentioned as a possibility. The explicit use of 3D manifolds to model 3D features is explored by Pigot [7, 8] and Pilouk [9] introduces the TEtrahedral irregular Network (TEN), in which the 3-simplex is used as a building block. However, in their work, a rigid mathematical foundation is missing. As far as can be deducted from their descriptions, the 3D simplices are explicitly represented by 2D simplices, specifically, triangles (which are in turn represented by edges and nodes). A topological data model based on 2D simplicial complexes (in 2D space) is introduced [10] and implemented in the PANDA system [11], an early object-oriented database. In applications polyhedrons are often used as 3D primitive [12, 13].

### 11.1.3 Overview of paper

Before describing the first implementation results and open issues, we will first describe the core characteristics of the previously introduced Poincaré-TEN approach in Section 11.2. After, the approach will be applied to modeling 3D Topography in Section 11.3, while Section 11.4 summarizes the implementation details. The preliminary implementation results with the 1,800 building data set are described in Section 11.5. This paper ends with discussing the current implementation and related open issues in Section 11.6.

## 11.2 The Poincaré-TEN approach

In this section, first three aspects of our Poincaré-TEN approach are further explained, before the full concept of the approach is used as the foundation for 3D topography modeling:

- It models the world as a full decomposition of 3D space
- The world is modelled in a Tetrahedronized Irregular Network (TEN)
- The TEN is modelled based on Poincaré simplicial homology

### 11.2.1 Characteristic 1: Full Decomposition of Space

As the Poincaré-TEN approach is developed with 3D topographic data in mind, two fundamental observations are of great importance [14]:

- Physical objects have by definition a volume. In reality, there are no point, line or polygon objects, only point, line or polygon representations exist (at a certain level of abstraction/generalization). The ISO 19101 Geographic information - Reference model [15] defines features as 'abstractions of real world phenomena'. In most current modeling approaches, the abstraction (read 'simplification') is in the choice of a representation of lower dimension. However, as the proposed method uses a tetrahedral network (or mesh), the simplification is already in the subdivision into easy-to-handle parts (i.e. it is a finite element method!).
- The real world can be considered a volume partition: a set of nonoverlapping volumes that form a closed (i.e. no gaps within the domain) modelled space. As a consequence, objects like 'earth' or 'air' are explicitly part of the real world and thus have to be modelled.

Although volume features are the basic elements in the model, planar features might still be very useful, as they mark the boundary (or transition) between two volume features. This approach allows for the existence of planar features, but only as 'derived features'. In terms of UML class diagrams, these planar features are modelled as association classes. For instance, the 'earth surface' is the result of the association between 'earth' and 'non-earth'. Such features might be labeled (for instance as 'grassland' or 'road surface' with additional attributes), but they do not represent or describe the volume object. For example, a road is represented by a volume (despite the appearance of planar features like the road surface), with neighboring volumes that might represent air, earth or other adjacent features.

The explicit inclusion of earth and air features is not very common, since these features are usually considered empty space in between topographic features. Based on following two arguments, we decided to deviate from common practice. First, air and earth features are often also the subject of analyses.

One can think of applications like modeling noise propagation or air pollution. Second, by introducing earth and air features future extensions of the model will be enabled (beyond Topography). Space that is currently labeled as air can be subdivided into air traffic or telecommunication corridors, while earth might be subclassified into geographic layers or polluted regions.

## 11.2.2 Characteristic 2: using a TEN

Despite initial ideas on a hybrid data model (an integrated TIN/TEN model, based on a pragmatic approach to model in 2,5D as much as possible and to switch to a full 3D model in exceptional cases only), the decision was made [14] to model all topographic features in a TEN. The preference for these simplex-based data structures is based on certain qualities of simplexes (a simplex can be defined as the simplest geometry in a dimension, regarded as the number of points required to describe the geometry):

- Well defined: a n-simplex is bounded by n + 1 (n - 1)-simplexes. E.g. a 2-simplex (triangle) is bounded by 3 1-simplexes (edges)
- Flatness of faces: every face can be described by three points
- A n-simplex is convex (which simplifies amongst others point-in-polygon tests)

Due to the use of simplexes, a 1:n relationship between features and their representations is introduced. The actual usability of the Poincaré-TEN approach depends on the actual size of this n and will be discussed later in this paper.

## 11.2.3 Characteristic 3: applying Poincaré simplicial homology

The new volumetric approach uses tetrahedrons to model real world features. Tetrahedrons consist of nodes, edges and triangles. All four data types are simplexes: the simplest geometry in each dimension, in which simple refers to minimizing the number of points required to define the shape. A more formal definition [16] of a $n$-simplex $S_n$ is: a $n$-simplex $S_n$ is the smallest convex set in Euclidian space $\mathbb{R}^m$ containing $n+1$ points $v_0, \ldots, v_n$ that do not lie in a hyperplane of dimension less than $n$. As the $n$-dimensional simplex is defined by $n+1$ nodes, it has the following notation: $S_n = <v_0, \ldots, v_n>$. The boundary of a $n$-simplex is defined by the following sum of $n-1$ dimensional simplexes [17] (the *hat* symbol indicates omitting the specific node):

# 11 Implementing the Poincaré-TEN data structure

$$\partial S_n = \sum_{i=0}^{n}(-1)^i <v_0,\ldots,\hat{v}_i,\ldots,v_n>$$

This results in the following boundaries (also see Figure 11.1):

$S_1 = <v_0, v_1>$  $\quad \partial S_1 = <v_1> - <v_0>$
$S_2 = <v_0, v_1, v_2>$  $\quad \partial S_2 = <v_1, v_2> - <v_0, v_2> + <v_0, v_1>$
$S_3 = <v_0, v_1, v_2, v_3>$  $\quad \partial S_3 = <v_1, v_2, v_3> - <v_0, v_2, v_3>$
$\quad\quad\quad\quad\quad\quad\quad\quad\quad + <v_0, v_1, v_3> - <v_0, v_1, v_2>$

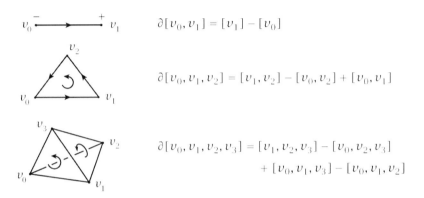

$\partial[v_0, v_1] = [v_1] - [v_0]$

$\partial[v_0, v_1, v_2] = [v_1, v_2] - [v_0, v_2] + [v_0, v_1]$

$\partial[v_0, v_1, v_2, v_3] = [v_1, v_2, v_3] - [v_0, v_2, v_3]$
$\quad\quad\quad\quad\quad\quad + [v_0, v_1, v_3] - [v_0, v_1, v_2]$

**Fig. 11.1** Simplexes and their boundaries (From [16])

All simplexes are ordered. As a simplex $S_n$ is defined by $n+1$ vertices, $(n+1)!$ permutations exist. All even permutations of an ordered simplex $S_n = <v_0,\ldots,v_n>$ have the same orientation, all odd permutations have opposite orientation. So edge $S_1 = <v_0, v_1>$ has boundary $\partial S_1 = <v_1> - <v_0>$. The other permutation $S_1 = -<v_0, v_1> = <v_1, v_0>$ has boundary $\partial S_1 = <v_0> - <v_1>$, which is the opposite direction. As a consequence operators like the dual of a simplex, that is the simplex with the opposite orientation, become very simple: it only requires a single permutation.

The direction of all oriented boundaries of a given simplex obtained with the above boundary operator formula is the same. In 3D this results in the favorable characteristic that with $S_3$ either all normal vectors of the boundary triangles point inwards or all normal vectors point outwards. This is a direct result of the boundary operator definition, as it is defined in such a manner that $\partial^2 S_n$ is the zero homomorphism, i.e. the boundary of the boundary equals zero (summing-up the positive and negative parts). For example, consider $\partial^2 S_3$, a tetrahedron. The boundary of this tetrahedron consists of four triangles, and the boundaries of these triangles consist of edges. Each of the six edges of $S_3$ appears twice, as each edge bounds two triangles. Since the zero homomorphism states that the sum of these edges equals zero, this is the case if and only if the edges in these six pairs have opposite signs. The edges of two neighboring triangles have opposite signs if and only if the triangles have similar orientation, i.e. either both are oriented outwards or both are oriented inwards. This characteristic is important in deriving the boundary of a simplicial complex (construction of multiple simplexes). If this identical orientation is assured for all boundary triangles of tetrahedrons (which can be achieved by a single permutation when necessary), deriving the boundary triangulation of a feature will reduce to adding up boundary triangles of all related tetrahedrons, as internal triangles will cancel out in pairs due to opposite orientation. Figure 11.2 shows an example in which all boundaries of the tetrahedrons are added to obtain the boundary triangulation of the building.

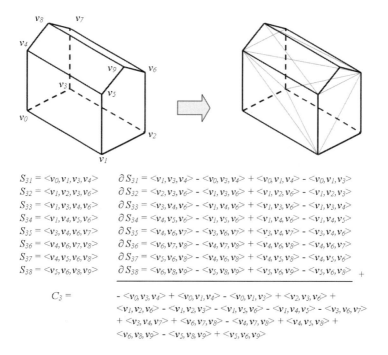

**Fig. 11.2** Deriving the boundary triangulation from the TEN

## 11.3 Poincaré-TEN approach to modeling 3D Topography

### 11.3.1 Conceptual model

Usually [8, 9], tetrahedrons are defined by four triangles, triangles by three edges and edges by two nodes. Geometry is stored at node level. As a result, reconstructing geometry, for instance a tetrahedron, becomes a relatively laborious operation. In simplicial homology, simplexes of all dimensions are defined by their vertices only, while relationships between other simplexes can be derived by applying the boundary operator. Due to the availability of this operator, there is no need for explicit storage of these relationships. This concept is illustrated in the UML class diagram in Figure 11.3. Tetrahedrons, triangles and edges are defined by an ordered list of nodes. The mutual relationships between tetrahedrons, triangles and nodes (the boundary/coboundary relationships) are derived and signed (i.e. oriented).

Figure 11.3 shows the concept of full space decomposition. The real world consists of volume features and features of lower dimension are modelled as association classes. As a result, instances of these classes are lifetime dependent on the relationship between two volume features.

### 11.3.2 Extending simplex notation: vertex encoding

In the Poincaré-TEN approach to 3D topographic data modeling, simplexes are defined by their vertices. Identical to the simplex notation from simplicial homology, where for instance a tetrahedron is noted as $S_3 = <v_0, v_1, v_2, v_3>$, simplex identifiers are constructed by concatenating the vertex ID's. In doing so, unique identifiers exist that contain orientation information as well, since the order of vertices determines the orientation. In an earlier paper [1], we suggested the use of x, y and z coordinate concatenation as node ID. Since geometry is the only attribute of a vertex, adding a unique identifier to each point and building an index on top of this table will cause a substantial increase in data storage. The geometry itself will be a unique identifier. Concatenating the coordinate pairs into one long identifier code and sorting the resulting list, will result in a very basic spatial index. In a way this approach can be seen as building and storing an index, while the original table is deleted.

Figure 11.4 [1] illustrates this idea of vertex encoding in a simplicial complex-based approach. A house is tetrahedronized and the resulting tetrahedrons are coded as the concatenation of their four vertices' coordinates. Each row in the tetrahedron encoding can be interpreted as $x_1y_1z_1x_2y_2z_2x_3y_3z_3x_4y_4z_4$. For reasons of simplicity, only two positions are used for each coordinate ele-

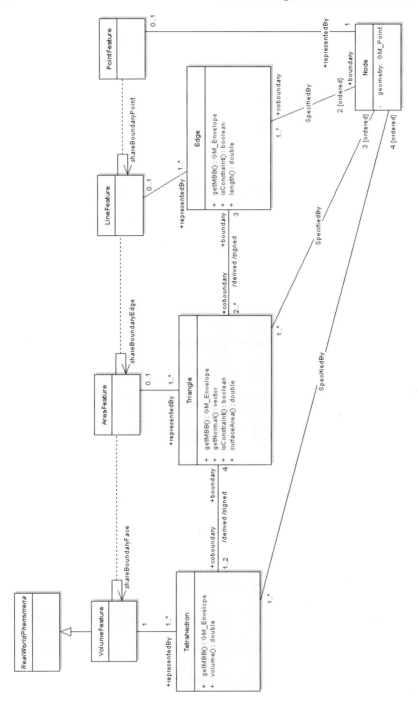

**Fig. 11.3** UML class diagram of the simplicial complex-based approach

ment. Therefore, the last row (1000000000600100600100608) should be interpret as the tetrahedron defined by the vertices (10, 00, 00), (00, 06, 00), (10, 06, 00) and (10, 06, 08), which is the tetrahedron at the bottom right of the house.

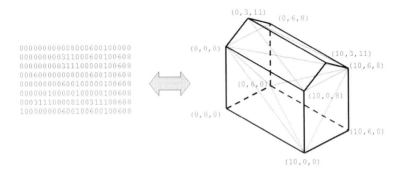

**Fig. 11.4** Describing tetrahedrons by their encoded vertices

## 11.4 Current implementation

To provide greater insight into the proposed new approach, the basic structure is implemented within the Oracle DBMS. This section will summarize the current status of the implementation, but one has to realize that this is still a work in progress. At this moment, the required tetrahedronization algorithms are not implemented within the DBMS, so TetGen [18] is used to perform an external batch tetrahedronisation. The input is a Piecewise Linear Complex (PLC), see Figure 11.5. A PLC [19] is a set of vertices, segments and facets, where a facet is a polygonal region. Each facet may be non-convex and hcontain holes, segments and vertices, but it should not be a curved surface. A facet can represent any planar straight line graph (PSLG), which is a popular input model used by many two-dimensional mesh algorithms. A PSLG is [20] a graph embedding of a planar graph (i.e. a graph without graph edge crossings), in which only straight line segments are used to connect the graph vertices.

Compared to a polyhedron, a PLC is a more flexible format. If one looks at the shaded facet in Figure 11.5, one can see that this facet cannot be described by a polygon because there are loose and dangling line segments. However, in our application, situations like these will be rare or not appear at all. Based on an input PLC, TetGen creates a constrained Delaunay tetrahedronisation. This tetrahedronization is loaded into the database and then converted into the Poincaré-TEN format. Figure 11.6 shows this concept with a small test dataset, from the input PLC (top), via the tetrahedronisation (mid) to the

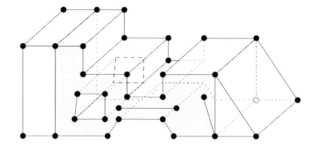

**Fig. 11.5** A Piecewise Linear Complex (PLC), input for the tetrahedronization algorithm (From [21])

output in which only the constrained triangles (the feature boundary faces) are drawn (bottom).

The tetrahedron table is the only table in the implementation. It consists of a single column (NVARCHAR2) in which the encoded tetrahedrons are described in the form $x_1y_1z_1x_2y_2z_2x_3y_3z_3x_4y_4z_4id$ (based on fixed length character strings). Note that besides the geometry, an unique identifier is added, which refers to a volume feature that is (partly) represented by the tetrahedron. The tetrahedrons are not signed, but are assumed to be a positive permutation, meaning that all normal vectors on boundary triangles are oriented outwards. This is checked and ensured during the initial loading proces. A consistent orientation is required to ensure that each boundary triangle appears twice: once with positive and once with negative orientation. The orientation simplifies determination of left/right and inside/outside relations. Based on the encoded tetrahedrons the boundary triangles can be derived by applying the boundary operator:

```
create or replace procedure deriveboundarytriangles(
    (...)
    a  := (SUBSTR(tetcode,1,3*codelength));
    b  := (SUBSTR(tetcode,1+3*codelength,3*codelength));
    c  := (SUBSTR(tetcode,1+6*codelength,3*codelength));
    d  := (SUBSTR(tetcode,1+9*codelength,3*codelength));
    id := (SUBSTR(tetcode,1+12*codelength));
    ordertriangle(codelength,'+'||b||c||d||id, tricode1);
    ordertriangle(codelength,'-'||a||c||d||id, tricode2);
    ordertriangle(codelength,'+'||a||b||d||id, tricode3);
    ordertriangle(codelength,'-'||a||b||c||id, tricode4);
    (...)
```

Note that the triangles inherit the object id from the tetrahedron, i.e. each triangle has a reference to the volume feature represented by the tetrahedron

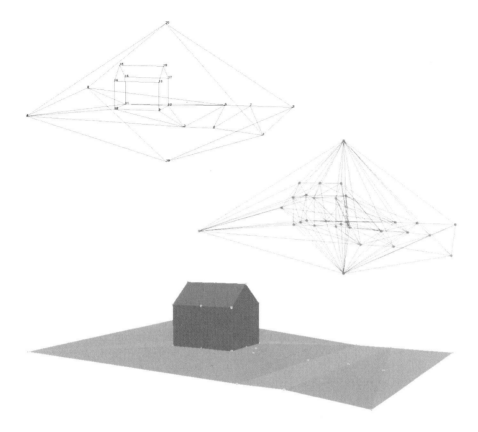

**Fig. 11.6** Input PLC (top), the resulting tetrahedronization (mid) and as output the constrained triangles (i.e. the feature boundaries)(bottom)

of which the triangle is part of the (internal) boundary. The reason for this will be introduced in the next section. Also, it can be seen that each boundary triangle is ordered by the `ordertriangle` procedure. The objective of this procedure is to gain control over which permutation is used. A triangle has six (= 3!) permutations, but it is important that the same permutation is used both in the positive and negative orientations, as they will not cancel out in pairs otherwise. The `ordertriangle` procedure always rewrites a triangle $<a,b,c>$ such that $a < b < c$ holds, which is an arbitrary criterion.

Based on a slightly altered version of the `deriveboundarytriangles` operator the triangle view is created. The resulting view contains all triangles (coded by their geometry and inherited object id's) and their coboundaries (the coboundary of a $n$-dimensional simplex $S_n$ is the set of all $(n+1)$-dimensional simplexes $S_{n+1}$ of which the simplex $S_n$ is part of their boundaries $\partial S_{n+1}$). In this case, the coboundary is a tetrahedron, of which the triangle is part of the boundary. This coboundary will prove useful in deriving topological relationships later in this section. The resulting view will contain four times the number of tetrahedrons and every triangle appears twice: once with a positive and once with a negative sign (and not in a permutated form, due to the ordertriangle procedure). However, it must be realized that this is just a view and no actual storage takes place:

```
create or replace view triangle as
    select deriveboundarytriangle1(tetcode) tricode,
    tetcode fromtetcode from tetrahedron
    UNION ALL
    select deriveboundarytriangle2(tetcode) tricode,
    tetcode fromtetcode from tetrahedron
    UNION ALL
    select deriveboundarytriangle3(tetcode) tricode,
    tetcode fromtetcode from tetrahedron
    UNION ALL
    select deriveboundarytriangle4(tetcode) tricode,
    tetcode fromtetcode from tetrahedron;
```

Features in the model are represented by a set of tetrahedrons. To ensure that these tetrahedrons represent the correct geometry, the outer boundary is triangulated and these triangles are used as constraints. This implies that these triangles will remain present as long as the feature is part of the model (i.e. they are not deleted in a update proces). To achieve this, the incremental tetrahedronization algorithm needs to keep track of these constrained triangles. In contrast with what one might expect, it is not necessary to store these constraints explicitly, as they can be derived. This derivation is based on the fact that although every triangle (in a geometric sense) appears two times (with opposite orientation) in the triangle view, not every triangle *code* appears twice. As stated before, the triangle code inherits the object id from the tetrahedron (its coboundary). This implies that for internal triangles (i.e. within an object) the triangle and its dual will have (apart from the sign) the exact same triangle code (geometry + object id), but in case of boundary triangles (i.e. constrained triangles) this code will differ due to the different inherited object id's. A view with constrained triangles can be derived:

```
create or replace view constrainedtriangle as
    select t1.tricode tricode from triangle t1
    where not exists (select t2.tricode from triangle t2
                      where t1.tricode = t2.tricode*-1);
```

Other views might be defined to simplify operations, for instance, a view with triangles stored by their geometry alone or a view without duals.

Similar to deriving the triangle views, views with edges, constrained edges and nodes can be constructed. Note that the views with edges contain no duals, i.e. edges are described only by their geometry:

```
create or replace view edge as
```

## 11 Implementing the Poincaré-TEN data structure

```
      select distinct deriveabsboundaryedge1(tricode) edcode
      from triangle
      UNION
      select distinct deriveabsboundaryedge2(tricode) edcode
      from triangle
      UNION
      select distinct deriveabsboundaryedge3(tricode) edcode
      from triangle;

   create or replace view constrainededge as
      select distinct deriveabsboundaryedge1(tricode) edcode
      from constrainedtriangle
      UNION
      select distinct deriveabsboundaryedge2(tricode) edcode
      from constrainedtriangle
      UNION
      select distinct deriveabsboundaryedge3(tricode) edcode
      from constrainedtriangle;

   create or replace view node as
      select distinct deriveboundarynode1(edcode) nodecode
      from edge
      UNION
      select distinct deriveboundarynode2(edcode) nodecode
      from edge;
```

In the current implementation edges are undirected and do not inherit object id's, as no application for this has been identified. However, strict application of the boundary operator results in directed triangles. With the tetrahedron table and triangle, edge and node views, the data structure is accessible at different levels. Due to encoding of the vertices, both geometry and topology are present at every level, thus enabling switching to the most appropriate approach for every operation.

## 11.5 Preliminary implementation results

The very small dataset from Figure 11.6 is now replaced by a larger dataset. It consists of 1796 buildings in the northern part of Rotterdam (see Figure 11.7) and covers an area of about seven square kilometres. At this moment other topographic features like the earth surface, roads, tunnels etc. are still missing due to lacking appropriate 3D data. Complex situations like multiple land use and highway interchanges are lacking as well. However, this dataset will provide more insight into the number of elements of a TEN.

**Fig. 11.7** Rotterdam test data set: 1796 buildings

## 11.5.1 An alternative model

Until now, simplex codes are obtained by concatenating the node coordinates, like $x_1y_1z_1x_2y_2z_2x_3y_3z_3x_4y_4z_4$. This approach is based on the idea that a node table only contains geometry and that adding an identifier would be a bit redundant, since the geometry is already an unique identifier. Nevertheless, one can question whether this approach actually reduces storage requirements, since each node is part of multiple tetrahedrons (the Rotterdam tetrahedronization shows an average of about fifteen tetrahedrons per node; see next subsection). Due to this result, the concatenated coordinate pair is used multiple times. As long as a node identifier requires considerably less storage space compared to this concatenated geometry, switching to a tetrahedron-node approach might be feasible. An additional node table containing a node identifier is required to create shorter simplex codes like $nid_1 nid_2 nid_3 nid_4$.

With this idea in mind tetrahedronization of the Rotterdam data set was performed. Since TetGen output consists of a comparable structure with both a tetrahedron and a node table, incorporating TetGen results in the Poincaré-TEN structure was easier in the tetrahedron-node version. Since obtaining a working implementation was strongly favored over minimizing storage requirements or optimizing performance at this point, the tetrahedron-node implementation was used. As a result, two tables with tetrahedrons and nodes are stored. All simplexes are identified by a concatenation/permutation of node id's instead of concatenated coordinate pairs. All pro's and con's regarding the choice between a tetrahedron-only and a tetrahedron-node implementation will be discussed into more detail in Section 11.6.2.

## *11.5.2 Preliminary results on storage requirements*

The Rotterdam data set is first converted into the input format for TetGen, the tetrahedronization software. This input format requires a list of nodes with geometry, a list of faces (described by their nodes) and a list of points inside each object to identify the object. Real volumetric 3D data is rare, so one has to convert data or integrate multiple sources. Modifying this into the topological format in which faces are described by their nodes is usually a very time-consuming task. The input dataset consists of 26,656 nodes, 16,928 faces and 1,796 points to identify the 1,796 buildings. Tetrahedronizing this input set with TetGen results in a TEN, consisting of 30,877 nodes, 54,566 constrained triangles and 167,598 tetrahedrons. One should note that the tetrahedronization results in one network, i.e. the space in between the buildings is tetrahedronized as well! The increase in the number of nodes is caused by the addition of Steiner points; additional points required to either enable tetrahedronization or improve tetrahedronization quality (in terms of avoiding ill-shaped triangles and tetrahedrons to avoid numerical instability).

The tetrahedronization results are loaded into the Poincaré-TEN structure. The tetrahedron table consists of 167.598 tetrahedrons. Based on these tetrahedron table, views are created with triangles, constrained triangles, edges and nodes by repeatedly applying the boundary operator. Since the Poincaré-TEN structure contains duals of all triangles as well, the numbers differ from the initial tetrahedronization. From the 167,598 tetrahedrons 670,392 (4 x number of tetrahedrons) triangles are derived, of which 109,120 are constrained triangles. This number slightly differs from multiplying the original 54.566 constrained triangles by two (because of inclusion of the dual), since the outer boundary of the TEN consists of twelve triangles without a dual. The edge view provides information for the 198,480 edges. Note that these edges are described by their nodes alone, so without inherited object id, dual or sign, i.e. each geometry is unique.

As stated before, the current implementation is very straightforward. Obtaining a working implementation was strongly favored over minimizing storage requirements or optimizing performance. However, improving these aspects is one of the most important tasks for the upcoming period. Nevertheless, storage requirements of the current approach are compared to requirements of a polyhedron approach. The polyhedrons are described in Oracle as a solid, defined by a set of polygonal faces, each described by their vertices. The TEN approach slightly differs from previously described implementations, as it consists of both a tetrahedron and a node table.

The Poincaré-TEN approach requires 1.44 and 19.65 MB, respectively, for the node and tetrahedron table, while the polyhedron tables requires 4.39 MB. This means that the current (absolutely not optimized!) implementation requires about 4.8 times more storage space. However, as will be discussed in Section 11.6.4, the feasibility of our approach should be assessed both based on storage requirements as well as performance. A simple storage reduction can be obtained by using bit string instead of character string representation of the coordinates. Not only will this save storage space (estimated between a factor 2 to 3), but it would also increase performance and no ascii to binary conversions are necessary when using the coordinates.

## 11.6 Discussion of open issues

The prototype implementations show that the Poincaré-TEN approach is indeed feasible and can be used for well defined representation, but is still usable in basic GIS functions: selection of relevant objects and their visualization. Further, basic analysis is very well possible: both using topology (e.g. find the neighbors of a given feature) and geometry (e.g. compute volume of a given feature by summing tetrahedron volumes). Finding neighbors of a given feature can be implemented by querying the constrained triangle view to find all boundary triangles with a specific feature identifier. Through a view with triangle duals, the neighboring features can be identified quickly. An alternative approach would be to traverse the TEN tetrahedron by tetrahedron and test for feature identifier changes. A function to find neighboring tetrahedrons can be defined:

```
create or replace function getneighbourtet1(
 (...)
   select fromtetcode into neighbourtet from triangle
   where removeobjectid(tricode)= -1 *removeobjectid(tricode);
 (...)
```

or the volume of a tetrahedron can be calculated using the Cayley-Menger determinant [22] (with $d_{ij}$ as length of edge $< v_i, v_j >$ ):

$$288V^2 = \begin{vmatrix} 0 & 1 & 1 & 1 & 1 \\ 1 & 0 & d_{01}^2 & d_{02}^2 & d_{03}^2 \\ 1 & d_{10}^2 & 0 & d_{12}^2 & d_{13}^2 \\ 1 & d_{20}^2 & d_{21}^2 & 0 & d_{23}^2 \\ 1 & d_{30}^2 & d_{31}^2 & d_{32}^2 & 0 \end{vmatrix}$$

Although these capabilities have been established, ongoing research is attempting to provide answers to a number of some open issues. These issues will be described in the final subsections of this paper.

### 11.6.1 Open issue 0. Spatial clustering and indexing

The large real world data set will require spatial organization of the data to enable the efficient implementation of spatial queries such as the rectangle (or box) selections. Spatial organization includes spatial clustering (things close in reality are also close in computer memory, which is tricky given the one dimensional nature of computer memory) and spatial indexing (given the spatial selection predicate, the addresses of the relevant objects can be found efficiently). If the current coding of the tetrahedrons (first x, then y, then z) is replaced by bitwise interleaving, the tetrahedron code itself may be used for spatial clustering (similar to the Morton code) and used for spatial indexing without using additional structures (such as quad-tree or r-tree, also requiring significant storage space and maintenance during updates) [23]. Only the coboundary references of the triangles might need functional indexes to improve performance.

### 11.6.2 Open issue 1. Minimizing redundancy: tetrahedron only vs. tetrahedron-node

In this paper, two variants of the implementation have been described. If a separate node table is used, with compact node id's, then the issue of realizing spatial clustering and indexing is relevant for both tables. As there is no direct geometry in the tetrahedron table, the bitwise interleaving approach of the coordinates cannot be used (and probably a more explicit technique has to be applied). At this time, no comparative results are available, but one can expect the tetrahedron-node variant to be cheaper in terms of storage than the tetrahedron-only approach. However, reducing data storage might deteriorate performance, as additional operations are necessary to perform geometrical operations on top of simplexes. If one thinks, for instance, of the operation that checks whether a tetrahedron is oriented positively or negatively, one needs the node coordinates to calculate a normal vector on one of the triangles and calculate the angle between this normal vector and a vector from a triangle opposite to the fourth node to determine whether the normal points inwards or outwards. To perform this operation in the tetrahedron-node implementation, one has to search the node table first to obtain the node geometries.

### 11.6.3 Open issue 2. Dealing with storage requirements: storing all coordinates vs. storing differences

Assuming that one opts for the tetrahedron only approach, storage requirements can be reduced by avoiding storage of the full coordinates. Since the four nodes are relatively close to each other, one might choose to store the coordinates of one node and only give difference vectors to the other three nodes: $x_1y_1z_1x_2y_2z_2x_3y_3z_3x_4y_4z_4$ would change into $xyz\delta x_1\delta y_1\delta z_1\delta x_2\delta y_2\delta z_2\delta x_3\delta y_3\delta z_3$. Similar to the choice between the tetrahedron only and the tetrahedron-node implementation, reducing data storage will come at a price. Again additional operators are required to reconstruct the four node geometries when necessary. However, if these can be implemented efficiently (and there is no reason why this can not be done), they could be used in a view translating the compact physical storage representation in a more verbose full representation (but as this is only a view, it is not stored so it does not matter that this size is larger). Also, the bitwise interleaving approach to provide spatial clustering and indexing may still work well with this approach (as it is sufficient to do only bitwise interleaving of the first coordinate).

### 11.6.4 Open issue 3. How to assess feasibility of the Poincaré-TEN approach

In the implementation of the theory, as indicated in this paper (first prototype and also the open issues described above for further improvement), care has to be taken so that the storage requirements are not excessive (compared to other approaches) as this would make the approach less feasible (storage requirements should be linear in the number of features represented). In general, bulky storage requires more time to retrieve data from the disk, as compared to compact storage. However, if very expensive computations are needed (e.g. joins which are not well supported), then bad response times could occur. It is important to implement the typical basic functionality effectively (both w.r.t. storage and time performance). At this moment, there seems to be no basic functions that cannot be implemented time efficiently (when proper clustering/indexing is applied). However, this assumption still has to be proven.

## 11.6.5 Open issue 4. Correct insertion of 3D objects: snapping to the earth surface

3D data sets are required to load into the current implementation. Although research efforts are made to increase availability of such datasets [24], dependence of the availability of such data sets seriously limits applicability of the data structure at this time. Therefore, additional functionality is required to switch from importing 3D data sets into importing 3D data from different sources. One can imagine that creation of a 3D topographic model starts with the creation of the earth surface, followed by inclusion of 3D buildings. In general, buildings are built on top of the earth surface. As the earth surface and building data originates from different sources, these objects are not likely to fit together perfectly. To cope with such situation, one needs a snap-to-earth-surface operator. Such an operator will project the buildings footprint onto the terrain and determine the distance between terrain and buildings underside. If this distance is smaller than a certain pre-set tolerance, the building will be placed on the terrain by applying a vertical displacement, thus ensuring a tight fit. Two options exist for this, as one can either adjust the buildings underside to fit the terrain or adjust the terrain to fit the (usually flat) underside of the building. The snapping operator can also be utilized for inclusion of infrastructural objects and land coverage objects.

## 11.6.6 Open issue 5. Incremental updating of existing structure in DBMS

The current implementation lacks any tetrahedronization algorithms. At this time, TetGen software is used and the resulting output is loaded into the database and subsequently converted into the Poincaré-TEN structure. With the intended application of 3D Topography in mind, bulk loading is useful for the initial model build, but updates should be handled incrementally. A theoretical framework of incremental updates in a TEN structure is presented in [2, 25]. However, these ideas still need further implementation and development. It will be most effective to devise incremental update procedures that act as local as possible, with the risk that quality parameters like the Delaunay criterion or shortest-to-longest edge ratios are temporarily not met. This could be compensated by a cleaning function that performs a local rebuild or even a full retetrahedronization. Obviously, such an operation needs to be performed every now and then, but not after every update, thus speeding up the update process.

# References

[1] Penninga, F., van Oosterom, P.: A Compact Topological DBMS Data Structure For 3D Topography. In Fabrikant, S., Wachowicz, M., eds.: Geographic Information Science and Systems in Europe, Agile Conference 2007. Lecture Notes in Geoinformation and Cartography, Springer (2007)

[2] Penninga, F., van Oosterom, P.: Updating Features in a TEN-based DBMS approach for 3D Topographic Data modeling. In Raubal, M., Miller, H.J., Frank, A.U., Goodchild, M.F., eds.: Geographic Information Science, Fourth International Conference, GIScience 2006, Münster, Germany, September 2006, Extended Abstracts. Volume 28 of IfGI prints. (2006) 147–152

[3] Penninga, F., van Oosterom, P., Kazar, B.M.: A TEN-based DBMS approach for 3D Topographic Data modeling. In Riedl, A., Kainz, W., Elmes, G., eds.: Progress in Spatial Data Handling, 12th International Symposium on spatial Data Handling, Springer (2006) 581–598

[4] Zlatanova, S., Abdul Rahman, A., Pilouk, M.: 3D GIS: Current Status and Perspectives. In: Proceedings of Joint Conference on Geo-Spatial Theory, Processing and Applications, Ottawa, Canada. (2002)

[5] Carlson, E.: Three-dimensional conceptual modeling of subsurface structures. In: Auto-Carto 8. (1987) 336–345

[6] Frank, A.U., Kuhn, W.: Cell Graphs: A provable Correct Method for the Storage of Geometry. In: Proceedings of the 2nd International Symposium on Spatial Data Handling, Seattle, Washington. (1986)

[7] Pigot, S.: A Topological Model for a 3D Spatial Information System. In: Proceedings of the 5th International Symposium on Spatial Data Handling. (1992) 344–360

[8] Pigot, S.: A topological model for a 3-dimensional Spatial Information System. PhD thesis, University of Tasmania, Australia (1995)

[9] Pilouk, M.: Integrated modeling for 3D GIS. PhD thesis, ITC Enschede, Netherlands (1996)

[10] Egenhofer, M., Frank, A., Jackson, J.: A Topological Data Model for Spatial Databases. In: Proceedings of First Symposium SSD'89. (1989) 271–286

[11] Egenhofer, M., Frank, A.: PANDA: An Extensible Dbms Supporting Object-Oriented Software Techniques. In: Datenbanksysteme in Büro, Technik und Wissenschaft. Proceedings of GI/SI Fachtagung, Zürich, 1989. Informatik Fachberichten, Springer-Verlag (1989) 74–79

[12] Zlatanova, S.: 3D GIS for urban development. PhD thesis, Graz University of Technology (2000)

[13] Stoter, J.: 3D Cadastre. PhD thesis, Delft University of Technology (2004)

[14] Penninga, F.: 3D Topographic Data modeling: Why Rigidity Is Preferable to Pragmatism. In Cohn, A.G., Mark, D.M., eds.: Spatial Infor-

mation Theory, Cosit'05. Volume 3693 of Lecture Notes on Computer Science., Springer (2005) 409–425
[15] ISO/TC211: Geographic information - reference model. Technical Report ISO 19101, International Organization for Standardization (2005)
[16] Hatcher, A.: Algebraic Topology. Cambridge University Press (2002) Available at http://www.math.cornell.edu/ hatcher.
[17] Poincaré, H.: Complément á l'Analysis Situs. Rendiconti del Circolo Matematico di Palermo **13** (1899) 285–343
[18] http://tetgen.berlios.de/: (2007)
[19] Miller, G.L., Talmor, D., Teng, S.H., Walkington, N., Wang, H.: Control Volume Meshes using Sphere Packing: Generation, Refinement and Coarsening. In: 5th International Meshing Roundtable, Sandia National Laboratories (1996) 47–62
[20] http://mathworld.wolfram.com/PlanarStraightLineGraph.html: (2007)
[21] Si, H.: TetGen, A Quality Tetrahedral Mesh Generator and Three-Dimensional Delaunay Triangulator. User's Manual. Technical report, Weierstrass Institute for Applied Analysis and Stochastics, Berlin, Germany (2006) Available at http://tetgen.berlios.de/files/tetgen-manual.pdf.
[22] Colins, K.D.: Cayley-Menger Determinant. From Mathworld – A Wolfram Web Resource. http://mathworld.wolfram.com/Cayley-MengerDeterminant.html (2003)
[23] van Oosterom, P., Vijlbrief, T.: The Spatial Location Code. In Kraak, M.J., Molenaar, M., eds.: Advances in GIS research II; proceedings of the seventh International Symposium on Spatial Data Handling - SDH'96, Taylor and Francis (1996)
[24] Oude Elberink, S., Vosselman, G.: Adding the Third Dimension to a Topographic Database Using Airborne Laser Scanner Data. In: Photogrammetric Computer Vision 2006. IAPRS, Bonn, Germany. (2006)
[25] Penninga, F., van Oosterom, P.: Editing Features in a TEN-based DBMS approach for 3D Topographic Data modeling. Technical Report GISt Report No. 43, Delft University of Technology (2006) Available at http://www.gdmc.nl/publications/reports/GISt43.pdf.

# Chapter 12
# Drainage reality in terrains with higher-order Delaunay triangulations

Ahmad Biniaz and Gholamhossein Dastghaibyfard

**Abstract**

Terrains are often modeled by triangulations, which ideally should have 'nice shape' triangles and reality of drainage in terrains (few local minima and drainage lines in the bottoms of valleys). Delaunay triangulation is a good way to formalize nice shape, and if higher-order Delaunay triangulations are used, drainage reality can be achieved. Two heuristics are presented, one for reducing the number of local minima and one for reducing the number of valley edges and components. The empirical results show how well they perform on real-world data; on average we see a 16% improvement over known algorithms.

## List of Abbreviations and Symbols

CHS    California Hot Springs
QP     Quinn Peak
SL     Sphinx Lakes
SM     Split Mountain
WP     Wren Peak
dl     Delaunay triangulation
of     Old flip heuristic
nf     New flip heuristic
h      Hull heuristic
fv     Flip plus valley heuristics
hv     Hull plus valley heuristics
fvvr   Flip plus valley plus valley-reduce heuristics

---

Department of Computer Science and Engineering, Shiraz University, Shiraz, Iran
biniaz@cse.shirazu.ac.ir, dstghaib@shirazu.ac.ir

hvvr   Hull plus valley plus valley-reduce heuristics

## 12.1 Introduction

Terrains are often modeled by triangulations. In nearly all applications where triangulations are used, the triangles must have a 'nice shape'. This is true for visualization, mesh generation [1], and terrain modeling [2]. Delaunay triangulation (*DT*) is a good way to formalize nice shape. Delaunay triangulation of a set *P* of $n$ points maximizes the minimum angle of its triangles, over all possible triangulations of *P*, and also lexicographically maximizes the increasing sequence of these angles.

For terrain modeling, there are criteria other than nice shape, such as reality of drainage and slope fidelity in terrains. Natural terrains do not have many local minima, because terrains are formed by natural processes and local minima would be eroded away by water flow [2]. When constructing a triangulated model of a terrain by Delaunay triangulation, some local minima may appear. This may be because Delaunay triangulation is defined for a planar set of points, and does not take into account the third dimension [3]. Another triangulation of the same points may not have these minima; it is therefore better to generate triangulated terrains with nice shape and few local minima. This leads us to use *higher-order Delaunay triangulations (HODT)* [3].

**Definition 12.1.** An edge in a point set *P* is order-$k$ if there exists a circle through its endpoints that has at most k points of *P* inside. A triangle in a point set *P* is order-$k$ if its circumcircle contains at most $k$ points of *P*. A triangulation of a set *P* of points is an order-$k$ Delaunay triangulation if every triangle of the triangulation is order-$k$ (see Fig. 1).

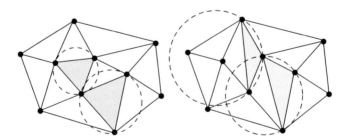

**Fig. 12.1** Left, order-0 Delaunay triangulation. Right, an order-2 Delaunay triangulation, with two triangles of orders 1 and 2.

Therefore a standard Delaunay triangulation is a unique order-0 Delaunay triangulation. For any positive integer $k$, there can be many different order-$k$

Delaunay triangulations. By the definition, any order-$k$ Delaunay triangulation is also an order-$k'$ Delaunay triangulation for all $k' > k$ [4]. The bigger $k$, the more freedom to eliminate artifacts like local minima, but the shape of the triangles may deteriorate. Higher order Delaunay triangulations have applications in realistic terrain modeling [5] and mesh generation [1].

Kok et. al. [5] showed that minimizing the number of local minima in order-$k$ Delaunay triangulations is *NP*-hard for larger values of $k$. For any $0 < \varepsilon < 1$ and some $0 < c < 1$, it is *NP*-hard to compute a $k$-th order Delaunay triangulation that minimizes the number of local minima of the polyhedral terrain for $n^{\varepsilon} \le k \le c.n$.

The terrain model is also influenced by the drainage lines, which are often known separately. Real terrains have these drainage lines in the bottoms of valleys. The contiguity of valley lines is a natural phenomenon. Valleys do not start and stop halfway a mountain slope, but the Delaunay triangulation may contain such artifacts [5]. So, the problem is to build a terrain model that has few valley edges [6] and valley components by choosing the correct triangulation, if it exists.

Therefore, optimization criteria for terrain modeling include minimizing the number of local minima and the number of valley line components [5].

This paper discusses the reality of drainage networks in terrains using higher-order Delaunay triangulations of a point set $P$, for which elevations are given. In section 12.2 we present a survey on the number of heuristics for generating realistic terrains, namely; flip and hull heuristics for reducing the number of local minima and the valley heuristic for reducing the number of valley edges and components. In section 12.3 a new efficient algorithm based on the flip heuristic is proposed, which generates better outcomes. In section 12.4 we propose a new heuristic called *valley reduce* to reduce the number of valley edges and components, as well an attractive method for removing isolated valley edge components. Section 12.5 presents experimental results and compares the heuristics on various terrains. Tables and visualizations show how well the two proposed algorithms perform on real-world data. Finally, we discuss our conclusions in section 12.6.

## 12.2 Background

There are two heuristics for reducing the number of local minima: hull and flip. The hull heuristic was firstly described by Gudmundsson et al. [3] with $O(nk^3 + nk \log n)$ time complexity, and has an approximation factor of $\Theta(k^2)$ of the optimum. The algorithm starts with the Delaunay triangulation and adds a useful order-$k$ Delaunay edge $e$ to the triangulation, if it reduces the number of local minima. This edge may intersect several Delaunay edges, which are removed; the two holes in the triangulation that appear are retriangulated with the constrained Delaunay triangulation [7] in $O(k \log k)$. The union of these two holes is called the *hull* of $e$. During the algorithm, the hull of the new inserted edge must avoid intersecting the hulls of previously inserted edges. Gudmundsson's own implementation uses the hull intersection graph for this purpose. Kok et. al. [5] improved the hull heuristic to run in $O(nk^2 + nk \log n)$ by marking the edges of the triangulation.; this is more efficient for larger values of $k$.

The flip heuristic proposed by Kok et. al. [5] flips the diagonal of a convex quadrilateral in the triangulation if certain conditions hold (for more details see section 12.3). The algorithm starts with the Delaunay triangulation and $k' = 1$, does all possible flips to obtain an order-$k'$ Delaunay triangulation, then increments $k'$ and repeats until $k' = k$. The run time of this algorithm is also $O(nk^2 + nk \log n)$.

To reduce the number of valley edge components and improve the drainage quality, Kok et. al. [5] proposed an $O(nk \log n)$ algorithm called the valley heuristic. They applied two methods: removing isolated valley edges and extending valley components downward. An isolated valley edge is removed by flip, if possible. To extend each valley component downward, the triangulation around its endpoint is changed locally and connected to another valley component if possible, reducing the number of valley components.

## 12.3 A More Efficient Flip Heuristic

In the flip heuristic proposed by Kok et. al. [5], any edge that satisfies two conditions is flipped. Conditions are: (i) the two new triangles are order-$k$ Delaunay triangles and (ii) the new edge connects the lowest point of the four to the opposite point. These conditions do not prevent the generation of new valley edges (valley edges are defined in the next section). However, they flip an edge without considering whether its end points are local minima or not. The tables and visualizations in [5] show that: 1) the number of valley edges and components are increased, since the method tries to connect edges to the points with steepest descent and 2) it generates an unrealistic drainage network with long edges, because it retriangulates the whole terrain surface instead of only around local minima.

As mentioned earlier, the flip heuristic reduces the number of local minima, but increases the number of valley edges and components. To overcome this problem, we modify the flip heuristic and call it *new flip*. In this new heuristic, the diagonal of a convex quadrilateral in the triangulation is flipped if two conditions hold simultaneously: (i) the two new triangles are order-$k$ Delaunay triangles and (ii) the new edge connects a local minimum to the opposite point. These conditions not only reduce the number of local minima, but also reduce the number of valley edges and valley components. The shape of the terrain is more realistic compared to the results of flip (see tables and visualizations in section 12.5). The time complexity of this algorithm is still

# 12 Drainage reality in terrains with higher-order Delaunay triangulations

$O(nk^2 + nk \log n)$, but in practice, it is more efficient and faster because it checks only the edges around local minima instead of all the edges of the triangulation. The concrete numbers for spent CPU time are given in Table 3 from section 12.5.

## 12.4 Reducing the Number of Valley Edge Components

Flow on terrains is usually assumed to take the direction of steepest descent; at any point, the steepest descent direction is unique. This is a common assumption in drainage network modeling [6, 8]. Yu et al. [6] distinguish three type of edges in triangulation: *cofluent*, *transfluent* and *difluent* edges. Cofluent or *valley* edges are edges that receive water from both adjacent triangles (because for both adjacent triangles, the direction of steepest descent is directed towards this edge); transfluent or *normal* edges receive water from one adjacent triangle, which continues down to another triangle; and difluent edges or *ridges* receive no water (because for both adjacent triangles, the direction of steepest descent is directed away from this edge) (see Fig. 2).

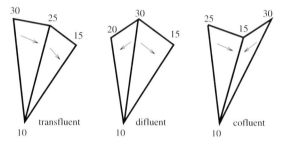

**Fig. 12.2** The three types of edges.

**Definition 12.2.** A valley (edge) component is a maximal set of valley edges such that flow from all of these valley edges reaches the lowest vertex incident to the valley edges.

So a valley component is a rooted tree with a single target that may be a local minimum. This target is tree root and is called the *end point*; the leaves of the tree are called *start points*.

Figure 3 shows an example of a terrain with three valley components. The valley edges are shown by the direction of the flow, numbers indicate the identity of each component, and squares show local minima. Component 2 ends at an inner local minimum and component 1 ends in a boundary local minimum (a local minimum that is on the boundary of the terrain). Component 3 ends in the inner vertex that is not a local minimum; flow proceeds over a triangle. In this example, the direction of steepest descent from vertex $q$ – where components 2 and 3 touch – is over the edge labeled 2 to the local minimum. Each component has an end point as the root of the tree and some start points as leaves. In our example, component 2 has three start points (leaves), while the other components have two.

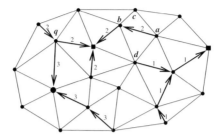

**Fig. 12.3** An example of a terrain with three valley edge components.

Kok et. al. [5] observed that the drainage quality of a terrain is determined by:

- the number of local minima, and
- the number of valley edge components that do not end in a local minimum.

The sum of these two numbers immediately gives the total number of valley edge components. They attempt to reduce this number with flip, hull and valley heuristics. We provide a new heuristic that reduces this number and consequently increases the drainage quality.

### 12.4.1 The Valley Reduce Heuristic

Just as isolated valley edges in the triangulation of a terrain are often artifacts, so are valley components with two or three edges (and in general valley components with few valley edges, i.e. small valley components). Berg et. al. [8] showed that the worst-case complexity of a river for triangulations with $n$ triangles, is $\Theta(n^2)$ and $\Theta(n^3)$ for all rivers, if drainage is allowed through the interiors of triangles, according to steepest descent. Thus, sequences of valley edges that do not end in local minima, where flow proceeds over a triangle, are also artifacts.

After removing single valley edges and extending valley components downhill by the valley heuristic, we reduce and possibly remove the valley components by using the valley reduce heuristic.

To reduce a valley component, we attempt to repeatedly flip (valley) edges from each starting point. Figure 3 shows the situation when trying to reduce component 2 from starting point $a$ by changing the flow situation on the valley edge $(a,b)$. Five candidate flips take care of this: the valley edge $(a,b)$ itself, and the four other edges incident to the two triangles incident to this valley edge: $\overline{ac}$, $\overline{ad}$, $\overline{bc}$ and $\overline{bd}$. A flip can potentially remove one valley edge but create another one at the same time; such a flip is not useful and termination of the heuristic would not be guaranteed [5]. Any flip changes the flow situation at four vertices. There are many possible choices for how to allow flips. We choose to flip only if no new valley edge or local minimum is generated, no valley edge except $(a,b)$ is removed, and the two new triangles are order-$k$ Delaunay.

**Fig. 12.4** Component 2 is reduced from starting point $a$.

Figure 4 shows the most common case, when the valley edge itself was flipped. The old starting point $a$ is removed from the list of starting points for this valley component. Point $b$ is added to the list as a new starting point if no valley edge is ascending from it. Then, this process is repeated on the new start point $b$. Some small valley components that are not isolated from the valley edge may be removed during this heuristic (see results in the next section).

The valley reduce heuristic may create single valley edge components, if all the edges of a component except for the final edge are removed. In this case, the number of isolated valley edges is increased and may influence the shape of drainage basins [9]. Therefore we relax the conditions that remove isolated valley edges. Kok et. al. [5] remove such edges if the flow situation of the four vertices of the convex quadrilateral does not change, and the two new triangles are order-$k$ Delaunay. We remove such edges only if no new valley edge or new local minimum is generated or removed except that edge, and the two new triangles are order-$k$ Delaunay. In this case another valley edge may be affected, but it remains a valley edge; e.g. one of its incident triangles may be changed but flow still goes through that edge. Figure 5 shows what happens when our method removes an isolated valley edge but the previous method does not. The edge $(a,b)$ is an isolated valley edge component (left figure). It becomes transfluent if edge $(b,c)$ is flipped (right figure). This flip causes valley edge $(c,d)$ to be affected, but it is still a valley edge. This type of flipping was not considered by the previous method.

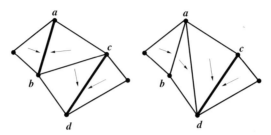

**Fig. 12.5** Removing an isolated valley edge component.

Algorithmically, there are at most $O(n)$ flips. The most expensive test is deciding if the two new triangles are order-$k$. This can be done in $O(\log n + k)$ time after $O(nk \log n)$ preprocessing time using an order-$(k+1)$ voronoi diagram [10].

**Theorem 12.1.** *The valley reduce heuristic to reduce the number of valley edges and components in order-k Delaunay triangulations on n points takes $O(nk\log n)$ time.*

## 12.5 Empirical Results

This section shows the experimental results of the proposed heuristics (new flip and valley reduce), and compares these heuristics with other heuristics for various terrains. The algorithms are tested on five real-world terrains: the California hot springs, Wren peak, Quinn peak, Sphinx lakes and Split Mountains. As in [5], the terrains have roughly 1950 vertices. The vertices were chosen by a random sampling of 1% of the points from elevation grids. We only examine order 8; higher orders are less interesting in practice since the interpolation quality deteriorates, and skinny triangles may cause artifacts in visualization [5]. Lower orders are also uninteresting because they limit our freedom to flip edges. Flip, hull and valley heuristics were written by Thierry de Kok, and we implemented the new flip and valley reduce heuristics. All the algorithms were written in C and use CGAL-3.1 library. We test the algorithms on an ASUS A3000 laptop with an Intel Pentium M+1.6 GHz processor and 504 MB of RAM running under Cygwin for the Microsoft Windows XP operating system.

The number of edges in the California hot springs, Wren, Quinn, Sphinx and Split are 5234, 5192, 5214, 5184 and 5147 respectively, for all triangulations. To define valley edges and flow when vertices have the same height, we

12 Drainage reality in terrains with higher-order Delaunay triangulations 207

treated height as a lexicographic number $(z,x,y)$, where $x$ and $y$ are the lesser significant components in the lexicographic order.

**Table 12.1** Statistics for five terrains. For each terrain, counts for the Delaunay triangulation are given, as well as for the outcome of the original flip, new flip, and hull heuristics for order 8.

| Terrain | CHS | | | | QP | | | | SL | | | |
|---|---|---|---|---|---|---|---|---|---|---|---|---|
| Heuristic | dl | of | nf | h | dl | of | nf | h | dl | of | nf | h |
| Valley edges | 860 | 955 | 827 | 803 | 755 | 864 | 684 | 703 | 670 | 855 | 648 | 627 |
| Local minima | 63 | 35 | 37 | 35 | 56 | 28 | 30 | 27 | 54 | 28 | 29 | 29 |
| Valley components | 245 | 256 | 224 | 230 | 250 | 302 | 227 | 231 | 246 | 301 | 230 | 227 |
| Not min.ending | 191 | 225 | 196 | 204 | 199 | 276 | 204 | 207 | 197 | 278 | 206 | 203 |

| Terrain | SM | | | | WP | | | |
|---|---|---|---|---|---|---|---|---|
| Heuristic | dl | of | nf | h | dl | of | nf | h |
| Valley edges | 692 | 862 | 668 | 651 | 783 | 902 | 725 | 740 |
| Local minima | 53 | 31 | 32 | 32 | 55 | 29 | 30 | 30 |
| Valley components | 268 | 290 | 253 | 255 | 230 | 286 | 216 | 212 |
| Not min.ending | 230 | 269 | 236 | 238 | 180 | 261 | 191 | 187 |

**Table 12.2** Statistics for five terrains when applying the valley reduce heuristic (order 8) to the outcomes of the hull plus valley and new flip plus valley heuristics

| Terrain | CHS | | QP | | SL | | SM | | WP | |
|---|---|---|---|---|---|---|---|---|---|---|
| Heuristic | hv | hvvr | hv | hvvr | hv | hvvr | hv | hvvr | hv | hvvr |
| Valley edges | 753 | 609 | 645 | 504 | 567 | 452 | 592 | 469 | 703 | 557 |
| Local minima | 35 | 35 | 27 | 27 | 29 | 29 | 32 | 32 | 30 | 30 |
| Valley components | 161 | 146 | 157 | 135 | 147 | 125 | 169 | 149 | 137 | 121 |
| Not min.ending | 136 | 121 | 132 | 110 | 124 | 102 | 153 | 133 | 114 | 98 |
| Single edge valley | 56 | 60 | 62 | 64 | 49 | 47 | 59 | 58 | 42 | 49 |

| Terrain | CHS | | QP | | SL | | SM | | WP | |
|---|---|---|---|---|---|---|---|---|---|---|
| Heuristic | fv | fvvr | fv | fvvr | fv | fvvr | fv | fvvr | fv | fvvr |
| Valley edges | 777 | 624 | 622 | 495 | 583 | 458 | 603 | 479 | 685 | 541 |
| Local minima | 37 | 37 | 30 | 30 | 29 | 29 | 32 | 32 | 30 | 30 |
| Valley components | 159 | 142 | 149 | 128 | 147 | 124 | 165 | 145 | 139 | 124 |
| Not min.ending | 132 | 115 | 123 | 102 | 124 | 101 | 149 | 129 | 116 | 101 |
| Single edge valley | 56 | 58 | 58 | 60 | 46 | 42 | 55 | 56 | 44 | 52 |

Table 1 shows the statistics obtained after applying Delaunay triangulation, flip, new flip, and hull heuristics to five different terrains. The number of local minima in the table represents the sum of inner local minima and boundary local minima. The last row shows the number of components that do not end in local minima; this number is the sum of components that end in the inner part of the terrain and components that end in the boundary of the terrain. The flip heuristic increased the number of valley edges considerably, while the new flip heuristic decreased this number. The same is true for the number of valley components.

The valley reduce heuristic was applied to the outcomes of new flip plus valley and hull plus valley heuristics, see results in Table 2. The results show that the valley reduce heuristic reduces the number of valley edges and valley components considerably in all cases. The reduction is between 19% to 22% for valley edges and between 9% to 16% for valley components for all triangulations. There is no considerable difference between fvvr and hvvr in the number of valley edges; in some cases new flip is better and in other cases hull is better. The number of valley components is lowest when applying the valley reduce heuristic to the outcome of new flip plus valley heuristics in all terrains except Wren Peak. The number of local minima is not changed by the valley reduce heuristic. A more careful look at Table 2 shows that the valley reduce heuristic removes between 11% and 19% of the valley components that do not end in local minima. The last rows of Table 2 show that the valley reduce heuristic may create single valley edge components that can not be removed.

**Table 12.3** Number of flips and CPU time spent for flip and new flip heuristics. Times are in seconds and do not include times of I/O.

| Terrain | CHS | | QP | | SL | | SM | | WP | |
|---|---|---|---|---|---|---|---|---|---|---|
| Heuristic | of | nf | of | nf | of | nf | of | nf | of | nf |
| No of points | 1967 | 1967 | 1957 | 1957 | 1947 | 1947 | 1938 | 1938 | 1951 | 1951 |
| No. of flips | 1696 | 137 | 1694 | 156 | 1758 | 165 | 1703 | 77 | 1734 | 172 |
| Spent CPU time | 50 | 3 | 51 | 4 | 51 | 4 | 53 | 2 | 51 | 4 |

Table 3 compares the flip and new flip heuristics on five terrains, regarding CPU time spent and the number of flips made to achieve 8-order Delaunay triangulations. We can see that the CPU time and number of flips for new flip is roughly one tenth of that for flip heuristic.

The visualization of heuristics applied at Sphinx Lakes and Quinn Peak terrains are given in Figs. 6 and 7 respectively. New flip has more realistic outcomes than its original version does. A close inspection shows that there is no significant difference between the outcomes of new flip and hull. However, when valley reduce is applied to their outcomes, the reality of the terrains increases significantly.

## 12.6 Conclusion

This paper deals with the reality of drainage networks on terrains using higher-order Delaunay triangulations. For drainage applications, it is important to have the drainage network coincide with the triangulation edges, and not go over the middles of triangles. Natural terrains have few local minima and valley edge components [5]. We presented two heuristics: new flip (for

12 Drainage reality in terrains with higher-order Delaunay triangulations 209

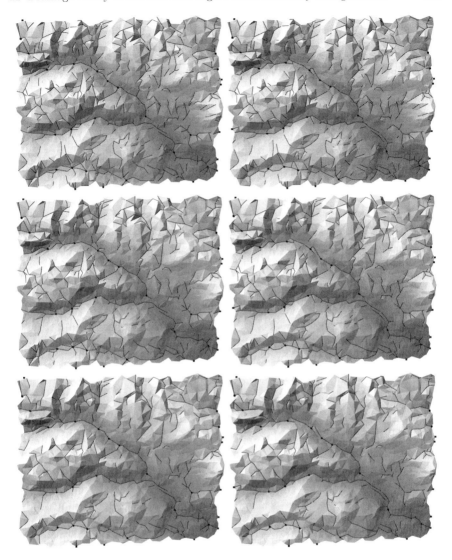

**Fig. 12.6** Visualization of the valley edges and local minima after applying the valley reduce heuristic to the Sphinx Lakes data set. Left column: outcome of new flip, outcome of new flip plus valley, outcome of new flip plus valley plus valley reduce. Second column: same, but for hull.

reducing the number of local minima) and valley reduce (for reducing the number of valley edge components in terrains). The empirical results of five real-world terrains shows on average a 16% reduction in the number of valley edges and components over known algorithms.

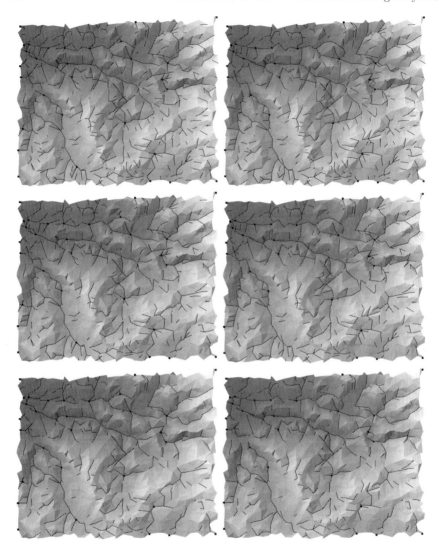

**Fig. 12.7** Visualization of valley edges and local minima after applying the valley reduce heuristic for the Quinn Peak data set. Left column: outcome of new flip, outcome of new flip plus valley, outcome of new flip plus valley plus valley reduce. Second column: same, but for hull.

# Acknowledgment

We would like to thank Mr. Maarten Löffler, a Ph.d student from Utrecht University, the Netherlands, for sharing his code and terrain data with us.

## References

[1] J. R. Shewchuk, Lecture notes on Delaunay mesh generation, University of California at Berkeley, (1999).
[2] Kreveld M van, GOGO project homepage. http://www.cs.uu.nl/centers/give/geometry/compgeom/gogo
[3] Gudmundsson J, Hammar M, Kreveld M van (2002) Higher order Delaunay triangulations. Computational Geometry: Theory and Applications 23:85–98
[4] Gudmundsson J, Haverkort H, Kreveld M van (2005) Constrained higher order Delaunay triangulations. Computational Geometry: Theory and Applications 30:271–277
[5] Kok T de, Kreveld M van, Löffler M (2007) Generating realistic terrains with higher-order Delaunay triangulations. Computational Geometry: Theory and Applications 36:52–67
[6] Yu S, Kreveld M van, Snoeyink J (1997) Drainage queries in TINs: from local to global and back again. In: Kraak MJ, Molenaar M (eds) Advances in GIS research II: Proc. of the 7th Int. Symp. on Spatial Data Handling, pp. 829–842
[7] Chew LP (1989) Constrained Delaunay triangulations. Algorithmica 4:97–108
[8] Berg M de, Bose P, Dobrint K, Kreveld M van, Overmars M, Groot M de, Roos R, Snoeyink J, Yu S (1996) The complexity of rivers in triangulated terrains. In: Proc. 8th Canad. Conf. Comput. Geom. 325–330
[9] McAllister M, Snoeyink J (1999) Extracting consistent watersheds from digital river and elevation data. In: Proc. ASPRS/ACSM Annu. Conf.
[10] Ramos EA (1999) On range reporting, ray shooting and k-level construction. In: Proc. 15th Annu. ACM Symp. on Computational Geometry, 390–399

## Chapter 13
# Surface Reconstruction from Contour Lines or LIDAR elevations by Least Squared-error Approximation using Tensor-Product Cubic B-splines

Shyamalee Mukherji

### Abstract

We consider, in this paper, the problem of reconstructing the surface from contour lines of a topographic map. We reconstruct the surface by approximating the elevations, as specified by the contour lines, by tensor-product cubic B-splines using the least squared-error criterion. The resulting surface is both accurate and smooth and is free from the terracing artifacts that occur when thin-plate splines are used to reconstruct the surface.

The approximating surface, $S(x,y)$, is a linear combination of tensor-product cubic B-splines. We denote the second-order partial derivatives of $S$ by $S_{xx}$, $S_{xy}$ and $S_{yy}$. Let $h_k$ be the elevations at the points $(x_k, y_k)$ on the contours. $S$ is found by minimising the sum of the squared-errors $\{S(x_k,y_k) - h_k\}^2$ and the quantity $\int\int S_{xx}^2(x,y) + 2S_{xy}^2(x,y) + S_{yy}^2(x,y)\,dy\,dx$, the latter weighted by a constant $\lambda$.

Thus, the coefficients of a small number of tensor-product cubic B-splines define the reconstructed surface. Also, since tensor-product cubic B-splines are non-zero only for four knot-intervals in the x-direction and y-direction, the elevation at any point can be found in constant time and a grid DEM can be generated from the coefficients of the B-splines in time linear in the size of the grid.

---

Centre of Studies in Resources Engineering
Indian Institute of Technology, Bombay
P.O.: Powai - I.I.T.
Mumbai - 400 076
INDIA.
shyamali@csre.iitb.ac.in

## 13.1 Introduction

The contour lines of a topographic map are sometimes the only source of information that we have about the terrain elevations in a region. Tensor-product B-splines have been successfully used to model a wide variety of surfaces in various fields, e.g., turbulence simulation, regional gravity field approximation, buckling of shells under loads etc.. In this paper, we reconstruct the surface from contour lines by approximating the elevations by tensor-product cubic B-splines using the least squared-error criterion.

Some solutions that have been proposed for the problem of reconstructing the surface from contour lines are given in Section 2. Then, the least squared-error approximation technique is described in Section 3. The results obtained by approximating the elevation data, derived from the contours, by tensor-product cubic B-splines are given in Section 4. In Section 5, we show that the elevation of the reconstructed surface can be computed efficiently and also give an efficient way of computing the DEM. In Section 6, we reconstruct surfaces from LIDAR elevations using the same technique of least squared-error approximation by tensor-product B-splines. We conclude the paper in Section 7, with a summary of the advantages of reconstructing the surface using B-splines.

## 13.2 Solutions Proposed in the Literature

Thin-plate splines have been used to reconstruct the surface from contour lines but the computing time of the algorithms and terracing artifacts in the reconstructed surface are major drawbacks of these methods [1].

The most accurate approach, for reconstruction of the surface, is to interpolate the elevations using Hardy's multiquadrics [2, 3]. The reconstructed surface is a linear combination of multiquadrics, each centered on a data point, and a polynomial of degree one. The coefficients of the multiquadrics and the polynomial are determined by solving a system of linear equations whose size is the number of data points plus three. Since the multiquadrics have global support, the coefficient matrix of the system of equations is a full matrix and a direct solution of the system is not acceptable for large data sets with thousands of points. The parameters that define the reconstructed surface are the centers of the multiquadrics along with the coefficients of the multiquadrics, and the coefficients of the polynomial. Therefore, the number of parameters required is thrice the number of data points plus three. Also, all the multiquadrics contribute to the elevation of the surface at any point.

To alleviate these problems, Pouderoux et al. proposed a scheme for fast reconstruction of a $C^1$-continuous surface [4]. The scheme also allows more control over the numerical stability of the solution. In this scheme, the global domain of interest is sub-divided into smaller overlapping sub-domains. The number of coefficients that define the reconstructed surface goes up as the sub-divisions overlap. The number of multiquadrics that contribute to the elevation (of the surface) at any point is greatly reduced but is still of the order of hundreds.

Franklin proposed a computation and storage-intensive algorithm for reconstructing the surface from contour lines [5].

Goncalves et al. used piecewise cubics to approximate the elevation data [6]. The piecewise cubics he used, however, are $C^1$-continuous.

Dakowicz et al. solved the problem of 'flat triangles' in TINs by inserting skeleton points in the TIN and assigning them elevations [7]. Various interpolation techniques were then used to interpolate the thus enriched contour elevation data and were compared. Slight breaks in slope at contour lines or oscillations are the artifacts that remain in the best surfaces obtained.

In this paper, we choose to approximate elevation data using cubic B-splines as the natural cubic spline is the solution to the least squared-error approximation problem in one dimension and the B-spline is smooth ($C^2$-continuous). We find that approximating contour elevation data by tensor-product cubic B-splines results in a smooth surface. The solution is numerically stable. The reconstructed surface is defined by the coefficients of a small number of tensor-product B-splines. The number of B-splines required has been observed to be half the number of parameters that would be required to specify a surface that has been reconstructed by multiquadrics without applying Pouderoux's sub-division scheme. Also, exactly 16 tensor-product B-splines contribute to the elevation of the surface at a point. So, the elevation at any point can be found in real-time.

## 13.3 Least Squared-error Approximation by Tensor-product Cubic B-splines [8]

A cubic B-spline, $B(x)$, with uniform knot-spacing $\Delta$ and centered at $x_i$ is given by $h(|x - x_i|/\Delta)$ where

$$h(p) = \begin{array}{ll} 3p^3/6 - p^2 + 4/6 & 0 \leq p < 1 \\ -p^3/6 + p^2 - 2p + 8/6 & 1 \leq p < 2 \\ 0 & 2 \leq p \end{array}$$

A tensor-product cubic B-spline, centered at a point $(x_1, y_1)$, is the product of a cubic B-spline $B(x)$ centered at $x_1$ and a cubic B-spline $B(y)$ centered at $y_1$.

Let $B_i(x)$ and $C_j(y)$ be cubic B-splines along the x and y-directions, respectively. Let us take $n_x$ B-splines along the x-direction and $n_y$ B-splines along the y-direction. Then, the spline surface $S(x,y)$ which is to be constructed is

$$S(x,y) = \sum_{i=1}^{n_x} \sum_{j=1}^{n_y} c_{ij} B_i(x) C_j(y)$$

Let $N$ be the number of points, $p_k$, at which the elevations, $h_k$, are known on the contour lines.

A smoothing term is added to the squared-error that is to be minimized and the function to be minimized is

$$F(c) = \sum_{k=1}^{N} \{S(p_k) - h_k\}^2 + \lambda J(c), \tag{13.1}$$

where $\lambda$ is a positive constant and

$$J(c) = \int_{a_1}^{b_1} \int_{a_2}^{b_2} S_{xx}^2 + 2S_{xy}^2 + S_{yy}^2 \, dy \, dx$$

where $S_{xx}$, $S_{xy}$ and $S_{yy}$ are the second-order partial derivatives of $S(x,y)$ and $[a_1, b_1] * [a_2, b_2]$ is the domain of $S$.

This integral can be expressed as

$$\sum_{i=1}^{n_x} \sum_{j=1}^{n_y} \sum_{r=1}^{n_x} \sum_{s=1}^{n_y} E_{ijrs} c_{ij} c_{rs},$$

where

$$E_{ijrs} = A_{ijrs} + 2B_{ijrs} + C_{ijrs}$$

and

$$A_{ijrs} = \int_{a_1}^{b_1} B_i''(x) B_r''(x) \, dx \int_{a_2}^{b_2} C_j(y) C_s(y) \, dy,$$

$$B_{ijrs} = \int_{a_1}^{b_1} B_i'(x) B_r'(x) \, dx \int_{a_2}^{b_2} C_j'(y) C_s'(y) \, dy,$$

$$C_{ijrs} = \int_{a_1}^{b_1} B_i(x) B_r(x) \, dx \int_{a_2}^{b_2} C_j''(y) C_s''(y) \, dy.$$

A minimum of $F(c)$ must occur at a point $c$ where all partial derivatives are zero.

Let

$$J(c) = c^T E c,$$

where $E$ is a square matrix of dimension $n = n_x * n_y$ whose elements are

$$E_{(j-1)n_x+i, (s-1)n_x+r} = E_{ijrs}$$

for $i, r = 1, \ldots, n_x$ and $j, s = 1, \ldots, n_y$.

By differentiating (1), we get

$$(B^T B + \lambda E) c = B^T h, \tag{13.2}$$

where $h = (h_1, \ldots, h_N)^T$ and $B$ is the $N \times n$ matrix

$$\begin{pmatrix} B_1(x_1)C_1(y_1) & B_2(x_1)C_1(y_1) & \cdots & B_{n_x}(x_1)C_{n_y}(y_1) \\ \cdot & \cdot & \cdot & \cdot \\ \cdot & \cdot & \cdot & \cdot \\ B_1(x_N)C_1(y_N) & B_2(x_N)C_1(y_N) & \cdots & B_{n_x}(x_N)C_{n_y}(y_N) \end{pmatrix}$$

where $p_k = (x_k, y_k)$.

Then, the solution to (1) is the solution $c$ to (2).

The matrix $G = B^T B + \lambda E$ is positive semi-definite. The matrix is strictly positive definite if the only solution to $c^T G c = 0$ is $c = 0$.

First observe that
$$c^T E c = J(c) = 0$$
implies that $S$ must be a linear polynomial $a + bx + cy$. Second, observe that
$$c^T B^T B c = ||Bc||^2 = 0$$
implies that $S(p_k) = 0$ for all $k = 1, ..., N$. Thus, we have that $c^T G c = 0$ implies that $S$ is a linear polynomial which is 0 at every point $p_k$. Clearly then, if there are at least 3 points $p_k$ which do not lie on a straight line, $S$ would have to be 0 and all the coefficients $c_{ij}$ would have to be 0. Since the points $p_k$ can never all be collinear, we deduce that $G$ is indeed nonsingular and the minimizer $c$ of (2) is unique.

## 13.4 Results

The contour lines in Fig. 1 correspond to elevations ranging from 220 metres at the bottom of the map to 660 metres at the top of the map; at intervals of 20 metres. The lines were digitized with a planimetric accuracy of 2.28 metres. $14 * 19$ tensor-product cubic B-splines (corresponding to 14 B-splines along the x-direction and 19 B-splines along the y-direction) and a value of 0.0001 for $\lambda$ were used to obtain the reconstructed spline surface in Fig. 2. The r.m.s. error (between the elevations of the reconstructed surface at the points on the digitized contour lines and the contour line elevations) achieved is 1.89 metres. The surface is smooth with minor artifacts at a few places at the edges of the surface. The surface obtained by interpolation using Hardy's multiquadrics is shown in Fig. 3 for comparison. The difference surface is shown in Fig. 4.

$32 * 30$ tensor-product cubic B-splines and a value of 0.01 for $\lambda$ were used to achieve an r.m.s. error of 2.01 metres for the contour lines in Fig. 5. The reconstructed surface is shown in Fig. 6. The surface obtained by interpolation using Hardy's multiquadrics is shown in Fig. 7. The difference surface is shown in Fig. 8.

Thus, the reconstructed surface is defined by a small number of B-splines. The surface is free from the terracing artifacts that occur when thin-plate splines are used to reconstruct the surface.

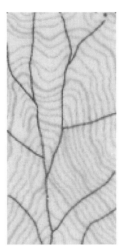

**Fig. 13.1** Contour lines.

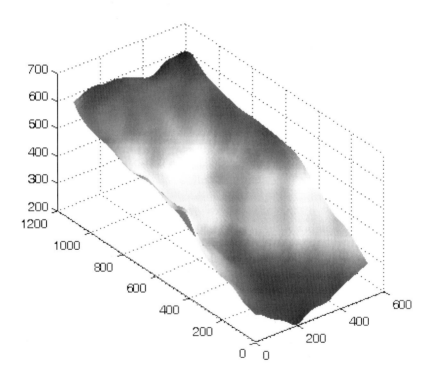

**Fig. 13.2** The reconstructed surface.

13 Surface Reconstruction from Contour Lines 219

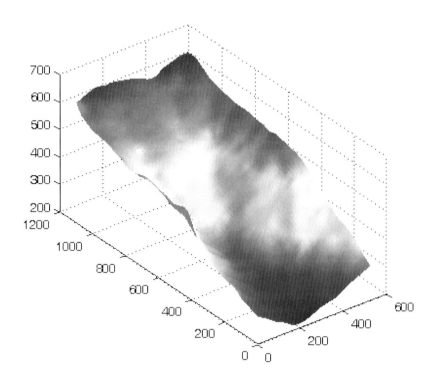

**Fig. 13.3** Surface obtained by interpolation using Hardy's multiquadrics.

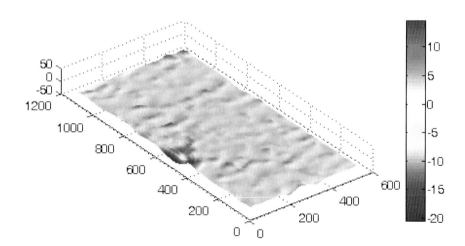

**Fig. 13.4** The difference surface.

**Fig. 13.5** Contour lines.

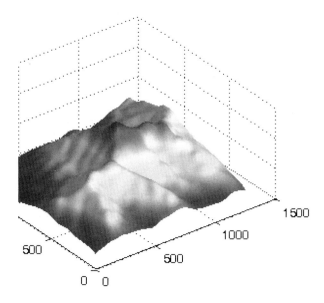

**Fig. 13.6** The reconstructed surface for the contour lines in Fig. 5.

## 13.5 Computing the DEM

In this Section, we show that the elevation of the reconstructed surface at a point can be computed efficiently. We then describe an efficient way of generating the DEM with low memory requirements.

Since tensor-product cubic B-splines are non-zero only for four knot-intervals in the x-direction and the y-direction, exactly 16 tensor-product cubic B-splines contribute to the elevation at a point.

Let us assume that the B-splines that are non-zero at a point $(x1, y1)$ are $B_k(x)$, $B_{k+1}(x)$, $B_{k+2}(x)$, $B_{k+3}(x)$, $C_l(y)$, $C_{l+1}(y)$, $C_{l+2}(y)$ and $C_{l+3}(y)$. The elevation at $(x1, y1)$ is found by evaluating the right-hand side of Eqn. 2 at $(x1, y1)$ as follows:

13 Surface Reconstruction from Contour Lines 221

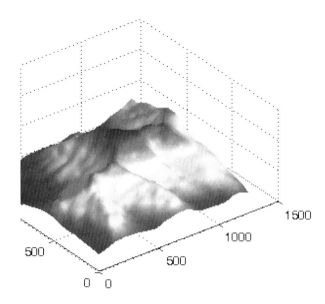

**Fig. 13.7** Surface obtained by interpolation using Hardy's multiquadrics.

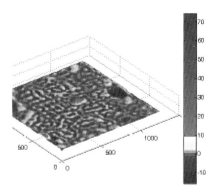

**Fig. 13.8** The difference surface.

We first evaluate the cubic polynomials $P_j(x) = \sum_{i=k}^{k+3} c_{ij} B_i(x)$, $j = l, \ldots, l+3$ at $x = x1$. To do this, we first find $p = |x1 - x_{k+1}|/\Delta_x$ where $x_{k+1}$ is the center of the B-spline $B_{k+1}(x)$ and $\Delta_x$ is the knot-interval for the $B_i(x)$. $P_j(x1)$ is, then,

$$c_{kj}(-p^3 + 3p^2 - 15p + 1) + c_{k+1,j}(3p^3 - 6p^2 + 4)$$
$$+ c_{k+2,j}(-3p^3 + 3p^2 + 3p + 1) + c_{k+3,j} p^3 \quad (13.3)$$

The quantities in the brackets require 17 operations {where an operation is either an addition (subtraction) or a multiplication}. So, finding the four $P_j(x1)$ requires $17 + 4*7 = 45$ operations.

Then, we find $q = |y1 - y_{l+1}|/\Delta_y$ where $y_{l+1}$ is the center of the B-spline $C_{l+1}(y)$ and $\Delta_y$ is the knot-interval for the $C_j(y)$. The elevation at $(x1, y1)$, which is $\sum_{j=l}^{l+3} P_j(x1) C_j(y1)$, can be expressed in a form that is analogous to the expression in (3) above. A rearrangement of the terms yields

$$q^3\{-P_l(x1) + 3(P_{l+1}(x1) - P_{l+2}(x1)) + P_{l+3}(x1)\}$$
$$+ 3q^2\{P_l(x1) + P_{l+2}(x1) - 2P_{l+1}(x1)\} + 3q\{-5P_l(x1) + P_{l+2}(x1)\}$$
$$+ P_l(x1) + P_{l+2}(x1) + 4P_{l+1}(x1) \quad (13.4)$$

the computation of which requires 21 operations.

Thus, the elevation at any point can be found with $45 + 21 = 66$ operations, or, in constant time. (In contrast, in Pouderoux's sub-division scheme, a small multiple of 800 operations would be required.)

When generating a DEM, we start with the distinct x co-ordinates, of the points of the DEM grid (Fig. 9), which lie between $x_2$ and $x_3$ ($x_1$ and $x_{n_x}$ lie at least $\Delta_x$ to the left of and $\Delta_x$ to the right of the first and last data points, respectively, in the horizontal direction; and there are no data points in the intervals $[x_1, x_2]$ and $[x_{n_x-1}, x_{n_x}]$), compute the respective $p = |x - x_2|/\Delta_x$ and compute and store the quantities in the brackets in expression (3), for these x co-ordinates. This takes $17 N_x$ operations where $N_x$ is the number of distinct x co-ordinates of the grid that lie between $x_2$ and $x_3$. We then compute and store the four $P_j(x1), j = 1, \ldots, 4$ for each of these x co-ordinates. This takes $N_x * 4 * 7 = 28 N_x$ operations.

We then take the y co-ordinates of the DEM grid that lie between $y_2$ and $y_3$ and find the corresponding $q = |y - y_2|/\Delta_y$. The elevation at the gridpoints, $(x1, y1)$, whose x co-ordinates lie between $x_2$ and $x_3$ and y co-ordinates lie between $y_2$ and $y_3$, is $\sum_{j=1}^{4} P_j(x1) C_j(y1) = P_1(x1)(-q^3 + 3q^2 - 15q + 1) + P_2(x1)(3q^3 - 6q^2 + 4) + P_3(x1)(-3q^3 + 3q^2 + 3q + 1) + P_4(x1) q^3$. Thus, the computation of the elevations at these grid-points takes a total of $17 N_y + N_y * N_x * 7$ operations where $N_y$ is the number of distinct y co-ordinates of the grid that lie between $y_2$ and $y_3$.

Next, we compute $P_5(x1)$ and use $P_j(x1), j = 2, \ldots, 5$ to find the elevations at the grid-points, whose x co-ordinates lie between $x_2$ and $x_3$ and y co-ordinates

# 13 Surface Reconstruction from Contour Lines

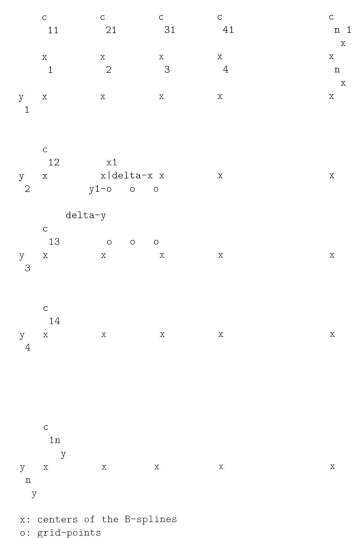

x: centers of the B-splines
o: grid-points

**Fig. 13.9** Computing the DEM

lie between $y_3$ and $y_4$. We proceed down the grid in this manner, computing the elevations at the grid-points, till we reach the bottom of the grid. Then we return to the top of the grid and repeat the entire procedure starting, this time, with the distinct x co-ordinates of the grid-points between $x_3$ and $x_4$. This process continues till the lower right corner of the grid is reached.

Thus, the computation of the elevations at the grid-points requires $17N + 7Nn_y + 17M(n_x - 3) + 7MN = 7MN + 17M(n_x - 3) + 7Nn_y + 17N$ operations for an $M*N$ grid. $n_x$ and $n_y$ can be atmost $N$ and $M$. So, the computation time is linear in the size of the grid and the constant is small, viz., 31. This is not the case with multiquadric-based DEMs wherein all multiquadrics contribute to the elevation at every grid-point.

## 13.6 Approximating LIDAR elevation data

We rotated the LIDAR scan-lines so that they are vertical as in Fig. 10. The surface obtained by approximating the elevations by tensor-product quadratic B-splines is shown in Fig. 11. This is a rural area scanned from an altitude of 1350 metres with a point spacing of 1.1 metres. The r.m.s. deviation of the surface from the LIDAR elevations is 1.8 cm.. Brovelli et al. approximates the LIDAR elevation data by bilinear splines [9]. The corresponding surface is shown in Fig. 12.

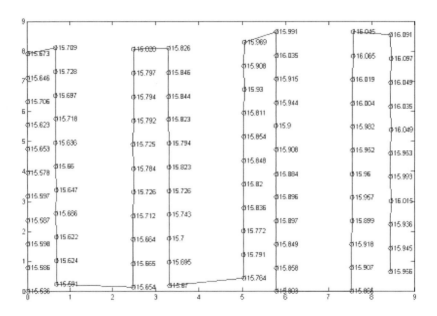

**Fig. 13.10** LIDAR elevations. (The values shown are in excess of 100 metres.)

LIDAR elevation data with gaps (Fig. 13) (rural area again) are approximated well by tensor-product cubic B-splines as is shown in Fig. 14. The value of lambda used is 0.000001 and the r.m.s. deviation of the surface from the LIDAR elevations is 2.6 cm..

13 Surface Reconstruction from Contour Lines 225

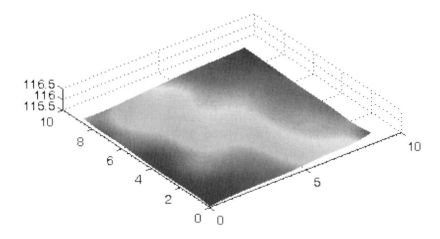

**Fig. 13.11** Surface approximating LIDAR data.

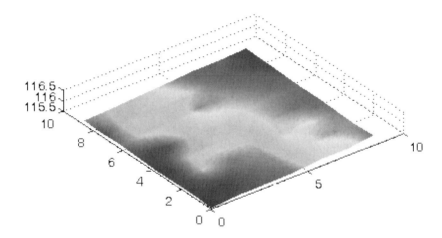

**Fig. 13.12** Approximation by bilinear splines.

## 13.7 Conclusions

We conclude, from the results in Section 4, that tensor-product cubic B-splines lead to a good reconstruction. The reconstructed surface is also free from the terracing artifacts that occur when thin-plate splines are used to reconstruct the surface. The smoothness of the surface results from the inherently smooth nature of cubic B-splines, which are $C^2$-continuous. An ad-

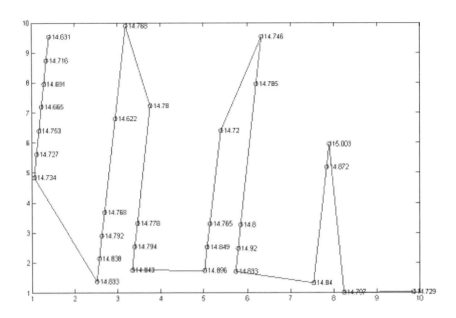

**Fig. 13.13** LIDAR elevations with gaps. (The values shown are in excess of 100 metres.)

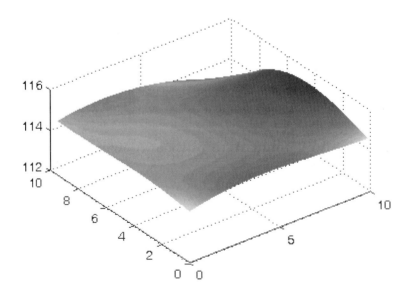

**Fig. 13.14** Surface approximating LIDAR data with gaps.

vantage of this method of reconstructing the surface is that the coefficients of a small number of tensor-product cubic B-splines define the surface.

Tensor-product cubic B-splines are non-zero only for four knot-intervals in the x-direction and y-direction. Therefore, exactly 16 tensor-product B-splines contribute to the elevation at a point. So, the elevation at any point can be found in constant time and a grid DEM can be generated from the coefficients of the B-splines in time linear in the size of the grid.

## 13.8 Acknowledgement

This work was supported by the Indian Space Research Organization.

# References

[1] Gousie MB (1998) Contours to digital elevation models: Grid-based surface reconstruction methods. PhD Thesis, Rensselaer Polytechnic Institute, Troy, New York
[2] Franke R (1982) Scattered data interpolation: tests of some methods. Math Comp 38(157):181-200
[3] Hardy RL (1971) Multiquadric equations of topography and other irregular surfaces. J Geophys Res 76:1905-1915
[4] Pouderoux J, Gonzato JC, Tobor I, Guitton P (2004) Adaptive hierarchical RBF interpolation for creating smooth digital elevation models. Proc 12th Ann ACM Intl Workshop GIS 2004, Washington, DC, USA, 232-240
[5] Franklin WR (2000) Applications of analytical cartography. Carto & GIS 27(3):225-237
[6] Goncalves G, Julien P, Riazanoff S, Cervelle B (2002) Preserving cartographic quality in DTM interpolation from contour lines. ISPRS J Photogram & Remote Sens 56:210-220
[7] Dakowicz M, Gold CM (2003) Extracting meaningful slopes from terrain contours. Intl J Comput Geom & Applns 13(4):339-357.
[8] Floater MS (2000) Meshless parameterization and B-spline surface approximation. In: Cipolla R, Martin R (eds) The mathematics of surfaces IX. Springer, Berlin Heidelberg NewYork
[9] Brovelli MA, Cannata M, Longoni UM (2004) LIDAR data filtering and DTM interpolation within GRASS. Trans in GIS 8(2):155-174

# Chapter 14
# Modelling and Managing Topology in 3D Geoinformation Systems[1]

Andreas Thomsen, Martin Breunig, Edgar Butwilowski, and Björn Broscheit

**Abstract**

Modelling and managing topology in 3D GIS is a non-trivial task. The traditional approaches for modelling topological data in 2D GIS cannot be easily extended into higher dimensions. In fact, the topology of real 3D models is much more complex than that of the 2D and 2.5D models used in classical GIS; in consequence there is a great number of different 3D spatial models ranging from constructive solid geometry to boundary representations. The choice of a particular representation is generally driven by the requirements of a given application. Nevertheless, from a data management point of view, it would be useful to provide a general topological model handling 2D, 2.5D and 3D models in a uniform way. In this paper we describe concepts and the realisation of a general approach to modelling and managing topology in a 3D GIS based on oriented d-Generalised Maps and the closely related cell-tuple structures. As an example of the applicability of the approach, the combination of a group of buildings from a 3D city model with the corresponding part of a 2D city is presented. Finally, an outlook to ongoing research is given in the context of topological abstraction for objects represented in multi-representation databases.

Institute for Geoinformatics and Remote Sensing, University of Osnabrück,
Seminarstr. 19 a/b, 49069 Osnabrück, Germany
{martin.breunig,andreas.thomsen,edgar.butwilowski,bjoern.broscheit}
@uni-osnabrueck.de

[1] This work is funded by the German Research Foundation (DFG) in the project 'MAT' within the DFG joint project 'Abstraction of Geoinformation', grant no. BR 2128/6-1.

## 14.1 Introduction

Topology and GIS belong together since the development of GIS. Already first GIS like GRASS and Arc/Info provided a topological data model storing relationships between points, lines, and areas of an area network. These traditional approaches for modelling topology in in 2D GIS were implemented by explicit links between geometric objects, e.g. from a line segment to its neighbouring left and right area. The more topologgical relationships the user required, the more complex the topological model became.

Unfortunately, there are no straightforward extensions into 3D space of the 2D topological data models used in traditional GIS. Instead, there is a number of different 3D spatial models ranging from constructive solid geometry to boundary representations, the choice of a particular representation being driven by the requirements of a particular application - from architecture and urbanism to numerical modelling, engineering and underground mining.

In a 3D Geoinformation System, objects of different dimension $d \leq 3$ are processed. The geometry of a geoscientific object in 3D GIS can be composed of sets of points, curves, surfaces and volumes, respectively. In a topology model of a 3D GIS, the components of the objects can be interpreted e.g. as a mesh of nodes, edges, faces and solids that describes both the interior structure of the geoscientific objects and their mutual neighbourhood relationships in 3D space.

To describe topology uniformly in 3D solid modeling [1] and 3D GIS [2], a general framework for topological data models has to be provided that abstracts from the dimension of the objects. Furthermore, it should be usable as a data integration platform for 2D, 3D and time-dependent 3D (sometimes termed "4D") topology.

In this paper, we investigate how oriented G-Maps and cell-tuple structures can be used to handle the topology of a digital spatial model in a more generic way, in order to support 2- and 3-dimensional and spatio-temporal models. For many 3D geo-applications not only the modelling, but also the management of topology in database management systems is relevant. Inspired by the work of GeoToolKit [3], that provides a geometric 3D library, we here present topological data structures and operations needed in 3D GIS.

After a short overview on related work, we recall Lienhardt's d-G-Map in section 3, and discuss basic topological operations in section 4. An object-relational representation based on Brisson's cell-tuple structures is introduced in section 5. The integration of triangular meshes is discussed in section 6, while the application to time-dependent topology is briefly presented in section 7. We conclude with an application example in section 8, and an outlook on future work.

## 14.2 Related work

Whereas basic relationships of point set topology, and in particular Egenhofer's *nine intersection model* have become standard in GIS, and lend themselves to a 3D generalisation [4, 5], 3D discrete topological structures stemming from algebraic topology have not gained comparable popularity, despite considerable development during the last decades. While [6] discussed the application of simplicial complexes to spatial databases, general approaches to representing topology in the context of 3D modelling have been examined by [7] and by other authors. [8] developed d-dimensional cell-tuple structures, and in parallel [9] developed d-dimensional Generalised Maps (d-G-Maps), to represent and manage the topological properties of cellular partitions of d-dimensional manifolds (d-CPM), cf. also [10]. [11] has shown that 3-G-Maps have comparable space and time behaviour as the DCEL and radial edge structures, but can be used for a wider range of applications, allowing a more concise and robust code. [12] used G-Maps to model architectural complexes in a hierarchy of multi-partitions. G-Maps and cell-tuple structures have been used to represent the topology of land-use changes [13], and are currently applied in the geoscientific 3D modelling software GOCAD[2] [14, 10], and in the topological modelling and visualisation tool Moka[15]. [16] give a concise overview of 2D and 3D topological models and propose the translation into geometric primitives for the integration of 2D and 3D topological models with 3D GIS based on relational databases. Recently, [17] describe the integration of 2D and 3D cadastral objects in a representation by regular polytopes based on pseudo-rational numbers. [18] presents the combination of 3D simplicial networks with Poincarié Algebra in a TEN-based spatial DBMS. Relations of our work with the work of [19], which has not been available at short notice, will be examined in our future work.

## 14.3 A general approach to modelling topology in 3D GIS

In the following, we use oriented d-CPM as a topological model for 3D GIS. d-CPM can be considered as a generalisation of simplicial complexes, but lack the algebraic properties of the latter. However, if a d-CPM is represented by a d-G-Map, the involution operations of the latter provide the cellular complex with the combinatorial structure of an abstract simplicial complex, where the cells and cell-tuples play the role of abstract nodes and abstract simplexes, while the involution operators define the neighbourhood relationships between the abstract simplexes [9]. Note that the abstract nodes $n, e, f, s$ of a 3-G-Map belong to 4 classes distinguished by different dimensions, whereas

---

[2] GOCAD is a registered trademark of Earth Decision Co.

all nodes of a simplicial complex belong to the same finite set of vertices in space.

According to [9], a *d-dimensional Generalized Map (d-G-Map)* is a d+2-tuple $G = (D, \alpha_0, \ldots, \alpha_d)$, consisting of a finite set $D$ of objects called *"darts"*, and $d+1$ permutations $\alpha_i, i = 0, \ldots, d$ that verify the following two conditions: the $\alpha_i$ are *involutions*, i.e. they verify for all $x$,

$$\alpha_i(\alpha_i(x)) = x, \qquad (14.1)$$

and for all $i, j$ with $0 \leq i < i+2 \leq j \leq d$, $\alpha_i \alpha_j$ is an involution, i.e.

$$\alpha_i(\alpha_j(\alpha_i(\alpha_j(x)))) = x, \qquad (14.2)$$

which implies $\alpha_i \alpha_j = \alpha_j \alpha_i$.

The G-Maps are *embedded* in space by a mapping that to each dart associates a unique combination $(n, e, f[, s])$ of a node $n$, an edge $e$, a face $f$, and in 3D a solid $s$.

The condition that a d-G-Map always represents a d-dimensional manifold ensures that to any cell-tuple, there exists at most one partner that differs from it only by one exchange operation $\alpha_i$.

A d-G-Map can be represented as a graph with cell-tuples as nodes (darts), and edges defined by the involution operations (Figure 14.1).

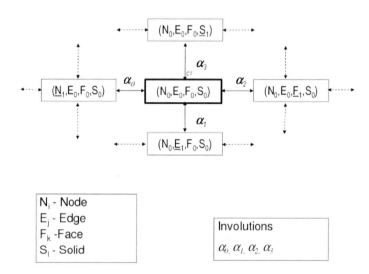

**Fig. 14.1** Representation of an oriented 3-G-Map as graph with symmetries determined by the combinatorial character of the involutions.

By assumption, the cellular complexes are orientable, and the corresponding G-Maps are oriented. This implies that there are two classes of darts or cell-tuples of the same cardinality, but carrying different polarity. Different from [9], we exclude the possibility that an involution attaches a cell-tuple to itself ($f(x) = x$), e.g. at the boundary of the cellular complex. Thus we ensure that involution operations always link pairs of cell-tuples of opposite sign. Instead, following [8], we introduce a special non-standard cell, the outside, or *universe*, which in general needs not be simply connected and may comprise holes and islands, and provide the universe also with cell-tuples and involutions on the boundary. This approach increases the number of objects to be handled. However, it simplifies some operations and algorithms.

## 14.4 Topological operations on oriented d-G-Maps and cell-tuple structures

In the following, we briefly present some topological operations on oriented G-Maps, a more extensive discussion can be found e.g. in [11, 20].

### *14.4.1 Orbits*

*Orbits* are defined as subsets that can be reached by any combination of involution operations of a given subset of $\alpha_0, \ldots, \alpha_d$, starting from a given dart or cell-tuple $cT_0$. They are noted $orbit^d(cT_0, \alpha_i, \ldots, \alpha_j)$, or shorter $orbit^d_{i \ldots j}(cT_0)$. Orbits that comprise all indices $0, \ldots, d$ with the exception of $k$ leave the $k$-th cell of cell-tuple $cT_0$ fixed and can be interpreted as the subset of all cell-tuples containing the same cell of dimension $k$ as $cT_0$. Such orbits can also be noted as $orbit^d(cT_0, k)$. Orbits of this type provide another way to describe cells of dimension $k$. While most types of orbits are implemented by single or double programming loops of fixed or variable size, [11] implements orbits of type $orbit^2_{012}()$, $orbit^3_{012}()$, $orbit^3_{123}()$, and $orbit^3_{0123}()$ recursively using a stack. Whereas loop implementations of orbits yield continuous closed paths in the G-Map graph, recursively implemented orbits in certain situations may produce discontinuities ("jumps"). Orbits of type $orbit^2_{012}()$, $orbit^3_{0123}()$ produce the complete connected component containing $cT_0$. Besides orbits, other *loops*, i.e. closed paths in the G-Map graph may be defined by an application or by a user. Orbits and loops are the main methods for the navigation on the G-Map graph. They are also indispensable for the implementation of some of the topological operations discussed below.

## 14.4.2 Topological operations on cells

Two classes of topological operations can be distinguished: Euler operations that conserve the Euler-Poincaré characteristic and thus the global connectivity properties of a G-Map ([1, 11]), and non-Euler operations that alter the connectivity of the structure. Examples of Euler operations are the subdivision of a cell of any dimension $k > 0$ by a newly created separating cell of dimension $k - 1$, e.g. the division of a face $f$ by a new edge $e$, and the corresponding inverse operations, under certain conditions ensuring the consistency of the resulting G-Map. An example of a Non-Euler operation is the attachment and subsequent *sewing* [11] of two previously disconnected cells, and the inverse operation. These operations constitute the most important methods for the building, transformation and in particular generalisation of d-G-Maps.

## 14.5 Data management for topological cell-tuple structures

In the following, we discuss some aspects of a different realisation of an oriented d-G-Map, namely as a cell-tuple structure in an object-relational database. This representation aims at providing a general topological access structure in 2D and 3D to existing GIS based on object-relational databases (ORDBMS). Whereas in the graph representation the main attention is given to the involutions $\alpha_i$, the relational representation uses Brisson's [8] cell-tuples as a realisation of the darts, while involutions are implemented using foreign keys and exchange operations. It is an interesting question, to what extent the functionality of orbits can be replaced by subset queries, join operations, and sorting, i.e. by standard operations of a relational DBMS.

### 14.5.1 Implementation of the topological data structures as database representation

The topological data structure presented here shall manage the topology of complex spatial objects in 2 and 3 dimensions. It is based on Lienhardt's [9] *d-Generalized Maps* and on Brisson's [8] closely related *cell-tuple structures*.

A d-G-Map can be represented in the relational model as follows (Figure 14.2): the set of cell-tuples is stored in tabular form, e.g. by two relations
$cTpos\,(node\_id, edge\_id, face\_id\,[, solid\_id]\,,' +', n\_inv, e\_inv, f\_inv\,[, s\_inv])$
$cTneg\,(node\_id, edge\_id, face\_id\,[, solid\_id]\,,' -', n\_inv, e\_inv, f\_inv\,[, s\_inv])$,
and the involution operations are modelled as symmetric 1:1 relationships

# 14 Modelling and Managing Topology in 3D GIS

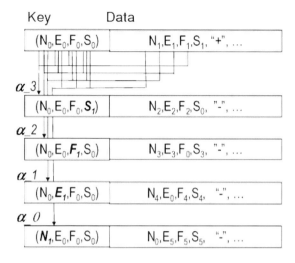

**Fig. 14.2** Representation of a 3-G-Map as relation with nodes $N_i$, edges $E_i$, faces $F_i$, solids $S_i$, and involutions $\alpha_i$

defined by the *switch operations* [8], linking e.g.
$cTpos(node\_id, \ldots, '+', n\_inv, \ldots,)$ to $cTneg(n\_inv, \ldots, '-', node\_id, \ldots,)$.

In a cell-tuple, the combination of cell identifiers, augmented by the positive or negative sign, $(node\_id, edge\_id, face\_is[, solid\_id], sign)$ is used as a unique *cell-tuple key*, while the identifiers of the cells to be exchanged by the involutions are stored as *data*. The data access by cell-tuple keys is enhanced by sorted indexes or hash indexes. The involutions are implemented in two steps: first, from a given cell-tuple entry, create a new cell-tuple key by exchanging exactly one cell_id. Second, use this key to retrieve the corresponding complete entry from the database.

The implementation of a d-G-Map is thus realised as a network of celltuples that is made persistent by relations of an Object-Relational Database Management System (ORDBMS). With the goal of a topological component for multi-representation databases [20], we implemented 2-G-Maps and 3-G-Maps with the ORDBMS PostgreSQL[3] [21] in combination with the open source GIS *PostGIS*[4] [22]. In our future work, we intend to implement the graph representation of a G-Map as a topological access structure for our object-oriented 3D/4D Geo-DBMS GeoDB3D [25, 26]. GeoDB3d uses sim-

---
[3] PostgreSQL ©1996-2005 by the PostgreSQL Global Development Group ©1994 by the Regents of the University of California is released under the BSD license
[4] PostGIS has been developed by Refractions Research and is released under the GNU General Public License

plicial complexes to represent geometry, and is based on the DBMS Object-Store[5].

## 14.5.2 Implementation of topological database operations

As the general topological data model is to be integrated into existent spatial ORDBMS, we focus on a clear translation of the G-Map into the relational model, and on the integration of the topological operations with the SQL-commands of a database server. In our view, optimization efforts should rather make use of RDBMS functionality, like sorting, indexing, clustering and caching, than perform the topological operations in client memory.

Orbits of the form $orbit^d_{0...\not{k}...d}(cT_0)$ comprise all celltuples that share with $cT_0$ a cell of dimension $k$. The corresponding cell-tuple subset can be retrieved by an appropriate relational query, though not in the same arrangement. For $orbit^2_{012}()$ and $orbit^3_{0123}()$, a corresponding relational query would yield all cell-tuples, regardless whether from the same connected component or not. For many purposes, this may be sufficient, but for the implementation of the two last-mentioned orbits, and for applications that require an identical arrangement, we can either explicitly model the orbit using the involution operations, or rearrange the subset on the client side after retrieval. Implementing orbit re-arrangement as an additional functionality of the server would be the best option, if this is supported by the ORDBMS. Loops can be implemented associating each cell-tuple with a selector variable that defines the involution to be performed next.

## 14.5.3 Example implementation of a database operation

A Basic Euler operation.

As an example of a basic topological operation, in a 2-G-Map comprising nodes $n...$, edges $e...$ and faces $f...$, consider the insertion of a new node $n$ that splits an edge $e(n_0, n_1, f_0, f_1)$ between nodes $n_0$, $n_1$ and faces $f_0, f_1$ into two edges $e_0(n_0, n)$ and $e_1(n, n_1)$ (Figure 14.3). At node $n$, four cell-tuples are inserted:

- $(n, e_0, f_0, -, n_0, e_1, f_1)$,
- $(n, e_0, f_1, +, n_0, e_1, f_0)$,
- $(n, e_1, f_0, -, n_1, e_0, f_1)$,

---

[5] ObjectStore is a registered trademark of Progress Software Co.

- $(n, e_1, f_1, +, n_1, e_0, f_0)$.

Eight cell-tuples at nodes $n_0$ and $n_1$ are transformed:

- $(n_0, e, f_0, +, n_1, e_a, f_1) \rightarrow (n_0, e_0, f_0, +, n, e_a, f_1)$,
- $(n_0, e, f_1, -, n_1, e_b, f_0) \rightarrow (n_0, e_0, f_1, -, n, e_b, f_0)$,
- $(n_1, e, f_0, -, n_0, e_c, f_1) \rightarrow (n_1, e_1, f_0, -, n, e_c, f_1)$,
- $(n_1, e, f_1, +, n_0, e_d, f_0) \rightarrow (n_1, e_1, f_1, +, n, e_d, f_0)$.
- $(n_0, e_a, f_0, -, \ldots, e, f_1) \rightarrow (n_0, e_a, f_0, -, \ldots, e_0, f_1)$,
- $(n_0, e_b, f_1, +, \ldots, e, f_0) \rightarrow (n_0, e_b, f_1, +, \ldots, e_0, f_0)$,
- $(n_1, e_c, f_0, +, \ldots, e, f_1) \rightarrow (n_1, e_c, f_0, +, \ldots, e_1, f_1)$,
- $(n_1, e_d, f_1, -, \ldots, e, f_0) (n_1, e_d, f_1, -, \ldots, e_1, f_0)$.

The translation into SQL is straightforward:

```
BEGIN TRANSACTION
INSERT INTO celltuples VALUE (n,e0,f0,-,n0,e1,f1);
INSERT INTO celltuples VALUE (n,e0,f1,+,n0,e1,f0);
INSERT INTO celltuples VALUE (n,e1,f1,-,n1,e0,f0);
INSERT INTO celltuples VALUE (n,e0,f0,+,n0,e1,f1);
UPDATE celltuples
 CASE
  WHEN edge=e AND node=n0 THEN SET edge=e0 SET node_inv=n
  WHEN edge=e AND node=n1 THEN SET edge=e1 SET node_inv=n
  WHEN edge_inv=e AND node=n0 THEN SET edge_inv=e0
  WHEN edge_inv=e AND node=n1 THEN SET edge_inv=e1
 END
 WHERE edge= e OR edge_inv= e;
COMMIT;
```

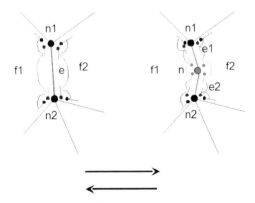

**Fig. 14.3** A simple Euler operation: splitting an edge by the insertion of a node

While this example is particularly simple, corresponding operations on cells of higher dimension, e.g. in a 3-G-Map the splitting of a face by the introduction of a separating edge initially follow a similar pattern. However, after the "sewing" i.e. adaptation of the $\alpha_i$ transitions at the new separating edge, additional operations are necessary to modify the links of all cell-tuples that refer the divided face. These operations are supported by two $orbit_{01}^2$ about the two newly created faces. For the splitting of a 3D solid by a new 2D face, a "loop" is required that defines the location where the new face is incident with the boundary of the existing solid to be split. Typically, a database client would provide a set of basic operations for the management, navigation and retrieval of topological information. These operations should be combined into short programs or scripts that fulfil more complex tasks.

### 14.5.4 Integrity checks for the relational representation of d-G-Maps

A spatial database has no a-priori knowledge about the way a newly introduced dataset has been constructed, nor on the order of update operations executed by a user or by a client application. It is therefore necessary to provide it with a set of tools to check the integrity of a stored G-Map at any time. In a relational representation of a G-Map, join operations can serve to implement some basic integrity checks. A possible test for $\alpha_0$ verifying condition (1) is the following operation which for a consistent G-Map returns zero:

```
SELECT COUNT(*)
   FROM cT_pos, cT_neg
   WHERE (cT_neg.node_id = cT_pos.node_inv)
     AND NOT (cT_pos.node_id = cT_neg.node_inv);
```

Condition (2) on $\alpha_0$, $\alpha_2$ is checked e.g. by the following SQL query involving a triple join, that must return zero, if the G-Map is consistent with condition (2):

```
SELECT COUNT(*)
   FROM cT_pos as cT_p1, cT_neg as cT_n1,
        cT_pos as cT_P2, cT_neg as cT_n2
   WHERE (cT_n1.node_id = cT_p1.node_inv)
     AND (cT_p2.face_id = cT_n1.face_inv)
     AND (cT_n2.node_id= cT_p2.node_inv)
     AND NOT (cT_p1.face_id = cT_n2.face_inv);
```

## 14.6 Mesh representation of geometry

By merging adjacent d-cells, hierarchies of G-Maps can be built, that consist of a sequence of nested subsets of cell-tuples and their involutions. The $\alpha_i$ transitions at a higher level correspond to a sequence of transitions at the lower, more detailed level. Such a nested hierarchy of G-Maps [11] can be used to integrate a more detailed geometric representation into the topological model. Suppose that the geometry of flat or curved surface patches is represented by triangle nets, which may have been generated by any modelling method, and are described by a set of vertices $(v_j, x_j, y_j, z_j)$ and a set of triangle elements $(tr_i, v_{i_0}, v_{i_1}, v_{i_2}, n_{i_1}, n_{i_2}, n_{i_3})$, where $tr_i$ is an identifier of the triangle element, $v_{i_0}, ..., v_{i_2}$ reference its vertices, and $n_{i_0}, ..., n_{i_2}$ reference the neighbour triangles (Figure 14.4).

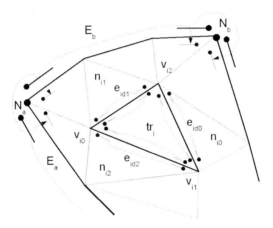

**Fig. 14.4** Representation of a face by a triangle mesh. Within each triangle element $tr_i$, celltuples and switch transitions are generated automatically (small darts). The cell-tuples of the complete face are situated at selected "corner" vertices $N_j$ of the mesh, while involutions follow the mesh boundaries $E_k$

The corresponding G-Map representation of topology of a mesh element comprises six cell-tuples $(v_{i_k}, edge_l, tr_i, sign)$, where the values $edge_l$ still have to be determined. It is possible to generate a new numbering of edges and a new cell-tuple representation, and to store it as a separate object, but this would greatly increase the size of the model, without adding any new information. In order to save space, we therefore suggest to generate the six cell-tuples of a triangle element on the fly when required, using an edge numbering scheme that can be reproduced as long as the triangle mesh is not altered: If an edge is situated at the triangle mesh boundary, we identify the edge by the triangle identifier $tr_i$ and by its relative position within the triangle, the latter being a number $p < 3$. For an interior edge separating two triangles $tr_i$ and $tr_j$, we choose the smaller of the two numbers, e.g. $tr_j$, and the corresponding relative position $p_j$. Clearly two bits are sufficient to store the local position, and the edge identifiers may be represented e.g. by $tr_j * 3 + p_j$. By restricting the admissible number of triangles, we can store triangle identifiers, vertex identifiers and edge numbers in fixed length fields, e.g. as long integers, and reproduce the corresponding cell-tuples at any time: for a given triangle $tr$ and its three neighbours $tr_0$, $tr_1$ and $tr_2$, determine first which neighbours have lower id, and second, the relative position $p_e$ of the corresponding edge $e$ within, then compose $tr$ or $tr_k$ and $p_e$ into the edge number $e_id$; finally, return the six cell-tuples, where the signs '+' and '-' are only given as an example:

$$(v_{i_0}, e_{id_2}, tr_i, +)(v_{i_1}, e_{id_1}, tr_2)$$
$$(v_{i_1}, e_{id_2}, tr_i, -)(v_{i_0}, e_{id_0}, tr_2)$$
$$(v_{i_1}, e_{id_0}, tr_i, +)(v_{i_2}, e_{id_2}, tr_0)$$
$$(v_{i_2}, e_{id_0}, tr_i, -)(v_{i_1}, e_{id_1}, tr_0)$$
$$(v_{i_2}, e_{id_1}, tr_i, +)(v_{i_0}, e_{id_0}, tr_1)$$
$$(v_{i_0}, e_{id_1}, tr_i, -)(v_{i_2}, e_{id_2}, tr_1)$$

As long as the triangle mesh topology is not altered, these cell-tuples can be reproduced at any time.

Thus, at the lowest, most detailed level of the hierarchy, the cell-tuple structure is represented implicitly by the triangle net. By a similar argument, any mesh consisting of elements with a bounded number of vertices can be integrated at the cost of a small number of bits for each element, e.g. a quadrangular mesh for boundary representation, or a tetrahedral mesh for solid modelling. Combined with a corresponding interpolation method, e.g. linear or bilinear interpolation for each mesh cell, at the lowest level topology and geometry representation can be integrated with the G-Map hierarchy.

## 14.7 Time-dependent topology

The objects of a city model, or of any other 3D GIS possess a "life span", i.e. a temporal interval of existence, that in turn can be decomposed into several intervals during which their structure is constant. As the G-Map representation of topology is a discrete structure, we consider intervals of constant topology as the smallest temporal units, separated by time instants at which the topology changes. Geometry and thematic properties, however may vary within these intervals, and using appropriate interpolation methods continuous variation of geometry and thematic properties at interval boundaries can be modelled.

We define a time-dependent d-G-Map as an application $\phi$ that to any temporal instant $t$ of a temporal interval $T$ attaches a d-G-Map $\phi(t) = G(D(t), \alpha_0(t), \ldots, \alpha_d(t))$. At each time instant $t$, $\phi(t)$ must verify the conditions (1) and (2) mentioned above. Given a sequence $t_0, T_1, \ldots, t_i - 1, T_i, \ldots, t_n$ of time interval composing a "life span" $[t_0, \ldots, t_n]$, we require $\phi(t)$ to be constant on each interval $T_i$. We do not in general impose continuity at the temporal interval boundaries $t_i$, but in some cases, a smoothness criterion at the transition between consecutive time intervals may be required. This can be achieved e.g. by associating with the time instant $t_i$ the common refinement of the topologies associated with time intervals $T_i$ and $T_{i+1}$ meeting at $t_i$. As a time instant or time inteval is attached to any component of a time-dependent G-Map, we may search, for a given cell-tuple or a given transition, to find a minimal subdivision of its "life-span". This can result in a bundle of a large number of different temporal interval sequences that may be more difficult to manage than their common subdivision. For the modelling of time-dependent d-G-Maps, we therefore propose a compromise between both approaches, by identifying larger groups of spatiotemporal elements, the so-called spatiotemporal components, that consist of the cartesian product of a temporal sequence with a constant subset of the time-dependent cellular complex. With each spatio-temporal component, a single sequence of time intervals is associated, thus reducing significantly the amount of storage required, while simplifying the management. This approach results in a hierarchical decomposition of the spatio-temporal G-Map into a number of component G-Maps, each of which is constant over its temporal interval of definition. The retrieval of a spatio-temporal cell then proceeds in two steps: first identify the temporal segment and the attached spatio-temporal component, and second, locate the cell within the ST-component.

## 14.8 Application example: combination of a 2D map with part of a 3D city model

In an ongoing application study, the first results of which are documented in [23], we examine the combination of 2D topology from a cadastral map of the city centre of Osnabrück with freely available 3D city model data of Osnabrück ([24] into a topological model of the environment of Osnabrück palace 14.5).

We used PostgreSQL ([21] as database platform, java[6] and pl/java [27] as programming languages, and the program Moka [15] as visualisation and editing tool. The cell-tuple structure, the involution operations, orbits and

---

[6] Java is a registered trademark of Sun Microsystems, Inc.

loops, as well basic Euler non-Euler operations on cellular complexes as described in [20] have been implemented. A certain gain in execution speed was achieved by server-side implementation as stored procedures [23], other optimisation attempts are still under way.

The application example consists of several connected buildings enclosing a courtyard, which is linked to streets and to a park by seven archways - a configuration which cannot be modelled without true 3D topology (14.5).

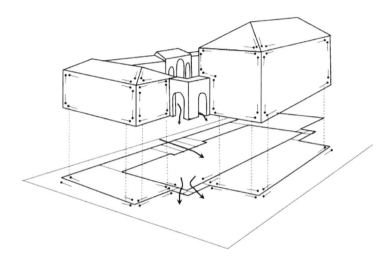

**Fig. 14.5** Topological representation (sketch): 3-G-Map of Osnabrück Palace with the 2D city map. Small pins symbolize a subset of the cell-tuples

From a database point of view, the goal is to provide topological and mixed database queries supported by a topological access structure. The queries contain adjacency queries and other useful queries for way finding, e.g.
*Can I pass through the courtyard on my way from the street to the park?*.

The use of G-Maps respectively of cell-tuple structures leads to a clear method:

1. Select a rectangular working area in the 2D cadastral map, and a set of buildings from the city model.
2. Correct the location of the vertices using digital elevation data.
3. Extract the topology of the 2D cadastral map as a 2-G-Map.
4. Convert the 2-G-Map of the working area into a 3-G-map:

    a. Extend the relational representation by addition of two columns.
    b. Duplicate the set of cell-tuples, inverting the orientation such that a "lower" and an "upper" side can be distinguished.

14 Modelling and Managing Topology in 3D GIS        243

   c. Add four nodes, eight edges and five faces, and introduce a solid - the "underground", resulting in a "sandbox" that carries the oroginal 2D map as upper surface.

5. Construct a 3-G-Map from the data of the 3D city model, which in fact is composed of 2D patches in 3D space:

   a. Extract simply connected surface patches, and represent them as faces, edges and nodes in the database.
   b. Build a 3-G-Map by composing the faces into boundaries of volume cells representing the topology of individual building parts.
   c. Transform the 3-G-Map by merging the common boundaries of adjacent building parts, representing the topology of the 3D city map.

6. Combine the models by defining faces on top of the "sandbox" corresponding to the ground faces of the solids composing the city model.

   a. Edit the two topology models to define the faces, edges and nodes to be matched
   b. Correct of vertex positions if necessary.
   c. Sew the cells of the two models at the contacts.

Most of these steps are either trivial or can easily be automated. The constructions of the topologies of the 2D map (step 3.) and of the 3D city model (step 5.), however are not simple. As the 2D cadastral map data are stored as polygons in a shape file, a spatial join on vertex and arc locations has to be used to establish the contacts between faces. The construction of a topologically consistent 3D model from the city model data involves considerable user interaction. In fact, the city model, derived from satellite data, consists of 2D surface patches suitable for a "virtual reality" visual representation, but is neither consistent nor complete, and does not comprise volume cells. Therefore available floor plans, elevations and vertical sections of the buildings, which belong to Osnabrück university, have to be consulted to control the construction of the 3D model.

After the two models are merged into one, further editing of the cells can be used to e.g. cut out cellars, windows, doors and archways, or to create interior walls and intermediate ceilings, in order to yield a more realistic consistent topological 3D model integrating indoor and outdoor spaces (Figure 14.6).

Topological database queries such as determining the neighbouring buildings of the palace can be directly answered using the topological 3-G-Map structure.

## 14.9 Conclusion and outlook

In this paper we have described a general approach for modelling and managing the topology of objects in a 3D GIS - based on oriented d-Generalised

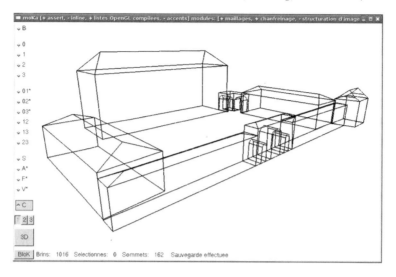

**Fig. 14.6** Screen snapshot during editing session with Moka [15]: Introduction of archways, and correction of vertex position nconsistencies. Edges are represented by straight lines, though the corresponding arcs may be curves or polylines

Maps. The topological data model and its realisation in a database management system have been presented in detail. The realisation of the approach in an Object-Relational Database Management System (ORDBMS) has been presented. An application example as part of the Osnabrück city map combined with a 3D model of Osnabrück Palace showed the applicability of the approach.

In our future work we will also deal with topological operations on objects with different levels of detail (LOD) based on hierarchical d-G-Maps. This approach shall be implemented in a Multi-Representation Database and evaluated with cartographic data of our project partners at Hannover University.

# References

[1] Mäntylä, M.: An Introduction to Solid Modelling. Computer Science Press (1988)
[2] Turner, A.K. (1992)(Ed.): Three-Dimensional Modelling with Geoscientific Information Systems, proc. NATO ASI 354, Kluwer, Dordrecht, 123–142.
[3] Balovnev, O., Bode, T., Breunig, M., Cremers, A.B., Müller, W., Pogodaev, G., Shumilov, S., Siebeck, J., Siehl, A., Thomsen, A.: The Story of the GeoToolKit – An Object-Oriented Geodatabase Kernel System. GeoInformatica 8(1) (2004) 5–47.

[4] Egenhofer, M.J.: A formal definition of binary topological relationships. In: Proc. 3$^{\text{th}}$ Int. Conf. on foundation of data organisation and algorithms (1989) 457–472
[5] Zlatanova, S.: 3D GIS for Urban Development. PhD dissertation, TU GRAZ, ITC Dissertation 69 (2000).
[6] Egenhofer, M.J., Frank, A.U. and Jackson, J.P.: A topological data model for spatial databases. In: Buchmann, A. P., Günther, O., Smith, T. R. and Wang, Y.-F.(eds.): Design and Implementation of Large Spatial, LNCS 409, Springer, Berlin (1990) 271–286 Zlatanova, S.: 3D GIS for Urban Development. PhD dissertation, TU GRAZ, ITC Dissertation 69 (2000).
[7] Pigot, S.: A topological model for a 3D spatial information system. 5$^{\text{th}}$ SDH, Charleston, (1992), 344–360.
[8] Brisson, E.: Representing Geometric Structures in d Dimensions: Topology and Order. Discrete & Computational Geometry 9 (1993) 387–426.
[9] Lienhardt, P.: N-dimensional generalized combinatorial maps and cellular quasi-manifolds. Int. Journal Comp. Geometry and applications 4(3) (1994) 275–324.
[10] Mallet, J.L.: Geomodelling. Oxford University Press (2002)
[11] Lévy, B.: Topologie Algorithmique – Combinatoire et Plongement. PhD Thesis, INPL Nancy (1999)
[12] Fradin, D., Meneveaux, D. and Lienhardt, P.: Partition de l'espace et hiérarchie de cartes généralisées. AFIG 2002, Lyon, décembre (2002), 12p.
[13] Raza, A., Kainz,W.: An Object-Oriented Approach for Modelling Urban Land-Use Changes. ACM-GIS (1999) 20–25.
[14] Mallet, J.L.: GOCAD: A computer aided design programme for geological applications. In: Turner, A.K. (Ed.): Three-Dimensional Modelling with Geoscientific Information Systems, proc. NATO ASI 354, Kluwer, Dordrecht, 123–142.
[15] MOKA· Modeleur de Cartes http://www.sic.sp2mi.univ-poitiers.fr/moka/ (2006).
[16] van Oosterom, P., Stoter, J.E., Quak, C.W., Zlatanova, S., The Balance Between Geometry and Topology. In: Richardson, D. and van Oosterom, P. (eds.), Advances in Spatial Data Handling, 10$^{\text{th}}$ SDH, Springer, Berlin (2002).
[17] Thompson, R.J.., van Oosterom, P.: Implementation issues in the storage of spatial data as regular polytopes. Information Systems for Sustainable Development-Part1 (2006) 2.33–2.46 (2006).
[18] Penninga, F., van Oosterom, P. and Kazar,B. M.: A tetrahedronized irregular network based DBMS approach for 3D topographic data modeling. In: Riedl, Andreas, Kainz, W., Elmes and Gregory A. (eds.): Progress in Spatial Data Handling, 12$^{\text{th}}$ SDH 2006, Springer, Berlin (2006) 581–598

[19] Saadi Mesgari, M.: Topological Cell-Tuple Structures for Three-Dimensional Spatial Data. PhD thesis University of Twente. ITC Dissertation 74 (2000).
[20] Thomsen, A., Breunig, M.: Some remarks on topological abstraction in multi representation databases. In: Popovich, V., Schrenk, M. and Korolenko, K. (eds.): 3rd workshop Inf. Fusion & GIS, Springer, Berlin (2007) 234–251.
[21] PostgreSQL.org: http://www.postgresql.org/docs (2006).
[22] PostGIS.org: http://postgis.refractions.net/documentation (2006).
[23] Butwilowski, E.: Topologische Fragestellungen bei der Kombination von 3D-Stadtmodellen mit 2D-Karten in einer Räumlichen Datenbank. Diplomarbeit, Fachgebiet Geographie, Universität Osnabrück, (2007).
[24] FRIDA: Free data from the city of Osnabrueck. http://frida.intevation.org/ueber-frida.html (2007).
[25] Breunig, M., Bär W. and Thomsen, A.: Usage of Spatial Data Stores for Geo-Services. $7^{th}$ AGILE Conf. Geographic Information Science, (2004) 687–696.
[26] Bär, W.: Verwaltung geowissenschaftlicher 3D Daten in mobilen Datenbanksystemen PhD Thesis, dept. of Mathematics/Computer Science, University of Osnabrück (2007).
http://elib.ub.uni-osnabrueck.de/cgi-bin/diss/user/catalog?search=sqn&sqn=693
[27] pl/java: http://wiki.tada.se/display/pljava/Home

# Chapter 15
# Mathematically provable correct implementation of integrated 2D and 3D representations

Rodney Thompson[1,2] and Peter van Oosterom[1]

## Abstract

The concept of the 'Regular Polytope' has been designed to facilitate the search for a rigorous closed algebra for the query and manipulation of the representations of spatial objects within the finite precision of a computer implementation. It has been shown to support a closed, complete and useful algebra of connectivity, and support a topology, without assuming the availability of infinite precision arithmetic. This paper explores the practicalities of implementing this approach both in terms of the database schema and in terms of the algorithmic implementation of the connectivity and topological predicates and functions. The problem domains of Cadastre and Topography have been chosen to illustrate the issues.

## 15.1 Introduction

One of the perennial problems in the spatial data industry is interchange of data. It is common for considerable outlay of time and effort (and funds) to be consumed in re-formatting and revalidating data, largely due to the lack of formal definition of spatial primitives and functions. For example, there is no agreed normative meaning of the 'equals' predicate when applied to geometric objects. Definitions of validity are in general defined by implementers – for example (Kazar *et al.* 2007). In addition, the language of spatial databases is couched in terms of the language of mathematics, with operations named

---

[1] Delft University of Technology, OTB, section GIS Technology,
Jaffalaan 9, 2628 BX the Netherlands
[2] Department of Natural Resources and Water,
Queensland, Australia
Rod.Thompson@nrw.qld.gov.au, oosterom@tudelft.nl

'union' and 'intersection' and using vector-like representations. This naturally leads to the impression that the representations form a topological space, and/or a vector space, which unfortunately is not the case. Generally speaking, the rigorous mathematics of the definition of spatial objects ends outside the database representation, which is only an approximation of the theoretical formalism used to define it. This leads to many cases where unexpected breakdowns in logic occur (Franklin 1984; Hölbling et al. 1998; Thompson 2004; Thompson 2007) due to the finite digital arithmetic implemented by computers and the necessary rounding or approximations.

By contrast, the Regular Polytope (Thompson 2005a) has been shown to be a promising candidate for the rigorous representation of geometric objects, in a form that is computable using the finite arithmetic available on digital computers. In order to explore practical issues in the Regular Polytope representation, a series of objects have been written in the Java programming language, and stored in a basic form using an Informix database.

The class of test data chosen was Cadastral property boundaries, since large volumes of data was available, and this topic presents some unique challenges, in particular, the mix of 3D and 2D data that is involved (Stoter 2004). The Regular Polytope representation provides a particularly elegant solution to this issue.

This paper describes alternative database schemas to support the implementation, and discusses some of the practical considerations that arise. This gives an indication of the requirements of a full implementation, and what further development is needed. Also discussed are some of the practicalities involved in converting geo-information to and from the regular polytope form from the conventional vertex representations. First, two different approaches to representation of geo-objects based on regions (resp. boundary-based and boundary-free) and two different application areas (resp. Cadastre and Topography) are discussed. This introductory section ends with an overview of the complete paper.

### *15.1.1 Boundary Representations and Mereology*

The conventional description of a geometric object partitions space into the interior, boundary and exterior of that object. The Egenhofer 9-matrix provides an exhaustive description of possible binary relationships (between two objects), and is frequently used in situations where a clear definition of a complex relationship is required (Egenhofer and Herring 1994). It consists of a $3 \times 3$ matrix of Boolean values representing the overlap of the interiors, boundaries and exteriors of the object pairs.

Alternately, there is an advantage in taking a Mereological approach to spatial logic, thus avoiding some of the distinctions between finite and infinite (smooth) sets. In this way, concepts such as 'set contacts set' and 'set includes set' move easily from the infinite to the finite realm, whereas the definition of a region as a collection of points defined by a boundary set of points (Smith 1997) raises the issue of the density of the representation. Briefly, the distinction is that point-set topology defines regions as sets of points, with boundaries being a separate set of points, either included or not depending whether the region is closed or open. The Mereological approach is to treat the region as the fundamental concept, with the boundary arising as a natural consequence of the region being limited in extent (Borgo *et al.* 1996).

Although the concept of a boundary as a point-set is useful in describing mathematical abstractions, it has no counterpart in the real world. '... it is nonsense to ask whether a physical object occupies an open or a closed region of space, or who owns the mathematical line along a property frontier' (Lemon & Pratt 1998a, page 10).

Similarly, in the computer representation, a boundary point set is problematic. On a line between two points in a 2D vector representation based on integers or floating point coordinates, it has been shown that in about 60% of cases, there will be no representable points at all lying on the line (apart from the end points)(Castellanos 1988; Thompson 2007). Thus, if a region is defined by a conventional polygon, the point set representing its boundary will consist of the vertices plus an insignificant, but highly variable number of points.

Thus, in summary, the concept of a boundary as a set of points does not sit well in the real world, or in the computer representation. It might be thought that a boundary would be needed to ensure a definition of such concepts as tangential contact, but this is not the case. An alternate approach comes from the Region Connection Calculus (RCC), defining such predicates without invoking boundary point-sets (Randell *et al.* 1992). Of the 512 possible relationships that can be defined in the Egenhofer 9 matrix, Zlatanova (2000) has shown that only 8 are possible between objects of the same dimensionality as the embedding space in 2D or 3D. These relationships can be directly modelled by the 8 relations of the RCC.

## *15.1.2 Application to Cadastre*

There are an interesting set of specific requirements in the realm of Cadastral data. Here, the fundamental item of interest is the land parcel. While these parcels are defined by measurement (survey) of their boundaries, there is no question of ownership of the boundary itself. Thus a boundary-free representation is ideal.

There is a growing need to represent 3D 'land' parcels (space) in a cadastre. These include strata parcels, unit rights, properties overhanging roads or rail, etc. (Thompson 2005b; Tarbit and Thompson 2006), but comprise a small minority of cadastral parcels in any jurisdiction, so that there is a strong argument for integrated 2D and 3D representations in the one database (Stoter and van Oosterom 2006). As pointed out by Stoter (2004), the so-called 2D parcels are really volumetric regions in space. It is merely the definition that is 2D (defining a prismatic volume, with undefined vertical limits), so it should be possible to evaluate any functions and predicates on mixed 2D/3D parcels. E.g. in figure 1, it can be asked whether $C$ intersects $D$ (which it does, since it encroaches on the volume above $D$.

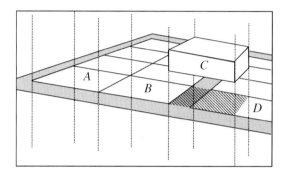

**Fig. 15.1** Mixing 3D and 3D Cadastre. $C$ is a volumetric parcel, with its projection onto the ground shown hashed

## *15.1.3 Topography*

The representation of the topography of the earth's surface has some similarity in requirements. The vast majority can be adequately represented by treating the elevation of the surface as a function of two variables (the $x$ and $y$ coordinates) – with only one elevation applying to each point, but this precludes the representation of vertical or overhanging cliffs, natural arches or many man-made structures (see figure 2). In the same way as with the cadastre, the vast majority of practical topographic data does not require full 3D, and could be modelled as a single-valued function of two variables - e.g. elevation($x$, $y$). Only minority of specific regions need true 3D; e.g. a Tetrahedral Network TEN (Verbree *et al.* 2005).

## *15.1.4 Structure of the Paper*

The remainder of the paper is structured as follows:

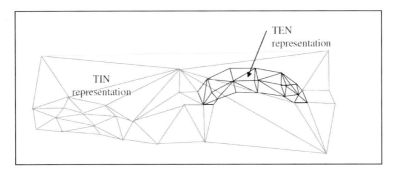

**Fig. 15.2** Combination of TIN and TEN representations

- Section 2 defines the concept of domain-restricted rational numbers, and the regular polytope concept.
- Section 3 discusses various database structures that could potentially be used to implement the approach.
- Section 4 discusses a set of Java language classes that have been written to assist in the investigation of, and demonstrate the algebra. This includes some test results with real data.
- Section 5 explores some issues that arise when converting from conventional vertex defined representations.

## 15.2 The Regular Polytope

In order to provide the required rigorous logic in 2D and 3D, within the computational arithmetic, the concept of a regular polytope has been proposed (Thompson 2005a; 2007) and will be summarized below. In this paper, the terminology and mathematics of 3D will be used, since the equivalent 2D case is usually obvious. Most of the illustrations are 2D for ease of drawing. This section contains a brief definition of the regular polytope representation (Sections 2.1 to 2.3), with some of its properties (Section 2.4), in comparison with more conventional approaches (Section 2.5).

### 15.2.1 Definition of domain-restricted rational numbers and points

Given two large integers $N'$ and $N''$, a domain-restricted (dr)-rational number $r$ can be defined with the interpretation $I/J$, with $I, J$ integers ($-N'' \leq I \leq N''$, $0 < J \leq N'$). The name 'domain-restricted rational' (dr-rational) is used because the values of $I$ and $J$ are constrained to a finite domain of values. Like floating point numbers, dr-rational numbers do not form a field[1] (in contrast to the true rational numbers) (Weisstein 2005), and therefore cannot span a vector space.

In 3D, a dr-rational point is defined as an ordered triple of dr-rational numbers $p = x,y,z$, with $x,y$ and $z$ representing the Cartesian co-ordinate values. Note that there are also counter-intuitive properties possessed by dr-rational points – e.g. it cannot be assumed that the mid-point between two dr-rational points is a dr-rational point. The advantage possessed by the dr-rational representation is that it is directly implementable in computer hardware.

## 15.2.2 Half Space Definition

In 3D a half space[2] $H(A,B,C,D)$ ($A$, $B$, $C$, $D$ integers, $-M < A,B,C < M$, $-3M^2 < D < 3M^2$ is defined as the set of dr-rational points $p(x,y,z)$: $-M \leq x,y,z < M$, for which computation of the following inequalities yields these results:

- $(Ax + By + Cz + D) > 0$ or
- $[(Ax + By + Cz + D) = 0$ and $A > 0]$ or [3]
- $[(By + Cz + D) = 0$ and $A = 0$ and $B > 0]$ or
- $[(Cz + D) = 0$ and $A = 0, B = 0$ and $C > 0]$,

where $M$ is the limit of values allowed for point representations. The choice of $M$ is dependant on the size of the region to be covered as a multiple of the unit of resolution. The values of N' and N" follow from M as:

- for 2D applications $N' = 2M^2$, $N'' = 4M^3$.
- for 3D applications use $N' = 6M^3$, $N'' = 18M^4$.

$H(0,0,0,0)$ is not a permitted half space.

The complement of a half space is defined as:

- $\overline{H} = (-A, -B, -C, -D)$ where $H = (A,B,C,D)$.

---

[1] The set of rational numbers $\mathbf{Q}$ obey the field axioms, including the closure axioms (e.g. $a \in \mathbf{Q}$, $b \in \mathbf{Q} \Rightarrow a.b \in \mathbf{Q}$). This is not the case for dr-rational or floating point numbers.

[2] The equivalent 2D object is known as a 'half plane' which (for convenience) is defined by the integer parameters A,B and D.

[3] This form of the definition with four parts, rather than just $(A.x + B.y + C.z + D) > 0$, is chosen so as to ensure a boundary-free representation. In effect, this eliminates all boundary points.

## 15.2.3 Regular Polytope Definition

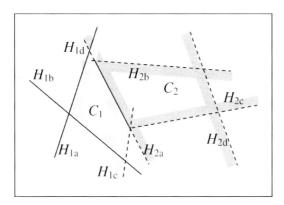

**Fig. 15.3** Regular polytope $O$ defined as the union of convex polytopes $C_1$ and $C_2$

A regular polytope $O$ is defined as the union of a finite set of (possibly overlapping) 'convex polytopes' ($C_1$ and $C_2$ in figure 3), which are in turn defined as the intersection of a finite set of half spaces ($H_{1a}$ to $H_{1d}$ and $H_{2a}$ to $H_{2d}$)(Thompson 2005a; Thompson 2007). Note – a regular polytope may consist of disconnected parts, and parts may overlap. The natural definitions of the union, intersection, and complement of regular polytopes is used, so that the meanings are exactly equivalent to the point set interpretations. E.g.:
$\forall p : p \in O_1 \cup O_2 \Leftrightarrow p \in O_1 \vee p \in O_2$ The concept of connectivity in this approach is based on two forms – so-called 'weak connectivity' $C_a$ in figure 4, and 'strong' or $C_b$ connectivity. The strongest form of connection is actual overlap, so that $OV \Rightarrow C_b \Rightarrow C_a$. This can also be expressed in dimensional terms (Clementini et al. 1993), such that in a 3D space, $Ca \equiv$ 0D or 1D meet, $Cb \equiv$ 2D meet, and $OV \equiv$ 3D meet.

## 15.2.4 Properties of the Regular Polytope

It is relatively simple to show that the space of regular polytopes is a topology (Thompson 2005c), based on the definition of regular polytope as an open set. It is readily apparent that for any regular polytope $O$, $\forall p : p \in O \Leftrightarrow p \notin \overline{O}$, and that[4] $O \cup \overline{O} = O_\infty$ and $O \cap \overline{O} = O_\Phi$. Thus no boundary points exist between $\overline{O}$ and $O$ ($\forall p : p \in O \vee p \in \overline{O}$), in contrast with most conventional approaches

---

[4] $O_\Phi$ and $O_\infty$ are the empty and universal regular polytopes respectively such that $\forall p : p \notin O_\Phi$ and $p \in O_\infty$.

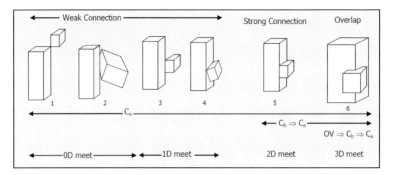

**Fig. 15.4** Modes of Connectivity in 3D

where (in the mathematical model) space is partitioned into a region's interior $R^o$, exterior $R^-$ and boundary $\partial R$, with $\forall p, p \in R^o \vee p \in R^- \vee p \in \partial R$. A further consequence of being a boundary free representation is that the axioms of a Boolean algebra (Weisstein 1999) are satisfied.

It has been shown (Thompson 2005c; Thompson 2007; Thompson and van Oosterom 2007) that the space of regular polytopes obeys the axioms of the Region Connection Calculus (Randell et al. 1992) based on the above definitions, and that it forms a Weak Proximity Space (Naimpally and Warrack 1970); and a Boolean Contact Algebra (Düntsch and Winter 2004). It is important to remember that it is the computational representation that satisfies the axioms, not an abstraction which is approximated by the computational representation. Thus it is possible to computationally apply the operations in any combination with complete confidence that no logic failure can result.

## *15.2.5 Vertex-based Representations*

In two-dimensional applications, the 'Point/Line/Polygon' paradigm for the representation of spatial features is well entrenched, albeit with some significant variations (van Oosterom et al. 2004), and provides a degree of comfort in the user. This is spite of some serious difficulties in terms of rigorous definitions of concepts such as validity, and equality (Thompson 2005a). The available 3D structures take various forms (Arens et al. 2003), with no one having proved to be the best in all circumstances (Zlatanova et al. 2004).

In this paper, the term 'vertex based' representation is used to cover all ways to model spatial data based on point coordinates of vertices as the major determinants of the shape and position of the objects. The vertices are defined as points with coordinates $x,y,z$, while all other geometric objects are defined in terms of sets of vertices or higher order constructive objects. This describes virtually all other two and three dimensional spatial data models, regardless of the level of topological encoding supported (Ellul and Haklay 2005). In contrast, but following a similar naming convention, the regular polytope in 3D would be called a 'surface based representation' and in 2D an 'edge based representation'. It can also be viewed as a restricted form of constraint-based representation.

## 15.3 The Proposed Database Structures

In implementing the concepts of the regular polytope, in a database management system, in order to manage large data volumes in a multi-user environment, several decisions need to be made. Principal amongst these is the decision as to how much redundant storage of information is to be tolerated, and more specifically whether 'topological encoding' is to be used. Section 3.1 discusses the basic data model, 3.2 briefly describes topological encoding, while 3.3 applies the concept to the regular polytope.

### 15.3.1 Discrete Polytope Encoding

This is perhaps the simplest structure, with some redundancy of storage, and no topological encoding. Each regular polytope is stored as a unit, containing its component convex polytopes and their defining half spaces.

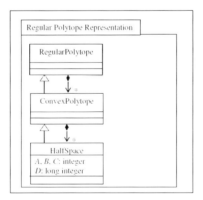

**Fig. 15.5** The Regular Polytope model. See also (Thompson 2005a; Thompson 2007)

Figure 5 shows, in the unified Modelling Language (UML) (OMG 1997), a possible implementation of the regular polytope representation discretely encoded (in 3D). Key points in the interpretation of this diagram is that:

- A convex polytope is a specialization of regular polytope.
- A half space is a specialization of convex polytope (which means it is also a regular polytope).
- A regular polytope consists of zero or more convex polytopes. Note that zero implies the empty regular polytope.
- A convex polytope consists of zero or more half spaces. Note that zero implies the infinite convex polytope.
- A half space has the integer attributes $A$, $B$, $C$ and $D$.
- No dr-rational numbers are stored in the database.
- A HalfSpace whose plane separates two convex polytopes will be stored twice (as an 'anti-equal' pair).

#### 15.3.1.1 Model Restrictions

This model is the most basic, intended for demonstration purposes. In practice, additional classes would be added to improve speed and responsiveness. For example, a convex polytope might be associated with an approximate bounding rectangle, which is the basis for a spatial index, however the bounding box could be computed with a function and the spatial index created on the return value of that function. That is, the bounding box need not be stored if a functional spatial index is used. There are a number of issues remaining that apply to this base level model:

- The Regular Polytope storage mechanism differs from the more familiar vertex representations, and requires non-trivial conversion routines to allow interoperability.
- The calculations require the use of large precision integer arithmetic (but these large precision integers are not stored in the database).
- The storage requirements are approximately double those required for simple polygon encoding.
- It is not trivial to map this storage form to/from the topological encoding form.
- Some analytic operations – such as those that require volumes, areas, distances, etc. of objects are inconvenient in the regular polytope representation.

#### 15.3.1.2 Model Advantages

- Data retrieval can readily be optimised since each regular polytope can be stored as an individual self-contained record on disc, and indexed using standard techniques such as r-tree (Guttman 1984).
- All RCC and topological predicates and functions can be rigidly supported in the computer-based representation.

### 15.3.1.3 The Disjoint Normal Form

A minor variant on the above strategy is to make the decomposition of the regular polytopes into convex polytopes more restricted. In the disjoint normal form (DNF), the convex polytopes that comprise a regular polytope are not permitted to overlap. In deciding whether or not to use DNF considerations include:

- Calculation of the volume of a regular polytope in DNF is simpler. The volume of each convex polytope can be calculated, and the results summed.
- Conversion of the regular polytope to a vertex defined polyhedron is simplified, since the individual convex polytopes can be converted, and the resultant polyhedra can be 'dissolved' together. Polyhedron dissolve is a simpler and faster operation than calculation of a union.
- It is not trivial to convert an arbitrary regular polytope to DNF.
- Some conversion algorithms from vertex representations to regular polytope produce DNF naturally.
- The number of convex polytopes in a conventional regular polytope (allowing overlap) can be less than in the case of DNF (see figure 6).
- The decomposition into disjoint convex polytopes is not unique (additional decision rules would be required to make it unique).

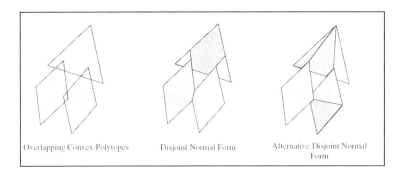

**Fig. 15.6** A regular polytope in overlapping, and in disjoint normal form

## 15.3.2 Topological Encoding

Topological Encoding is a traditional form of storage of spatial data in vertex representation, which provides two major advantages over discrete polygon storage of coverages:

- It gives the option of fast neighbour searches (find adjacent polygons).
- It reduces the redundancy of storage of boundary details.

There are several variants on space partitioning using topological encoding, but all are based on the common storage of boundary details, with links between the storage location of the boundary, and the details of the region(s) delimited by that boundary. It is in the definition of a coverage[5] that this approach is most significant, where every boundary is used in the definition of at least two regions (apart from those few boundaries surrounding the entire coverage).

In figure 7, the line string between node 1 and node 2 defines region $A$ to its left and region $B$ to its right. It is in cases such as this, where there is some complexity in the definition of the common boundary, that the greatest advantages of the approach are realised. (Since the definition of the line string from 1 to 2 contains many points, which do not need to be stored twice as would be the case if $A$ and $B$ were defined as discrete polygons).

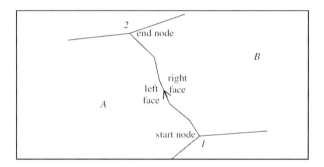

**Fig. 15.7** Two regions delimited by a common boundary line

## 15.3.3 Topological Encoding of Regular Polytopes

In the storage schemes that are appropriate to the Regular polytope representation, there are several possible analogues to topological encoding, but one in particular is quite promising for use in the field of cadastral data. This approach treats each half space as a common object, stored once only (in the way a common boundary is as described above), with links to each convex polytope that it participates in the definition of (either as a direct definition or as a definition by the complement of the half space); see figure 8.

---

[5] In this context, 'coverage' is used to mean that the entire area of interest is partitioned into non-overlapping regions (with no gaps between regions).

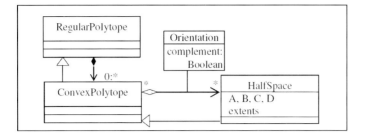

**Fig. 15.8** Regular polytope schema with common storage of half spaces

As an example from the cadastral domain, consider a series of property parcels with a common road frontage as depicted in figure 9. The single half space XY participates in the definition of the road, and its complement in the convex polytopes A, B1, C1, D and E1 (see figure 10. This ensures that:

- There are no gaps or overlaps between parcels and the road.
- The road frontages are straight.

Since the true definition of the parcels from the survey plan was probably in terms of a bearing and distance measurement from point X to point Y, this is a particularly appropriate representation, and allows the option of storing such measurement metadata within the half space record. Thus half space XY would be linked by the direct connection to the road section 1, and via the 'complement = true' link to convex polytopes A, B1, C1, D and E1. Note that half spaces can be used more than twice in contrast to the traditional encoding of topology based on edges, where a common edge is used twice (positive and negative).

Even where straight sections of frontage are non-contiguous, the half space record can be used in common. For example, the half space marked Y-Z defines the road section 2, with its complement defining E1, F, road section 3, G etc.

In full 3D parcels, the same is possible, with a half space being able to define a number of parcels in strata, as well as defining a non-stratum (2D) parcel adjoining it. In figure 11, the half space marked as XY, is the boundary of '2D parcel' A, and its complement is the boundary of strata parcels B1 to B5.

The HalfSpace record also should carry attributes defining its extents of use. This would probably be in the form of a minimum enclosing rectangle, and is used for two purposes:

- To distinguish between half spaces which are only co-incident by chance (in which case individual representation of the HalfSpaces is more practical). For example, it is possible that two boundaries many kilometres apart have the identical A,B,C,D values, but are not in any way related, and should not be linked.

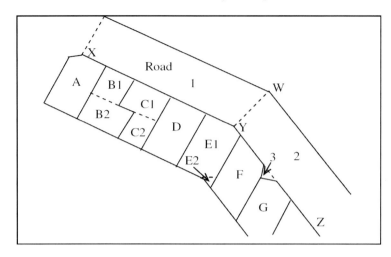

**Fig. 15.9** Cadastral data in the topologically encoded regular polytope form

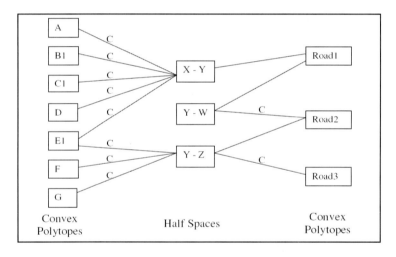

**Fig. 15.10** Object diagram showing some of the connections in figure 9 (The linkages marked 'C' are links with complement = true as in figure 8)

- To allow easy application of adjustments such as datum changes. Where an adjustment can be approximated by a 'block shift', the new definition of the halvers in a block can be calculated using the localisation provided by the extents.

The advantages that are created by using conventional topological encoding apply to the topologically encoded regular polytopes well as the rigorous logic of the regular polytope, so that:

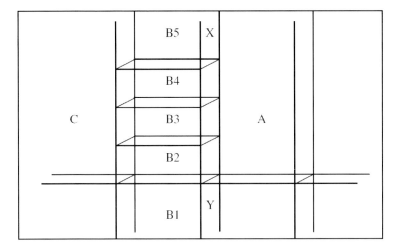

**Fig. 15.11** 3D parcels encoded using topologically encoded regular polytopes

- Some redundant storage is eliminated.
- Fast neighbour searches are facilitated.
- Accidental creation of overlaps and gaps is prevented.
- Frontages are kept straight.

It is unlikely that there will be much saving in storage requirements using this structure, since the cost of storing a halver redundantly is quite low, and largely offset by the keys and indexing needed to support the encoding. This is also true of the conventional form of topological encoding, and in both approaches the advantages are to be found in the ease of update, while retaining correct adjacency and consistency within the model.

## 15.4 The Java Demonstration Classes

A set of Java classes have been developed and tested on approximately a thousand cadastral parcels from the Queensland Cadastre, over a semi-urban region of average density and complexity. The region chosen contains primarily base (2D) parcels, but also has a smaller number of easements (secondary interests), and several 3D parcels. It consists of properties associated with residential, light commercial, light industrial, and recreational land usage.

The Java objects as developed parallel the definitions of the components of the regular polytope, and are divided into categories:

- The half space (or half plane),
- The convex polytope,

- The regular polytope.

This object model is intended for manipulation purposes in the processing software, and so differs from the various models given in section 3, which were intended primarily to illustrate a data storage strategy. The Java classes are set up to facilitate the mixing of 2 and 3 dimensional data.

This implementation models 3D and 2D objects, with no extensions to either lower of higher dimensionality. Since this is intended to explore practical issues associated with Cadastral data, no attempt has been made to produce a fully general n-dimensional model. Also there is no provision made for lower-dimensional objects such as lines, points and surfaces to be embedded in the space.

### 15.4.1 Description of the Java Objects

These classes and interfaces are defined for the calculation of the functions that have been defined on the regular polytopes. They contain redundant information and constructs that assist with these calculations, but are not necessarily stored permanently. Likewise, links that are described below may not be of a permanent nature. In contrast to the earlier class diagrams, which were intended to illustrate possible database table structures, the following diagrams document a series of Java classes that implement the RCC functions.

### 15.4.2 Classes and Relations

The half space/plane object is characterised by classes based on the Halver and the Face as in figure 12 described below:

- **HalfSpace**: defines a 3D half space and carries the parameters $A, B, C$ and $D$
- **HalfPlane**: defines a 2D half plane and carries the parameters $A, B$ and $D$. Parameter $C$ is not needed in 2D.
- *Halver*: a virtual class abstracting the HalfSpace and HalfPlane classes (this is restricted at present to 2D or 3D, since this is what the problem domain requires – see sections 1.2 and 1.3).
- **Point2R** and **Point3R** are domain-restricted rational point classes. They consist of a tuple of rational numbers (2 or 3 respectively), each consisting of a pair of Java BigIntegers (see 4.4.9).

The association in figure 12 is:

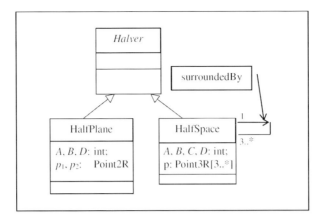

**Fig. 15.12** The Half Space, Half Plane and Halver Classes

- **Surrounded By**: a redundant one-way linkage, from a halfSpace, to the halfSpaces that adjoin and define it. It is needed during the calculation of vertices, and each time a new *halver* is added to a convex polytope. In modifying a halfSpace to take account of a new halfSpace, it may be necessary to calculate two new dr-rational points. These are the points of intersection of this halfSpace, the new halfSpace, and existing halfSpaces that surround the face. The relation is not needed in the 2D case, since only two half planes are needed to define a point.

The **ConvexPoly** (figure 13) contains a collection of *Halver* objects. In the prototype, a convex poly must contain all 2D or all 3D halvers (and is sub-classed as ConvexPoly2 or ConvexPoly3 respectively). The MBR in the ConvexPolytope and in the RegularPolytope is a 3D box defined by integer coordinates which is guaranteed to contain the whole convex (or regular) polytope.

The **Polytope** contains a collection of *ConvexPoly* objects. In this implementation, all ConvexPoly objects in a particular Polytope object must be 2D or all must be 3D. The nrUnitsA, and unitNrA attributes are used for the calculation of $C_a$ (weak) connectivity, nrUnitsB, and unitNrB attributes for $C_b$ (strong) connectivity. A unit in this context is a connected set of convex polytopes, so that that $C_a$ connectivity is defined as nrUnitsA $= 1$. Since $C_b$ is a stronger form of connectivity, $C_b \Rightarrow C_a$, and therefore nrUnitsA $\leq$ nrUnitsB.

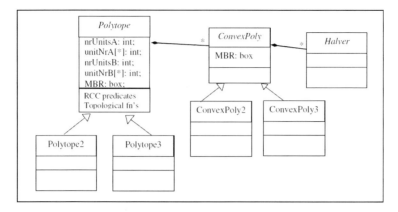

**Fig. 15.13** The Regular Polytope and Convex Polytope Classes

## 15.4.3 Methods

Only the more important methods are described in this section. The main classes based on Polytope, ConvexPoly, and Halver, have methods which convert them to and from database form. In this demonstration suite, only the bare minimum is stored in the database – the $A, B, C$ and $D$ values of the halvers, the structure of the regular polytope and a bounding box. For the purpose of the demonstration, and to assist with development, encoding was via a simple text string, but in a final system, a more sophisticated binary storage mechanism would be used. Vertices could also be stored for speed of processing.

### 15.4.3.1 Polytope Methods

A regular polytope is constructed by creating an empty regular polytope $O_\Phi$, with no convex polytopes, and extending it using Polytope.addConvexPoly(C). The methods provided in the regular polytope classes provide the full implementation of the RCC theory (Randell *et al.* 1992), extended for strong and weak connectivity:

- $C_a$ (p,p$_1$) Polytope.connectsToA(Polytope)
- $C_b$ (p,p$_1$) Poyltope.connectsToB(Polytope)
- $DC_a$ (p,p$_1$) ¬ Polytope.connectsToA(Polytope)
- $DC_b$ (p,p$_1$) ¬ Polytope.connectsToB(Polytope)
- P(p,p$_1$) Polytope.isWithin(Polytope)
- PP(p,p$_1$) Polytope.properPartOf(Polytope)
- EQ(p,p$_1$) Polytope.equals(Polytope)
- OV(p,$_1$) Polytope.intersects(Polytope)

# 15 Mathematically provable correct integrated 2D/3D representations

- $EC_a$ (p,p$_1$) Polytope.externallyConnectedToA(Polytope)
- $EC_b$ (p,p$_1$) Polytope.externallyConnectedToB(Polytope)
- $TPP_a$ (p,p$_1$) Polytope.tangentialProperPartOfA(Polytope)
- $TPP_b$ (p,p$_1$) Polytope.tangentialProperPartOfB(Polytope)
- $NTPP_a$ (p,p$_1$) Polytope.nonTangentialProperPartOfA(Polytope)
- $NTPP_b$ (p,p$_1$) Polytope.nonTangentialProperPartOfB(Polytope)
- PO(p,p$_1$) Polytope.properOverlap(Polytope)

Note that by RCC theory, all of these relations can be generated from the 'connects' relation. In practice, some are directly calculated (such as 'intersects' – for reasons as given in Chapter 5), but most are simply implemented as their definition suggests. e.g.:

```
/** Determines if this regular polytope is within the other
  * @param other The other Regular Polytope
  * @return True if this regular polytope is within the
    other */

public boolean isWithin(Polytope other) {
  Polytope otherM = other.inverse();
  otherM = otherM.intersection(this);
  return (otherM.convexPolys.size() == 0); }

/** Determines if this regular polytope is equal to the other
  * @param other The other Regular Polytope
  * @return True if every point in this regular polytope is
  * within the other and visa versa. */

public boolean equals(Polytope other) {
    return (this.isWithin(other) && other.isWithin(this)); }
```

## 15.4.4 Results

Approximately one thousand parcels were selected from the Queensland Cadastre – see figure 14. The area chosen was the region surrounding the 'Gabba' cricket grounds in Woolloongabba Brisbane, because this area contains some 3D parcels of non-trivial shape. The parcels obtained from the database are 2D only, but do include secondary interests (such as easements). Thus overlapping 2D register objects exist. There were several 3D parcels in the region. Two associated with the cricket stadium (figure 15), and one with a restaurant (figure 16) were hand encoded.

In the original data, some inaccuracies had been introduced through rounding, so there are slight overlaps and mismatches between the edges. This will be discussed further in section 5.

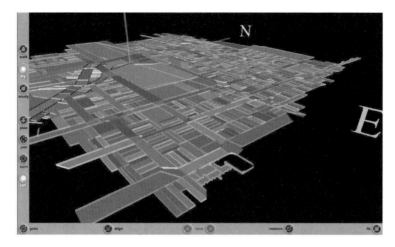

**Fig. 15.14** An overview of the test region

Figure 14 shows the data in question. The 2D parcels have been represented by colouring a plane (at $z = 0$) with a randomly selected colour. To further show the division between parcels, a vertical 'fence' has been drawn, of the same shade as the horizontal surface. Since the colour on each side is different, some interfering visualization effects can occur.

Figure 15 shows a detail of some of the 3D parcels (which abut without overlapping); see also (Stoter 2004) pages 269-272 for a view of these same parcels.

**Fig. 15.15** Detail of two 3D contiguous parcels with a third in the far background (the vertical grey cylinder is the Z axis)

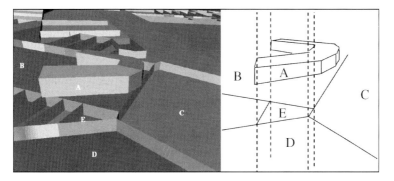

**Fig. 15.16** 3D parcel amongst 2D parcels. (Parcel A and 2D parcel E together comprise a restaurant. Parcel A overhangs the roadway represented by parcels B, C and D). Figure 16 depicts a 3D parcel A overhanging the footpath, and exactly abutting the 2D parcel E directly below the open space partially enclosed by it

#### 15.4.4.1 Data Quantities

One of the reasons for conducting this investigation was to determine the storage requirements of this approach. Had a more rural region been chosen, the averages below may have been less attractive (because there are more vertices per shared boundary on average, resulting in more halvers and convex polytopes), and this could be the subject of further investigation. The parcels in the test region required the following representation (Table 1). For comparison, an indication of the conventional polygon/polyhedron complexity is shown, but note that in the 3D case, the number of corners is estimated only, as a polyhedral model was not constructed as part of this investigation:

|  | 2D Case | 3D Case |
|---|---|---|
|  | 1044 parcels | 3 parcels |
| Average Convex Polytopes per Regular Polytope | 1.3 | 3.3 |
| Average Halvers per Regular Polytope | 5.3 | 23.6 |
| Average Halvers per Convex Polytope | 4.0 | 7.1 |
| Worst Case Convex Polytopes per Regular Polytope | 44 | 5 |
| Worst Case Halvers per Regular Polytope | 81 | 17 |
| Worst Case Halvers per Convex Polytope | 11 | 8 |
| Average Corners per Conventional Parcel | 6.3 | 36 |
| Maximum Corners per Conventional Parcel | 100 | 42 |

**Table 15.1** Average complexity of semi-urban parcels

### 15.4.4.2 Algorithmic Complexity

Java is a difficult language to obtain clear timing tests, since it is interpretive and uses various strategies of partial compilation. It also uses a 'garbage collector' form of memory management, leading to unpredictable timings of operations. For this reason, no strict timing tests were done. On the other hand, the actual algorithms are available for complexity analysis, and this leads to the suggestion that a practical implementation is possible. In the following, only the critical and potentially complex routines of the demonstration implementation are discussed.

### 15.4.4.3 ConvexPoly.compareWith(ConvexPoly)

This determines the relationship between two convex polytopes, returning the possible results: DISJOINT, CONTACTSa, CONTACTSb, INTERSECTS, CONTAINS, CONTAINED or EQUAL, and is probably the most critical method, since it is used in nearly all other operations. Inspection of the code shows that this operation will execute in $O(f_1 f_2 p^2)$ time, where $f_1$, $f_2$ are the number of half spaces or planes in the convex polytope, and $p$ is the average number of vertices in a face. In 2D, $p = 2$, so this becomes $O(f_1 f_2)$.

In 3D, it could be expected that the number of vertices on a face would be fairly constant in the range of about 3 to 6, so this also becomes $O(f_1 f_2)$. Thus it is important that in a practical system, the complexity of a convex polytope be kept limited. Fortunately, this is possible simply by dividing any highly complex convex polytopes into multiple smaller ones.

Thus, if the convex polytope is restricted to a specified maximum complexity, this routine is $O(1)$ (i.e. constant) in complexity. The cost of this simplification is an increase in the complexity of the regular polytope, so that more convex polytopes will be needed.

### 15.4.4.4 Constructing a Regular Polytope

As a regular polytope is constructed, each convex polytope must be compared with the convex polytopes previously added (to determine connectivity). This operation is thus of $O(n^2)$ where n is the number of convex polytopes in the regular polytope[6]. Since each convex polytope is a well defined geometric object, convex, and contains a MBR, it is relatively easy to optimise this operation. For example an in-memory spatial index could be used to reduce the search-time from $O(n^2)$ to $O(n \log n)$.

---

[6] This is assuming that the convex polytope complexity has been controlled as described above. Otherwise it would be $O(n^2 f^2)$ in 3D.

### 15.4.4.5 Polytope.intersection(Polytope)

This operation involves the calculation of the intersection of the Cartesian product of the convex polytopes. Thus it is by nature a $O(nm)$ operation, however the construction of the resultant polytope from this Cartesian product raises this in theory to $O(nm \log nm)$.

### 15.4.4.6 Polytope.inverse()

For regular polytope $O = \bigcup_{i=1..n} C_i$, with $C_i = \bigcap_{j=1..m_i} H_j$ the first step is to calculate: $O_i = \overline{C_i} = \bigcup_{j=1..m_i} \overline{H}_j$ for $i = i \ldots n$.

Thus, since each $m_i \leq c$ (by the assumption of the limited complexity of half spaces), this results in $n$ regular polytopes, each of up to $c$ convex polytopes. Each convex polytope consists of one half space only. Thus, this first part of the operation is $O(nc) = O(n)$ (because $c$ is constant). Note that the inverse of a convex polytope consists of a regular polytope with up to $c$ convex polytopes, each defined by one half space.

The second phase consists of forming the intersection of the $n$ regular polytopes $\overline{O} = \bigcap_{i=1..n} O_i$. If approached without any optimisation, this would be disastrous – leading to an operation of order $c^n$.

Fortunately, at each step in the algorithm, a large number of convex polytopes that are generated by the intersection operation are discarded. At the end of the process, assume there are $l$ convex polytopes left. If it is assumed that the number of convex polytopes in $R$ remains fairly constant during the process, this means that the cost of processing each $O_i$ in the intersection operation will be $O(l \log l)$. Since there are n polytopes, this gives $O(nl \log l)$. Note, this is an algorithm which could well repay some optimisation effort beyond the simple version used in the demonstration software.

### 15.4.4.7 Other Regular Polytope Operations

All of the other regular polytope operations (as shown in section 4.3.1) are simple combinations of other operations. So that the worst cases will be of no higher complexity that Polytope.inverse or Polytope.intersection.

### 15.4.4.8 Indexing and Searches

The programs as developed as a proof of concept do not use any database spatial indexing, and so are not efficient for doing spatial searches. On the other hand, they do generate a minimum bounding rectangle (or solid) surrounding the vertices of each regular polytope and each convex polytope, and so a standard R-Tree algorithm can be used.

#### 15.4.4.9 BigInteger Arithmetic

One of the advantages of implementing these routines in Java was the easy availability of the BigInteger object class. This provides a complete set of arithmetical operations on an integer representation with (effectively) no limit on the size of operands. The disadvantage of BigInteger is the slow speed of the operations, and the fact that the speed of operations is dependant on the size of the numbers involved.

In order to implement this software in a language other than Java (e.g. C), some of this functionality will need to be implemented. This is not a difficulty, since the algorithms are well known and documented. Further, not all functionality is needed. It is not necessary to allow for potentially infinite operands so memory allocation is not an issue. Although large numbers are involved, they are constrained. Further, not all arithmetic operations need be implemented – negation, addition and multiplication are needed, but division is not (this is a considerable simplification).

It is important to note that the use of BigInteger arithmetic is not an attempt to increase the accuracy of the data. The resolution of numeric forms such as 8 byte floating point is easily sufficient to cover the level of accuracy of virtually all spatial data in practical databases. The use of extended arithmetical types is to ensure repeatability and consistency in operations.

### *15.4.5 Optimizing the Model*

Optimising techniques would benefit from control of the complexity of the individual convex polytopes. The calculation of the vertices of a convex polytope is a significant process, strongly dependant on the cardinality of the set of half-spaces in a Convex Polytope. Restricting this cardinality can control this complexity, even at the cost of increasing the cardinality of the set of Convex Polytopes in a Regular Polytope. It is significantly easier to optimize the operations between convex polytopes.

In calculating the intersection of region $B$ with region $A$ in figure 17 (shown as two convex polytopes $A_1$ and $A_2$), even though all the half planes which define $A_1$ intersect all the half planes that define $B$ (since the half-plane definition is theoretically infinite), it can be determined by a conventional bounding box overlap test that all the vertices of $A_1$ are completely separated from the vertices of $B$ – therefore $A_1 \bigcap B$ is empty. This logic can be used to pre-eliminate large numbers of the partial intersections, and could be augmented by an in-memory spatial index – e.g. an R-tree based on the bounding boxes (shown as dashed lines) to further improve the calculation speed.

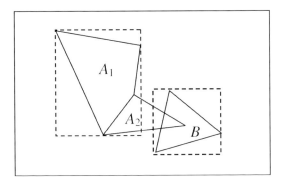

**Fig. 15.17** Calculating the intersection of two regular polytopes

While not necessary to the theory, it may improve the efficiency of many operations if the disjoint normal form (DNF) is used in representing regular polytopes (see 3.1.3). The indexing and comparison of convex polytopes can be improved thereby, since the disjoint convex polytopes will have smaller minimum bounding rectangles.

## 15.5 Data Load Issues

A major issue in the practical implementation of a regular polytope based storage mechanism is that of data conversion. Once the geometry is expressed in regular polytope form the operations between geometric regions are guaranteed to be rigorously correct, but the quality of the source data must be considered. Approximations may well have been made, and inaccuracies introduced to allow the data to be stored in the previous form, and this may create difficulties in data uptake.

### 15.5.1 Inaccuracy from Previous Systems

In many systems, calculation of the point of intersection between lines will have introduced rounding errors, as in the road frontage in figure 18 which was intended to be a straight line connection $A$ and $B$. In addition, to avoid later topological failures, a further displacement of the calculated point may have been applied, as in the case pictured in figure 19.

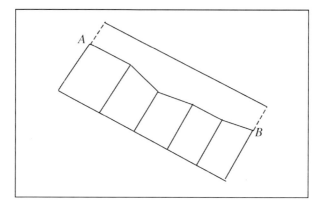

**Fig. 15.18** Points moved slightly in the calculation of intersections (shown exaggerated)

Note – this assumes that the data is to be loaded from an existing spatial database. In some cases, it may be possible to capture from the original source – e.g. the survey data. Unfortunately, while it would have been ideal to have captured original data in its uncompromised form, this is rarely the case, and much processing has been done to data before it reached the database. For example, bearings and distances will have been adjusted to 'close' and elevations of 3D points will have been calculated from the raw field notes. Note that there is a trend in which the original observations and measurements are more often stored in the (cadastral) database, in addition to the resulting interpretations (parcels).

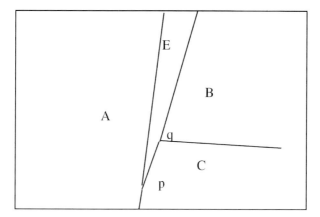

**Fig. 15.19** Polygon point $q$ moved (to the right) to prevent topological failure

15 Mathematically provable correct integrated 2D/3D representations    273

Ideally, before such parcels are converted into regular polytope form, the lines which were once straight, and were intended to be straight should be identified, and be represented as a single half plane (or at least half planes having the same $A,B,D$ values). As was mentioned above (section 4.4), this was not done in the demonstration software, resulting in small overlaps and mismatched edges. These can, however be detected by the rigorous operators available in the regular polytope representation.

### 15.5.2 2D Data Conversion to Regular Polytope Form

In the 2D version of the regular polytope, there is no difficulty generating a half plane whose edge passes exactly through any two points with integral coefficients. In the same way, any 2D data that is currently encoded using integer coefficients will create a 3D regular polytope with vertical walls with no loss of precision. In summary, it is possible to generate a half plane in 2D, or a half space parallel to the z axis through any two points with integral coefficients.

For example in figure 20, the incoming data is based on lines $(x_1,y_1)(x_2,y_2)$ and $(x_2,y_2)(x_3,y_3)$. The planes can be defined as $A_1 = y_1 - y_2$, $B_1 = x_2 - x_1$, $C_1 = 0$, and $D_1 = x_1 y_2 - x_2 y_1$, and $A_2 = y_3 - y_2$, $B_2 = x_2 - x_3$, $C_2 = 0$, and $D_2 = x_3 y_2 - x_2 y_3$. Clearly any point on the intersection of these planes will have $x = x_2$ and $y = y_2$.

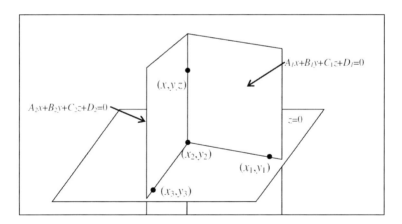

**Fig. 15.20** 3D planes based on incoming 2D data

Thus, any 2D data currently in a conventional format should easily be converted into regular polytope form with no loss of resolution, and no movement of vertices provided that integer representations are used for each. That is to say, if no attempt is made to straighten road frontages as part of the data load (as described above in 5.1), existing 2D cadastral data can be loaded unchanged. This is not necessarily the case with 3D data.

### 15.5.3 3D Data Conversion to Regular Polytope Form

Where 3D data is to be converted to regular polytope form, some care is needed. In general given any three non-colinear points, the best that can be asserted is that a half space can be generated whose boundary plane passes within one unit of resolution of each of the points. (In many special cases – specifically where the half space is parallel to any of the axes, much better results can be expected). If a situation such as that of figure 19 occurs in a 3D situation (the figure should be interpreted as a 'slice' through the 3D coverage), and the spurious bend at point $q$ is straightened, the actual position of point $p$ (as a point of intersection) is subject to a large variation.

In figure 21, where the half spaces that define region E have a possible imprecision of one unit, their point of intersection $p$ has a much larger possible error (shown shaded). If this is a critical issue, it may be solved by introducing a deliberate bias to the approximation, and an additional half space to limit the position of $p$. The bias is needed because the additional half plane can limit the error to the south (in figure 22) but not to the north. The bias is created by ensuring that all half planes that meet at acute angles are moved away from the convex polytope they define.

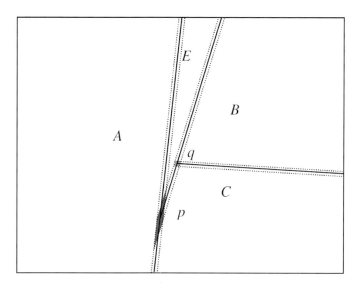

**Fig. 15.21** Imprecision in the placement of the point of intersection

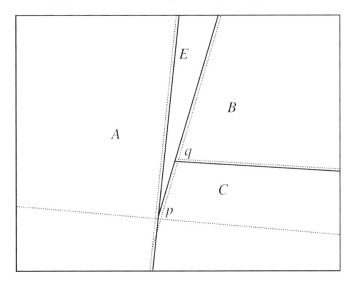

**Fig. 15.22** Half space introduced to constrain the point position

For example, in figure 22, region $E$ has been extended (still within one unit of resolution), and the acute angle at p has been truncated. This procedure – of truncating acute dihedral angles could be used as a general procedure in converting geometric objects to regular polytope form. In any case, objects with very acute angles need to be treated with caution in any representation to avoid the possibility of failures in algorithms such as buffer generation, generalisation, etc.

## 15.6 Conclusions

The Regular Polytope approach is practical, and could be rigorously implemented as a large-scale database (with proven functionality and without unpleasant surprises due to the mismatch of infinite real number mathematics and the finite digital computer). While some more optimization in the area of the regular polytope algorithms could yield speed improvements, for the sort of data used in this pilot system, acceptable results were obtained. In the test region, it was possible to run any combination the standard RCC and topological functions, and the combination and nesting of functions gave completely predictable results. The indication is that, a full implementation could be developed with query and analysis, and edit/update functionality.

It is expected that, as described above, restricting the complexity of convex polytopes will ensure practical speeds. In the case of 2D polytopes, several thousand half planes per complex polytope should be practicable. In 3D, the number is probably several hundred. This is appropriate particularly for cadastral data, whereas the parcels with large numbers of points (more than 2000) required in their definitions generally occur in rural areas, and are all 2D. Overly complex convex polytopes can be subdivided into a number of simpler convex polytopes.

The overwhelming advantage of the Regular Polytope approach is in the rigorously correct logic that it supports, and this justifies some additional data storage requirements, and the potentially slower processing times, but there is still much potential to improve the implementation of some of the operations – in particular, Polytope.intersection, and Polytope.inverse.

Possible future extensions of the Regular Polotype approach may include: non-linear half-spaces (e.g. circular arc, or polar coordinates defining a parcel boundary) and time as an additional dimension.

It has been shown that the definition of a half space is unique, and that the definition of a convex polytope in terms of half spaces is also unique (Thompson 2005c). If a decomposition of a regular polytope into a unique set of disjoint convex polytopes can be defined, it would be of great algorithmic value, among other things, providing a simple determination of equality.

# References

Arens, C., J. Stoter and P. van Oosterom (2003). Modelling 3D Spatial Objects in a Geo-DBMS Using A 3D Primitive. Association Geographic Information Laboritories Europe, Lyon, France.

Borgo, S., N. Guarino and C. Masolo (1996). A Pointless Theory of Space Based On Strong Connection and Congruence. 6th International Conference on Principles of Knowledge Representation and Reasoning (KR96), Morgan Kaufmann.

Castellanos, D. (1988). 'The Ubiquitous pi (Part II)'. Mathematics Magazine 61(3): 148-161.

Clementini, E., P. Di Felice and P. van Oosterom (1993). A Small Set of Formal Topological Relationships Suitable for End-User Interaction. Third International Symposium on Advances in Spatial Databases, Singapore.

Düntsch, I. and M. Winter (2004). 'Algebraization and Representation of Mereotopological Structures.' Relational Methods in Computer Science 1: 161-180.

Egenhofer, M. J. and J. R. Herring (1994). Categorising binary topological relations between regions, lines, and points in geographic databases. The nine intersection: formalism and its use for naturallanguage spatial predicates. M. J. Egenhofer, D. M. Mark and J. R. Herring, University of California.

Ellul, C. and M. Haklay (2005). Deriving a Generic Topological Data Structure for 3D Data. Topology and Spatial Databases Workshop, Glasgow, UK.

Franklin, W. R. (1984). 'Cartographic errors symptomatic of underlying algebra problems'. International Symposium on Spatial Data Handling, Zurich, Switzerland: 190-208.

Guttman, A. (1984). 'R-Trees: A Dynamic Index Structure for Spatial Searching' ACM SIGMOD 13: 47-57.

Hölbling, W., W. Kuhn and A. U. Frank (1998). 'Finite-Resolution Simplical Complexes.' Geoinformatica 2:3: 281-298.

Kazar, B. M., R. Kothuri, P. van Oosterom and S. Ravada (2007). On Valid and Invalid Three-Dimensional Geometries. In this book '2nd International Workshop on 3D Geo-Information: Requirements, Acquisition, Modelling, Analysis, Visualisation, 12-14 December 2007, Delft, the Netherlands'.

Naimpally, S. A. and B. D. Warrack (1970). Proximity Spaces. University Press, Cambridge.

OMG. (1997). 'UML 1.5'. Retrieved 2004 from http://www.omg.org/ technology/documents/formal/uml_2.htm

Randell, D. A., Z. Cui and A. G. Cohn (1992). A spatial logic based on regions and connection. 3rd International Conference on Principles of Knowledge Representation and Reasoning, Cambridge MA, USA, Morgan Kaufmann.

Smith, B. (1997). Boundaries: An Essay in Mereotopology. The Philosophy of Roderick Chisholm. L. Hahn, LaSalle: Open Court: 534- 561.

Stoter, J. (2004). 3D Cadastre. PhD Thesis. Delft, Delft University of Technology.

Stoter, J. and P. van Oosterom (2006). 3D Cadastre in an International Context. Taylor & Francis, Boca Raton FL.

Tarbit, S. and R. J. Thompson (2006). Future Trends for Modern DCDB's, a new Vision for an Existing Infrastructure. Combined 5th Trans Tasman Survey Conference and 2nd Queensland Spatial Industry Conference. Cairns, Queensland, Australia.

Thompson, R. J. (2004). 3D Topological Framework for Robust Digital Spatial Models. Directions Magazine.

Thompson, R. J. (2005a). 3D Framework for Robust Digital Spatial Models. Large-Scale 3D Data Integration. S. Zlatanova and D. Prosperi. Boca Raton, FL, Taylor & Francis.

Thompson, R. J. (2005b). 3D Cadastral Issues Within NR&M. Brisbane, Department of Natural Resources and Mines (Internal Report).

Thompson, R. J. (2005c). 'Proofs of Assertions in the Investigation of the Regular Polytope'. Retrieved 2 Feb 2007 from http://www.gdmc.nl/publications/reports/GISt41.pdf

Thompson, R. J. (2007). Towards a Rigorous Logic for Spatial Data Representation. Geo Database Management Centre. Delft, Delft University of Technology. PhD Thesis.

Thompson, R. J. and P. van Oosterom (2007). 'Connectivity in the Regular Polytope Representation.' submitted to GeoInformatica.

van Oosterom, P., W. Quak and T. Tijssen (2004). About Invalid, Valid and Clean Polygons. Developments In Spatial Data Handling. P. F. Fisher. New York, Springer-Verlag: 1-16.

Verbree, E., A. van der Most, W. Quak and P. van Oosterom (2005). Overlay of 3D features within a tetrahedral mesh: A complex algorithm made simple. Auto Carto 2005, Las Vegas.

Weisstein, E. W. (1999). 'Boolean Algebra'. MathWorld – A Wolfram Web Resource Retrieved 20 Jan 2007 from http://mathworld.wolfram.com/BooleanAlgebra.html

Weisstein, E. W. (2005). 'Rational Number'. MathWorld – A Wolfram Web Resource Retrieved 23 May 2005 from http://mathworld.wolfram.com/RationalNumber.html

Zlatanova, S. (2000). 3D GIS for Urban Development. Graz, Graz University of Technology.

Zlatanova, S., A. A. Rahman and W. Shi (2004). 'Topological models and frameworks for 3D spatial objects'. Journal of Computers & Geosciences 30(4): 419-428.

# Chapter 16
# 3D Solids and Their Management In DBMS

Chen Tet Khuan, Alias Abdul-Rahman, and Sisi Zlatanova

## Abstract

3D spatial modeling is one of the most important issues in 3D GIS research. It involves the definition of spatial objects, data models, and attributes for visualization, interoperability and standards. Real world complexity leads to different modeling approaches, as seen in different GIS applications. This paper provides some review of the problems, challenges and issues pertaining to the 3D GIS problems, especially in the handling and managing of 3D solids in DBMS. The paper also describes 3D spatial operators in DBMS and presents results using a simulation dataset. At the end of the paper, we provide and highlight requirements and recommendations for future research.

## 16.1 Introduction

'True' 3D GIS require extensive effort, as revealed from the recent research output and workshop (see Abdul-Rahman, et al. 2006). It is interesting to note that work on fundamental aspects, like 3D spatial analysis, has not been addressed to the level where an operational 3D system could be realized. The aim of this paper is to review recent research on 3D spatial data modeling and describe our recent work on the management of 3D solids in geo DBMS. Recent research development shows that 3D spatial modeling is becoming very important for many advanced GIS applications and the scenario is being enhanced by the advancement of computer graphics (hardware

---

Department of Geoinformatics, Faculty of Geoinformation Science and Engineering, Universiti Teknologi Malaysia, Skudai, Malaysia
Delft University of Technology, OTB, section GIS Technology,
Jaffalaan 9, 2628 BX the Netherlands
kenchen, alias@fksg.utm.my, s.zlatanova@tudelft.nl

and software), visualization, etc. and also influenced by developments in the OpenGIS consortium. 3D visualization environments such as Google Earth or 3D navigation software have already made some contribution and enabled more and more users to utilize visualization technology. Until recently, only specialized applications were able to manage and analyze 3D spatial data. The third dimension was used primarily for visualization and navigation. However, users are looking for applications that have one or more 3D GIS functionality. Due to the complexity of real-world spatial objects, various types of representations (e.g. vector, raster, constructive solid geometry, etc.) and spatial data models (topology, and geometry) have been investigated and developed, including e.g. Pilouk, 1996; Zlatanova, 2000; and Kada et al, 2006.

A universal and practical spatial data model that is capable of addressing more than one application is not available. This is due to the complexity of real world objects and situations. On the other hand, different disciplines emphasize different aspects of information e.g. including different requirements and output. Thus, a data model could be considered appropriate for a certain application but not so appropriate for other tasks. Different aspects and characteristics of real objects have led to the existence of several variations in object definition. The solution for these problems has directly referred to GIS standardization.

Current 3D GIS offer 2D functionality with 3D visualization and navigation capability. However, promising developments were observed in the DBMS domain where more spatial data types, functions and indexing mechanism were supported. In this respect, DBMS are expected to become a critical component in developing of an operational 3D GIS. However, extensive research and development are needed to achieve native 3D support at DBMS level.

This paper reviews works on 3D DBMS, especially on the aspect of managing volumetric objects. It is organized in the following order – Section 2, a short discussion on the standard specifications for 3D GIS spatial data modeling by Open GIS Consortium. Based on the specifications, Section 3 discusses the implementation of maintaining 3D spatial objects in DBMS. Section 4 describes the previous research works on 3D spatial data modeling. A brief discussion for 3D visualization is given in Section 5 and the paper concludes with recommendations for future work in Section 6.

## 16.2 The OGC Abstract Specifications for 3D Solids

The Open Geospatial Consortium (OGC 1999) is a non-profit organization that deals with the development of standards for modelling real-world objects. These standards deal with conceptual schemes for describing and manipulating the spatial characteristics of geographic features. The desire to provide a standard specification for GIS was initially driven by the developers - due to

# 16 3D Solids and Their Management In DBMS

the difficulty in GIS interoperability. The specification, in short, defines three important areas, namely:

- Data types: the need to have data types that represent real world object is obvious. Different kinds of data types and different kinds of objects could be modelled within DBMS.
- Functions/operations: there must be functions and operators to support the management of multi-dimensional objects that work for spatial analysis in DBMS, e.g. objects intersection.
- Spatial index: the main purpose is to deal with spatial searching (query), and sometimes it implements in different operators to speed up the query process.

According to the Spatial Schema, spatial characteristics are described by one or more spatial attributes whose value is given by a geometric object (GM_Object) or a topological object (TP_Object). Geometry provides means for the quantitative description, by means of coordinates and mathematical functions, of the spatial characteristics of features, including dimension, position, size, shape, and orientation. The mathematical functions used to describe the geometry of an object depend on the type of coordinate reference system used to define the spatial position. Geometry is the only aspect of geographic information that changes when information is transformed from one geodetic reference system or coordinate system to another.

Topology deals with characteristics of geometric figures that remain invariant when space is deformed elastically and continuously – for example, when geographic data is transformed from one coordinate system to another. Within the context of geographic information, topology is commonly used to describe the connectivity of an n-dimensional graph, a property that is invariant under continuous graph transformation. Computational topology provides information about the connectivity of geometric primitives that can be derived from the underlying geometry.

This paper will further concentrate on Geometry.

## *16.2.1 GM_Solid*

OGC defines 3D object as GM_Solid (OGC 2001) and it is a subclass of GM_Primitive and is the basis for 3-dimensional geometry. The extent of a solid is defined by the boundary surfaces. The boundary defines a sequence set of GM_Surfaces that limit the extent of this GM_Solid (see Fig. 1). These surfaces shall be organized into one set of surfaces for each boundary component of the GM_Solid. Each of these shells shall be a cycle (closed composite surface without boundary). In general, a solid in a bounded 3-dimensional manifold has no distinguished exterior boundary. In cases where 'exterior' boundary is not well defined, all shells of the GM_SolidBoundary shall be listed as 'interior'. The GM_OrientableSurfaces that bound a solid shall be oriented outward – that is, the 'top' of each GM_Surface as defined by its orientation shall face away from the interior of the solid. To represent a 3D solid as a volumetric object, GM_Solid is the best abstract spesification defined by OGC. Other than the GM_Solid, some feature geometry such as GM_Composite also involves a 3D solid object with other primitives, e.g. point, line, and polygon.

There are some functions or operations that could be implemented using GM_Solid. The function/operations are:

- Area: the operation shall return the sum of the surface areas of all of the boundary components of a solid. For example: GM_Solid::area() : Area
- Volume: the operation shall return the volume of this GM_Solid. This is the volume interior to the exterior boundary shell minus the sum of the volumes interior to any interior boundary shell. For example: GM_Solid::volume() : Volume
- GM_Solid (constructor): since this standard is limited to 3-dimensional coordinate reference systems, any solid is definable by its boundary. The default constructor for a GM_Solid is from a properly structured set of GM_Shells organized as a GM_SolidBoundary. For example: GM_Solid:: GM_Solid(boundary : GM_SolidBoundary) : GM_Solid

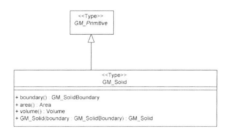

**Fig. 16.1** GM_Solid data type defined by OGC

Although the OGC does not discuss some operations that refer to 3D solid e.g. 3D intesection between 2 solids, to extend to the third dimension, similar specifications could be given to the 3D operations, if the z-coordinate is considered. The notion for operations provided by OGC are as provided:

    return-type type-1::operation(type-2, type-3 ... )

Example:

    Double Precision Geometry 1::Distance(Geometry 2, Geometry 3)

    operation(name-1 : type-1, name-2 : type-2, name-3 :
              type-3 ...) : return-type, ...

# 16 3D Solids and Their Management In DBMS

Example:

```
3D Intersects(A1:Geometry 1, A2:Geometry 2) : Geometry 3
```

There are other 3D objects being considered in the OGC specification, i.e. cone, sphere and, etc. Some 3D object are not considered volumetric solids, but still appear in 3D space, i.e. free-form curve and surface. Fig.2 denotes the complete list of 3D objects (with highlighted part) considered in OGC specification.

The OGC abstract specifications deal with geometry and functions. Spatial index is not mentioned in the abstract specification. Therefore, rule or specifications about developing spatial indexing is unavailable. However, the OGC provides for the implementation specification of R-Tree indexing according to the existing DBMS format, i.e. Oracle Spatial. The following section will discuss the basic idea of R-Tree index implemented within DBMS.

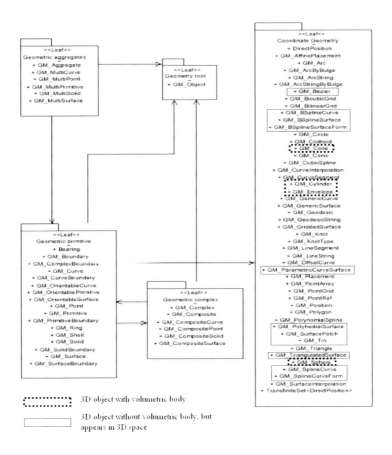

**Fig. 16.2** Geometry package in OGC abstract specification

## 16.2.2 The OGC Implementation Specifications for DBMS

The GM_Solid has been defined by the OGC as a general 3D primitive in abstract specification (OGC 1999a). However, the existing implementation (for SQL) of 3D solid (e.g. polyhedron, tetrahedron) is not available due to the absent of 3D data type (as 3D primitive) within existing DBMS. A volumetric object could be modeled using a multi-collection of similar or different geometries. OpenGIS implementation specification for 3D solid objects can be referred to as *PolyhedralSurface* and *MultiPolygon*. A *PolyhedralSurface* is a contiguous collection of polygons that share common boundary segments. It is a subtype of *Surface*. The primitive of *PolyhedralSurface* and *MultiPolygon* are referred to as *Polygon* (see Fig. 5). The difference between these two geometries is that the polygons that construct *PolyhedralSurface* must share boundaries with the neighboring polygons. The *MultiPolygon* is flexible, i.e. share boundary may not exist for certain polygon(s). For each pair of polygons that 'touch', the common boundary shall be expressible as a finite collection of LineStrings. Each LineString shall be part of the boundary of at most 2 polygon patches. A TIN (triangulated irregular network) is a PolyhedralSurface consisting only of Triangle patches. For any two polygons that share a common boundary, the 'top' of the polygon shall be consistent. This means that when two LinearRings from these two Polygons traverse the common boundary segment, they do so in opposite directions. Since the Polyhedral surface is contiguous, all polygons will be consistently oriented. This means that a non-oriented surface shall not have single surface representations. Fig. 3 shows an example of such a consistently oriented surface (from the top). The arrows indicate the ordering of linear rings from the polygon boundary in which they are located. The methods of implementing the polyhedral surface in DBMS is given as below (see Fig. 4):

```
NumPatches (): Integer - Returns the number of including
                        polygons

PatchN (N: Integer): Polygon - Returns a polygon in this
                        surface, the order is arbitrary.

BoundingPolygons (p: Polygon): MultiPolygon - Returns the
                        collection of polygons in this
                        surface that bounds the given
                        polygon 'p' for any polygon 'p'
                        in the surface.

IsClosed (): Integer - Returns 1 (True) if the polygon
                        closes on itself.
```

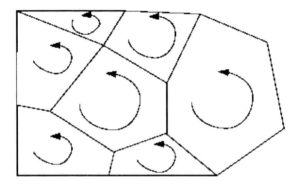

**Fig. 16.3** *PolyhedralSurface* with consistent orientation

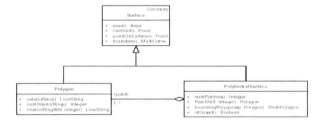

**Fig. 16.4** Implementation specification for *PolyhedralSurface*

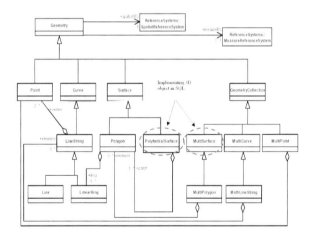

**Fig. 16.5** SQL Geometry type hierarchy

In the implementation specification, OGC provides the geometry function that is not limited to any dimension. Only DBMS itself decides the implementation of the standard functions (specified by OGC) that considers the third dimension or not. Some of the standard functions given by OGC (Simple Feature Specification for SQL, Revision 1.1) are:

```
Intersection (g1 Geometry, g2 Geometry): Geometry
```

Return a Geometry that is the set intersection of geometries g1 and g2.

```
Difference (g1 Geometry, g2 Geometry): Geometry
```

Return a Geometry that is the closure of the set difference of g1 and g2.

```
Union (g1 Geometry, g2 Geometry): Geometry
```

Return a Geometry that is the set union of g1 and g2.

```
SymDifference(g1 Geometry, g2 Geometry): Geometry
```

Return a Geometry that is the closure of the set symmetric difference of g1 and g2 (logical XOR of space).

```
Buffer (g1 Geometry, d Double Precision) : Geometry
```

Return as Geometry defined by buffering a distance d around g1

```
ConvexHull(g1 Geometry) : Geometry
```

Return a Geometry that is the convex hull of g1.

Implementing the spatial index that follows the standard specification is not available with the OGC document. This is because the spatial index deals with the method of searching, which often involves mathematical algorithms, e.g. the implementation of R-Tree indexing. A R-Tree is a depth-balanced tree extending the B-tree for n-dimensions. The index stores the minimum bounding boxes as representations, not the objects themselves. It is equally referred to as a minimum bounding rectangle (MBR). A detailed documentation about the R-Tree could be found in Rigaux et al. (2002). There is no standard syntax/command/structure stated by OGC that enables any DBMS to be implemented. Only the DBMSs themselves provide their own syntax/command/structure that establishes the spatial index. The following examples are provided:

(For Oracle Spatial)

```
CREATE INDEX [index_name] on
<table_name>(geometry_column)

INDEXTYPE IS mdsys.spatial_index

PARAMETERS('sdo_indx_dims=3');   -- Dimension = 3
```

(For PostGIS)

CREATE INDEX [index_name] ON <table_name>

USING GIST <geometry_column>
GIST_GEOMETRY_OPS);

The concept of sample R-tree structure is given in Fig. 6, Fig. 7, & Fig. 8 in two and three-dimensions. The impact of the z-coordinate on 3D spatial indexing will influence the execution time because the indexing mechanism will search each of the (x, y) elements that relate to its z-coordinate. For example, 7 (x, y, z) points will search 7 times greater than 7 (x, y) elements.

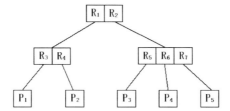

**Fig. 16.6** Directory of R -Tree indexing

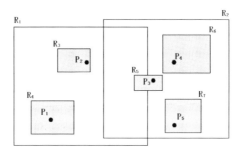

**Fig. 16.7** A planar representation of an R-tree

Note that the Oracle Spatial provides the spatial index up to 4D and the dimensionality should be defined in the syntax. However, the GiST index is widely used for 2D data. The implementation of GiST is rather limited for 3D data. The research and application of 3D GiST is expected in the near future. The next section discusses some implementations of spatial indexes for the third dimension in DBMS.

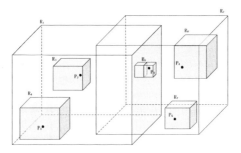

**Fig. 16.8** A 3D representation of an R-tree

## 16.3 Some Implementations of 3D Solid In DBMS

Since the Implementation specifications do not recommend a 3D data type, most of the DBMS (except Oracle Spatial) have not implemented volumetric data types. However, 3D data can be stored in the database since the data types are embedded in 3D space, i.e. point, line and polygon can be represented with their 3D coordinates. The next section describes how 3D real-world objects can be stored in the DBMS using 3D multipolygon.

### 16.3.1 Modeling 3D Solid Using MultiPolygon

In the Oracle Spatial object-relational model, a 3D solid object from 3D primitive is not possible. However, it could be done by implementing the MultiPolygon that bounds a solid. The geometric description of a spatial object is stored in a single row and in a single column of object type SDO_GEOMETRY in a user-defined table. Any tables that have a column of type SDO_GEOMETRY must have another column, or set of columns, that define a unique primary key for that table. Tables of this sort are referred to as geometry tables.

Oracle Spatial defines the object type SDO_GEOMETRY as:

```
CREATE TYPE sdo_geometry AS OBJECT (
   SDO_GTYPE NUMBER,
   SDO_SRID NUMBER,
   SDO_POINT SDO_POINT_TYPE,
   SDO_ELEM_INFO MDSYS.SDO_ELEM_INFO_ARRAY,
   SDO_ORDINATES MDSYS.SDO_ORDINATE_ARRAY);
```

An example of implementing a 3D multipolygon (where the geometry can have multiple, disjoint polygons in 3D) is provided below:

```
CREATE TABLE Solid3D (
```

```
      ID number(11) not null,
      shape mdsys.sdo_geometry not null);

   INSERT INTO Solid3D (ID, shape) VALUES (
      1 SDO_GEOMETRY(3007,   -- 3007 indicates a 3D multipolygon
      NULL,                  -- SRID is null
      NULL,                  -- SDO_POINT is null
      SDO_ELEM_INFO_ARRAY(   -- the offset of the polygon
         1, 1003, 1,
         16, 1003, 1,
         31, 1003, 1,
         46, 1003, 1,
         61, 1003, 1,
         76, 1003, 1),
      SDO_ORDINATE_ARRAY(
         4,4,0, 4,0,0, 0,0,0, 0,4,0, 4,4,0,   -- 1st polygon
         4,0,0, 4,4,0, 4,4,4, 4,0,4, 4,0,0,   -- 2nd polygon
         4,4,0, 0,4,0, 0,4,4, 4,4,4, 4,4,0,   -- 3rd polygon
         0,4,0, 0,0,0, 0,0,4, 0,4,4, 0,4,0,   -- 4th polygon
         0,0,0, 4,0,0, 4,0,4, 0,0,4, 0,0,0,   -- 5th polygon
         0,0,4, 4,0,4, 4,4,4, 0,4,4, 0,0,4    -- 6th polygon
   )));
```

For PostGIS, the 3D solid as a primitive object is also not available. To create a 3D object that implements existing primitives, then a MultiPolygonM could be used. The three dimensions simply allow a z-coordinate to be stored for each point. The geometry column in PostGIS differs from Oracle Spatial. The description of geometry column is given below:

```
AddGeometryColumn(<table\_name>, <column_name_of_geometry>,
                  <srid>, <geomery_type>, <dimension>)
```

An example of implementing the MultiPolygonM is given below:

```
CREATE TABLE Solid3D (ID integer primary key,
                     NAME varchar (20) not null);}

SELECT AddGeometryColumn('Solid3D', 'shape',
                        423, 'MULTIPOLYGONM', 3);
```

Note that the table name, Solid3D, is given a geometry column named 'shape', with MULTIPOLYGONM type in third dimension. The following example denotes a real multipolygon stored in PostGIS.

```
INSERT INTO Solid3D (ID, shape) VALUES (
   2, -- ID
   GeometryFromText('MULTIPOLYGONM(
```

```
      (4,4,0, 4,0,0, 0,0,0, 0,4,0, 4,4,0)  -- 1st lower polygon
      (4,0,0, 4,4,0, 4,4,4, 4,0,4, 4,0,0)  -- 2nd side polygon
      (4,4,0, 0,4,0, 0,4,4, 4,4,4, 4,4,0)  -- 3rd side polygon
      (0,4,0, 0,0,0, 0,0,4, 0,4,4, 0,4,0)  -- 4th side polygon
      (0,0,0, 4,0,0, 4,0,4, 0,0,4, 0,0,0)  -- 5th side polygon
      (0,0,4, 4,0,4, 4,4,4, 0,4,4, 0,0,4)  -- 6th upper polygon
)));
```

The advantage of implementing the multipolygon in DBMS is that the integration between CAD and GIS is possible for 3D visualization, i.e. Oracle (or called Spatial) spatial schema is supported by MicroStation and Autodesk Map 3D. This is due to the geometry column provided by Spatial directly accesses the 3D coordinates of the object, which allow the display tools to retrieve spatial information from the geometry column. However, problems occur if the data volume is huge, i.e. more polygons are stored for a single 3D solid body. Data size will directly affect data retrieval and yield a slow dataset loading within visualization environment. This weakness could be overcome with the approach of implementing polyhedron as 3D data type in DBMS as proposed by Arens (2003), see Section 4.

Although the implementation of MultiPolygons and Multipatch could be done for 3D visualization, these objects do not represent real 3D objects. They define only a set of bounding surfaces that construct a 3D object. Thus, it is not suitable for 3D analysis. This is one of the main reasons why 3D analytical functions are limited.

### *16.3.2 Spatial Indexing*

Another important aspect of 3D data management is spatial indexing. Spatial indexes are used in DBMS for fast search especially when spatial functions are applied. Without indexing, any searches for a feature would require a sequential scan of every record in the database. Indexing speeds up searching by organizing the data into a search tree that could be quickly traversed to find a particular record. There are several types of indexes within DBMS, e.g. PostGIS and Oracle Spatial: they are B-Tree indexes, R-Tree indexes, and GiST indexes.

- B-Trees are used for data, which can be sorted along one axis; for example, numbers, letters, dates. GIS data cannot be rationally sorted along one axis (which is greater, (0,0) or (0,1) or (1,0)?) so B-Tree indexing is of no use for GIS user.
- R-Trees break up data into rectangles, and sub-rectangles, and sub-sub rectangles, etc. R-Trees are used by some spatial databases to index GIS data, but the PostGIS R-Tree implementation is not as robust as the

GiST implementation. Oracle Spatial will implement the 3D R-Trees in the coming version 11g.
- GiST (Generalized Search Trees) indexes break up data into 'things to one side', 'things which overlap', 'things which are inside' and can be used on a wide range of data-types, including GIS data. PostGIS (2006) uses an R-Tree index implemented on top of GiST to index GIS data.

GiST indexes have two advantages over R-Tree indexes in PostGIS. First, GiST indexes enable the null value in the index columns. Secondly, GiST indexes could easily deal with GIS objects larger than the PostGIS 8K page size. The important part of an object in an index will only be considered within DBMS, e.g. in the case of GIS objects, just the bounding box. GIS objects larger than 8K will cause R-Tree indexes to fail in the process of being built. It could take a long time to create a GiST index if there is a significantly large amount of data in a table. Moreover, 3D indexing is not available within PostGIS.

Other DBMS, e.g. Oracle Spatial, are able to provide 3D indexing for 3D object (MULTIPOLYGON). For Spatial, the metadata that maintains the lower and upper bounds and tolerance of 3D object needs to be created. Later, a spatial index (R-tree in 3D) could be created on tables to speed up spatial queries. The following example denotes the sample in creating a 3D spatial index within Spatial.

```
-- Inserting metadata for 3D object: MULTIPOLYGON

INSERT INTO user_sdo_geom_metadata VALUES
   ('Solid3D', 'shape',
   mdsys.sdo_dim_array(
      mdsys.sdo_dim_element('X', 0, 100, 0.1),
      mdsys.sdo_dim_element('Y', 0, 100, 0.1),
      mdsys.sdo_dim_element('Z', 0, 100, 0.1))
   , NULL);

-- Creating 3D Spatial Index

CREATE INDEX Solid3D_I on Solid3D(shape)
   INDEXTYPE IS mdsys.spatial_index
   PARAMETERS(sdo_index_dims=3);        -- Dimension = 3

ANALYZE TABLE Solid3D COMPUTE STATISTICS;
```

## 16.3.3 Functions and Operations In DBMS

The 3D functions/operations in DBMS are mainly based on 2D objects that appear in 3D space, i.e. point, line, and polygon (in 3D). Most of the functions consider only the x,y coordinates of the data types although, and they may be given with 3D coordinates. However, there are some exceptions. Some of the 3D functions provided in PostGIS are:

- length3d(geometry): Returns the 3-dimensional length of the geometry if it is a linestring or multi-linestring.
- length3d_spheroid(geometry,spheroid): Calculates the length of of geometry on an ellipsoid, taking the elevation into account. This is just like length_spheroid except vertical coordinates (expressed in the same units as the spheroid axes) are used to calculate the extra distance vertical displacement adds.
- perimeter3d(geometry): Returns the 3-dimensional perimeter of the geometry, if it is a polygon or multi-polygon.
- MakeBox3D(<LLB>, <URT>): Creates a BOX3D defined by the given point geometries. LLB denotes lower left bottom, whereas URT denotes upper right top.
- xmin(box3d) ymin(box3d) zmin(box3d): Returns the requested minimum of a bounding box.
- xmax(box3d) ymax(box3d) zmax(box3d): Returns the requested maximum of a bounding box.

3D operations in existing DBMSs are hardly available. For example, due to the third dimension, Oracle Spatial is not considered in any function and operation, thus the 3D function and operation are not available. Maintaining objects with 3D coordinates are possible but the functions available within DBMS still do not consider the third-dimension. Some exceptions are only limited to geometry calculations, e.g. 3D length and 3D perimeter. The existing spatial functions are only based on the native geometry model, i.e. buffer for 2D polygon. The 3D operation for DBMS must focus on two directions:

- The existing operations have to be extended to the third-dimension, in which the z-coordinate must be involved, i.e. 3D intersection, 3D buffer, and etc.
- New 3D operations have to be developed based on topological models, i.e. 3D overlap, 3D meet that extended from 9-intersection model.

In the coming Oracle Spatial 11g, the 3D coordinate system will be implemented in DBMS environment. The 3D coordinate systems are all based on European Petroleum Survey Group (EPSG) specifications. The supported coordinate systems are: Vertical coordinate systems, Geocentric (3D Cartesian), Geographic (3D ellipsoidal), and Compound coordinate System.

## 16.4 Problems and Issues on 3D Data Modeling in DBMS

A number of works attempt to address the problem of spatial data modeling for 3D GIS where most of these efforts focused on polyhedron, tetrahedron, triangulated tetrahedron and even free-form curves and surfaces as a mechanism to formalize 3D spatial data modelling. The following section discusses some recent works on data modeling in DBMS.

### *16.4.1 Modeling 3D Solid in DBMS*

#### 16.4.1.1 Polyhedron

The modelling 3D spatial object and corresponding operations in a spatial DBMS has been investigated quite successfully by Arens (2003), and Arens et al. (2005). The basic idea was that a 3D polyhedron could be defined as a bounded subset of 3D space enclosed by a finite set of flat polygons, such that every edge of a polygon is shared by exactly one other polygon. The polygons are in 3D space because they are represented by vertices that appear in 3D space. The 3D primitive implemented by Arens was in a geometrical model with internal topology. The polyhedron was realized by storing the vertices explicitly (x,y,z) and describing the arrangement of these vertices in the faces of the polyhedron. This yields a hierarchical boundary representation (Aguilera 1998; Verbree and Zlatanova 2004). The sample of a polyhedron is illustrated in Fig. 9a, and the polyhedron storage is depicted in Fig. 9b.

The functions/operations given by Arens includes validation for polyhedron, spatial conversion, topological operation, and metric functions. To visualize 3D objects, it is necessary to use programs that actually show the third dimension. There are two options as proposed by Arens:

- GIS/CAD programs make a DBMS connection, for instance Microstation (Bentley 2007). These programs can only handle 3D objects that consist of multiple 2D objects. The 3D data stored as a 3D type needs a conversion before it can be visualised, e.g. splitting up the 3D object in multiple 2D polygons.
- VRML (Virtual Reality Modelling Language). When using VRML, there needs to be translation between the 3D type in the database and the VRML syntax.

These two representations have advantages and disadvantages. Displaying 3D objects using VRML require an extra step for 3D visualization. The polyhedron needs to be converted into a VRML file. First, the VRML file is stored as an SQL-loader file. Then, SQL-loader (from Oracle tool) load this file into DBMS environment to construct a table. The object's geometry will

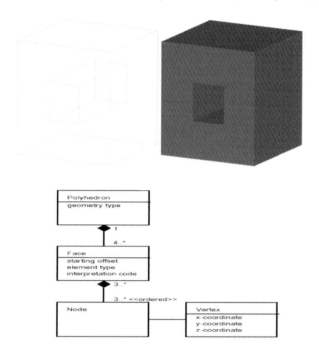

**Fig. 16.9** Sample of polyhedron, and UML diagram of polyhedron storage (after Arens 2003)

be added into the table and the VRML file can be browsed on the Internet. Taking advantage of web display, the data exchange/transfer could be done easily by extracting the VRML file. However, the VRML file is not part of the DBMS environment. The 3D visualization becomes inefficient if the data volume is huge – this happens when the conversion of geometry (from DBMS) to VRML file is carried out. However, the weakness could be overcome by integrating DBMS and display tool directly, i.e. GIS/CAD integration. In this case, a CAD system, such as Microstation, could be connected directly to the DBMS and retrieve the 3D data for 3D visualizaton.

### 16.4.1.2 TEN

Another attempt to define 3D object has been reported by Penninga, 2005. The 3D object, i.e. tetrahedron, is used to represent 3D volumetric shapes. The tetrahedron is the simplest possible geometry in the 3D domain. The conceptual design was intended for implementation of both geometrical and topological models in topographic modeling.

Initially, Penninga (2005) attempted to implement the TIN/TEN(2.5D /3D) model approach for topographic modeling. The idea is that the earth's terrain can be modelled in 2.5D TIN. The complex object will be mapped on top or below this terrain. This leads to the combination of TIN/TEN model (TIN: Triangulated Irregular Network / TEN: Tetrahedronized Irregular Network). However, since problems appear at both the conceptual and implementation level, an alternative model was suggested, i.e. the full TEN model. The shift to the full 3D model avoids the complication of designing multiple data structures in both TIN and TEN models for different spatial objects (Penninga et al. 2006; Penninga and van Oosterom 2007).

In the TEN model, four types of topographic features can be determined in this integration: 0D (point features), 1D (line features), 2D (area features) and 3D (volume features). For each type, feature simplexes of corresponding dimension are available to represent the features with nodes, edges, triangles and tetrahedrons (see Fig. 10).

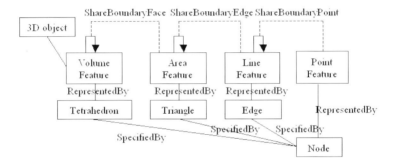

**Fig. 16.10** Logical design of 3D TEN (after Penninga & van Oosterom 2007)

With this TIN/TEN integration, a minor drawback would occur if an object became more complex, such as a complex building block. The entire building could be modeled using triangles as a whole to complete the geometry. An undesirable side effect is that the data size may become rather large, because more faces have to be stored in the data structure. The triangulation approach produces more storage, as compared to the polyhedron approach depending on the complexity of 3D objects. Since the space is completely subdivided into tetrahedrons, the interiors of objects (e.g. buildings), as well as the open space, are also decomposed into tetrahedrons. These tetrahedrons, however, require additional algorithms to be developed as a whole building block. This leads to database size expansion (see Fig. 11) and longer response time for visualization. More information on this comparison (polyhedron and TEN) can be found in Zlatanova 2000.

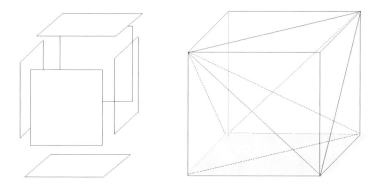

**Fig. 16.11** Comparison of the total faces/triangles between polyhedron and TEN (after Zlatanova 2000)

### 16.4.1.3 Triangulated Polyhedron

The triangulated polyhedron was proposed by Ledoux and Gold (2004) and it was based on 3D Voronoi Diagram (VD) and Delaunay Tetrahedralization (DT). The Voronoi diagram for a set of points (in a given space, $R$ is the partitioning of that space into regions such that all locations within any one region are closer to the generating point than to any other.

In 2D, this structure is defined by partitioning the plane into triangles (where the vertices of the triangles are points that generate each Voronoi cell) that satisfy the empty circumcircle test (a circle is empty when no points are in its interior, but more than three points can be directly on the circle). In any dimensions, the VD has a geometric dual structure called the Delaunay Triangulation. The two-dimensional DT is illustrated in Fig. 12 by the dashed lines. The Delaunay Triangulation is appropriate for modelling surfaces because among all the possible triangulations of a set of points, it creates one where the minimum angle in each triangle is maximized (triangles are as equilateral as possible), thus being useful for interpolation.

In three-dimension, a Voronoi cell generalizes to a convex polyhedron formed by convex faces, as shown in Fig. 13. The generalization to three dimensions of the Delaunay Triangulation is the Delaunay tetrahedralization: each triangle becomes a tetrahedron that satisfies the empty circumsphere rule. The DT is unique for a set of points, except when there are degenerate cases in the set (if five or more points are cospherical in 3D). In these cases, an arbitrary choice must be made among all the possible solutions. The number of tetrahedra in a DT constructed with n points depends on the configuration of these points.

It can be realized that triangulated polyhedron could be utilized for generating 3D spatial objects and eventually into DBMS.

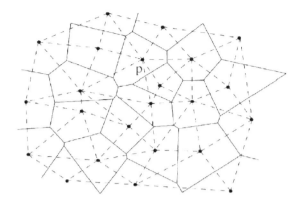

**Fig. 16.12** Two-dimensional VD (bold lines) and DT (dashed lines) (after Ledoux & Gold 2004)

**Fig. 16.13** The Voronoi cell in 3D (after Ledoux & Gold 2004)

### 16.4.1.4 Modeling 3D Freeform Curves and Surfaces

Complex geometry types such as freeform curves and surfaces can be implemented in DBMS. Many shapes in the real world are freeforms, i.e. not only contain points, linestrings and polygons, but also curves and curved surfaces, e.g. roads, building surfaces, and etc. Pu (2005) has created complex geometry data types that describe freeform curves and surfaces. Although freeform shapes can be simulated by tiny line segments/triangles/polygons, it is quite unrealistic and inefficient to store all these line segments/triangles/polygons into a DBMS especially when shapes are rather huge or complex. The freeform shapes discussed by Pu 2005, Pu and Zlatanova, 2006 are Bezier (Fig. 14), B-spline and NURBS.

A B-Spline surface is an expansion of B-spline curves (Fig. 15a) and B-spline curves are a generalization of Bezier curves, and the same applies for

**Fig. 16.14** (a) Bezier curve, and (b) bi-cubic Bezier surface (Pu, 2005)

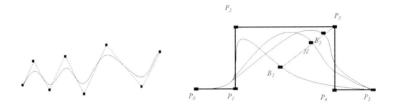

**Fig. 16.15** (a) B-spline, and (b) NURBS curve (Les, 1991)

surfaces. NURBS (Non-Uniform Rational B-Splines) curve is generalized from B-Spline curve (Fig. 12b). The implementation was done in Oracle Spatial, where new data types for each Bezier curve, B-spline curve and NURBS curve were created separately. An alternative approach was to create a data type for NURBS curves, which also represented Bezier curves and B-spline curves by leaving some parameters of NURBS curve empty - because NURBS curve is actually the generalization of Bezier curve and B-spline curve. The final freeform datatypes could include:

- Three curve types: GM BezierCurve, GM BSplineCurve and GM NURBSCurve (see Fig. 16a),
- One surface type: GM NURBSSurface (see Fig. 16b), and
- Four supplementary types: GM PointArray, GM WeightArray, GM Knot Vector and GM Trim.

The freeform curve and surface developed by Pu 2005 could not represent a 3D solid object. Although these data types consider the z-coordinate, the objects do not bound a volumetric body. The research could be extended from freeform surface that able to envelope a solid body, but it yields greater complexity due to more complex mathematical algorithms are required. As a result, it will slow down the process of 3D visualization and spatial operation.

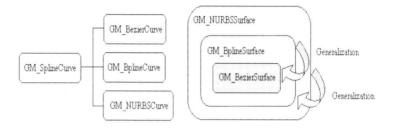

**Fig. 16.16** (a) Freeform curve, and (b) Freeform surface datatypes

## 16.4.2 3D Spatial Indexing

Spatial searching is a fundamental primitive in non-traditional databases such as GIS, CAD/CAM and multi-media applications. With the rapid proliferation of these databases in the past decade, extensive research has been conducted on the design of efficient data structures to enable fast spatial searching. Several data structures have been developed in this context, including Quadtrees (Wang, 1991), R-trees (Guttman, 1984), hB-trees (Lomet and Salzberg 1990), and TV-trees (Lin et al. 1984). Subsequent research has improved these basic structures further by proposing new techniques for query processing (Berchtold et al. 2000; Ferhatosmanoglu et al. 2001), faster and better index creation (Garcia et al. 1998), and better split-strategies in dynamic updates (Beckmann et al. 1990; Berchtold et al. 1996). These techniques are especially effective for low-dimensional spatial data such as those in GIS and CAD/CAM applications.

For indexing low-dimensional spatial data, certain DBMSs allow users to choose between one of two spatial indexes: a (Linear) Quadtree or an R-tree. The Oracle implements these two kinds of spatial indexes and incorporates and enhances some of the best proposals from existing spatial indexing research. The PostGIS implements the GISt indexing for spatial query.

Most of the spatial indexes are extended from these two kinds of indexing methods. The Linear Quadtree (or Quadtree for short) computes tile approximations for geometries and uses existing B-tree indexes to perform spatial searches. This approach results in simpler index creation, faster updates and inheriting a built-in B-tree concurrency control protocol. The R-tree is implemented logically as a tree and physically using tables inside the database and the search involves recursive SQL for traversing the tree from root to relevant leaves. This approach may be more efficient for queries due to the enhanced preservation of spatial proximity but may be slow in updates or index creation and implements its own concurrency protocols on top of spatial DBMS table level concurrency mechanisms.

The conventional approach to support similarity searches in high-dimensional vector spaces can be broadly classified into two categories:

The first approach uses data-partitioning index trees. Neighbouring vectors are coveredby MBRs (minimum bounding rectangles) or MBSs (minimum bounding spheres), which are organized in a hierarchical tree structure. Many index tree schemes have been proposed. They include the R-tree, the R*-tree (Beckmann et al. 1990, the Hilbert R-tree (Kamel and Faloutsos 1994), and the SS-tree (White and Jain 1996). In addition, nearest neighbour search methods using such indices have been proposed (Henrich 1994; Hjaltason and Samet 1995). Two recently proposed indices, the X-tree (Berchtold et al. 1996) and the SR-tree (Katayama and Satoh 1997), are reported to offer good performance. The X-tree introduces the notion of a supernode and outperforms the R*-tree. The SR-tree has a unique feature in that it uses both MBRs and MBSs and is reported to outperform both the R*-tree and the SS-tree.

The second approach is the use of approximation files. Among the others, the VA-file (vector approximation file) (Weber et al. 1998) is a simple yet powerful scheme. The VA-file divides the data space into cells and allocates a bit-string to each cell. The vectors inside a cell are approximated by the cell and the VA-file itself is simply an array of these geometric approximations. When searching vectors, the entire VA-file is scanned to select candidate vectors. Those candidates are then verified by visiting the vector files. Weber et al. (1998) report that the VA-file outperforms both the R*-tree and the X-tree when dimensionality is high. In the field of spatial search of high-dimensional data, this problem looms larger and larger. Search methods that present an approximate answer (Arya et al.1994; Gionis et al. 1999), have been proposed to avoid the problem. Although these methods are useful, to overcome this problem, an A-tree index was proposed by Sakurai et al. 2002. Introduction of the A-tree is motivated by a comparison and analysis of the SR-tree and VA-file. The basic idea of A-tree is the introduction of virtual bounding rectangles (VBRs), which contain and approximate MBRs or data objects. The A-tree indexing is based on the following design structures:

- Tree structure: It adopts a tree index to limit the searching result from one phase to the next phase.
- Relative approximation: to overcome the problem of tree indices identified in evaluation results, a new notion (i.e. relative approximation) was introduced, which is a simple yet powerful approximation method utilizing the hierarchy of tree indices. In relative approximation, bounding regions or data points are approximated by their relative positions in terms of the parent's bounding region.
- Partial usage of MBSs: since the SR-tree is one of the best indices among the tree indices proposed so far, the SR-tree is used as the starting point in designing the A-tree. However, the effect of MBSs is limited when searching high dimensional data. Hence, MBSs are not stored in the A-tree. As a result, the centroid of data objects in a subtree is used only for insertion and deletion. The A-tree is a new index that applies the notion of relative approximation to the hierarchical structure of the SR-tree.

However, this application is not naive; A-tree's configuration is unique in that: i) each node contains an MBR and a representation of the relative approximation of its children; and ii) the centroid of data objects is used only for updating purposes.

### *16.4.3 3D Operations*

Since the subject of implementing 3D topological operations for geometrical structure in a relational DBMS is a fairly unexplored area, some approaches will be considered in developing 3D spatial operation for DBMS:

- The 3D spatial operation will cover all necessary topological structures that define a complete solid object. In certain cases, not all primitives are needed, e.g. a polyhedron is defined by an ordered set of coordinate triplets for each polygon that bound a volumetric body, line will not be used in the data structure.
- Implementation of the 3D spatial operations will be tested within the DBMS environment.
- The results from 3D topological operations return to a Boolean form (TRUE/FALSE). It involves two spatial objects, polyhedron and polyhedron.

The topological operations presented here are based on the body-body relation (Zlatanova, 2000). Typically, the results given by this operation are in Boolean type, i.e. either TRUE or FALSE. The related operations include Overlap, Meet, Disjoint, Inside, Covers, CoveredBy, Contain, and Equal (see Fig. 17).

For topological operations in a geometrical model, a coordinate triplet of the vertex is used. Similar to computational-geometry operation from previous sections, the binary operation is divided into base and target object. However, the vertices from base object and polygons from target object will be discussed (see Fig. 18a). This topological operation involves vertices (from base object) and polygon (from target object). Therefore, the relation between these two objects will be examined. The location of base vertices relative to target polygon will be either outside, touch, or inside as has been implemented and discussed in Chen and Abdul-Rahman (2006). These relations will be used to determine how these two polyhedrons intersect each other.

The following table (Fig. 18b) denotes the complete relationship between base and target object. The 'X' sign represents the impossible intersection between two objects, whereas the 'check' sign represents the possible intersection for geometrical models.

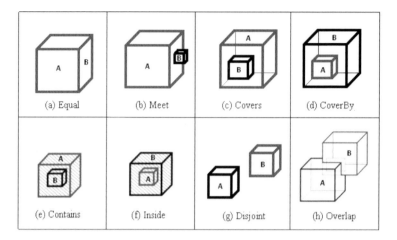

**Fig. 16.17** Body and body relation (after Zlatanova, 2000)

**Fig. 16.18** Vertices (base) touch the target polygon (3D Meet), and Conditions for topological operations (geometrical model)

The implementations of 3D topological operations involve two intersecting polyhedrons. This implementation was performed within the PostGIS environment. Below is the structure of polyhedron:

SELECT * FROM Solid3D WHERE PID = 1;

1,POLYHEDRON(PolygonInfo(6,24),SumVertexList(8),
SumPolygonList(4,4,4,4,4,4),
VertexList(100.0,100.0,100.0,400.0,100.0,100.0,400.0,
400.0,100.0,100.0,400.0,100.0,100.0,100.0,400.0,400.0,
100.0,400.0,400.0,400.0,400.0,100.0,400.0,400.0),
PolygonList(1,2,6,5,2,3,7,6,3,4,8,7,4,1,5,8,5,6,
7,8,1,4,3,2)) o,400.0,100.0,100.0,100.0,400.0,400.0,100.0,
400.0,400.0,400.0,400.0,100.0,400.0,400.0),
PolygonList(1,2,6,5,2,3,7,6,3,4,8,7,4,1,5,8,5,6,7,8,1,4,3,2))}

1. PolygonInfo(6,24) denotes 6 polygons and 24 IDs of polygon arrange in PolygonList,
2. SumVertexList(8) denotes the total vertices,

3. SumPolygonList(4,4,4,4,4,4) denotes total vertices for each of polygon (total polygon is 6, referred to (1)),
4. VertexList() denotes the list of coordinate-values for all vertices (with no redundant), and
5. PolygonList() denotes the information about each polygon from sets of ID.

The following SQL statement runs the 3D Overlap (see Fig. 19):

```
SELECT GMOVERLAP3D(a.POLYHEDRON,b.POLYHEDRON) AS GM_OVERLAP3D
FROM test a, test b where a.PID=1 and b.PID=2;
```

The result:

```
GM_OVERLAP3D
------------
   (TRUE)
```

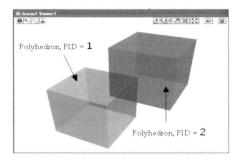

**Fig. 16.19** 3D Overlap

## 16.5 3D Visualization

Without visualization, any queries from database would be just numbers and characters – thus hard for users to decipher the meaning of the generated information. DBMS only provides a medium for data set management, and it certainly requires a front-end tool for visualizing the information as it is perceived in the real world. The data from DBMS needs to be integrated into a visualization tool so that it could be viewed as graphic. The 3D spatial data stores in the spatial column (within DBMS), and a connection needs to be built so that a display tool manages to access the spatial column and retrieve the data for 3D visualization.

It is also important to note that 3D objects need to be visualized in realism. With the benefit of the computer graphic technology, GIS could provide a good display with textures and colours. Some web application, e.g. Google Earth (GE) maintains the texture of spatial object over the Internet. GE is a dynamic 3D virtual globe application that contains high-resolution satellite and airborne images streamed through the Internet. GE uses a specific standard for external data sources called the Keyhole Markup Language (KML and KMZ is the zip version). KML is a file format used to display geographic data in an earth browser. A KML file is processed in much the same way that HTML (and XML) files are processed by web browsers. Like HTML, KML has a tag-based structure with names and attributes used for specific display purposes. Thus, Google Earth and Maps act as browsers for KML files. In Fig. 20, elements to the right on a particular branch in the tree are extensions of elements on the left. For example, Placemark is a special kind of Feature. It contains all of the elements that belong to Feature and adds some elements that are specific to the Placemark element.

Other than visualization in 3D, it is important to have spatial query and data updating based on the 3D data from the display tool. Some of the software, e.g. ArcGIS, and Microstation could provide data editing/update, and perform spatial query. Another advantage of integration between DBMS and the visualization tool is that posting data could be done from the display tool. Furthermore, the data will be stored and converted into the DBMS enviroment.

Another important element of 3D visualization is Level-Of-Detail (LOD). The concept of Levels of Detail (LOD) has been introduced to facilitate visualization of large scenes (see Clark 1976). The idea is to represent spatial objects that are compatible with the pixel size of the screen, relative of the observer's distance. This permits the original geometric representation to be replaced with a new low-resolution represebtation. Low-resolution representations require less memory and processing time for rendering and hence speeding-up the visualization process. The different representation is used by the visualisation system only if the object is far enough from the user. Closer objects are still represented in their full resolution. Moreover, if the distant object gets closer (as a result of the user's navigation through the model), the high-resolution representation is restored. The intentions are an unnoticeable switch between low and high levels of detail.

Currently, the CityGML (Kolbe et al. 2006) supports different LOD. It requires independent data collection processes with differing application requirements. In a CityGML dataset, the same object may be represented in different LOD simultaneously, enabling the analysis and visualization of the same object with regard to different degrees of resolution. Furthermore, two CityGML data sets containing the same object in different LOD may be combined and integrated. The CityGML provide multiple kinds of LOD as given in below (see Fig. 21):

# 16 3D Solids and Their Management In DBMS

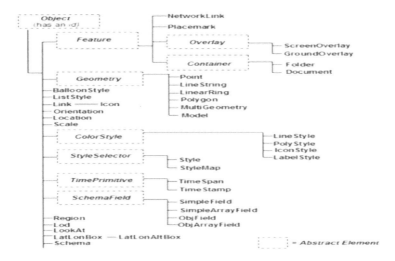

**Fig. 16.20** Sample KML file format

- LOD0 is essentially a two and a half dimensional Digital Terrain Model. An aerial image or a map may be draped on the DTM.
- LOD1 is the blocks model representing the buildings with flat roofs.
- LOD2 is used to differentiate the roof structures among different building. It also used in differentiating surfaces thematically.
- LOD3 provide a building model with detailed wall and roof structures, balconies, bays and projections. Vegetation objects may also be represented in this level. High-resolution textures can be mapped onto these structures too.
- LOD4 completes a LOD3 model by adding interior structures for 3D objects. For example, buildings are composed of rooms, interior doors, stairs, and furniture.

LOD are used not only to speed up visualisation but also for different applications. For example LOD1 (CityGML) is perfect for air pollution analysis; LOD3 is good for realistic visualisation; LOD4 can be used for evacuation from buildings.

To maintain the LOD and colour/texture attributes could be performed in both the DBMS and display tool. Certain display tools, e.g. VRML browser, ArcGIS, and Microstation are able to maintain the colour and texture well. This is not necessary to maintain the texture and colour attribute within DBMS since these display tool manage to maintain these features with better interactive functions. However, the LOD is required to store within DBMS. This is because different LOD represent different kinds of geometry. For example, from CityGML, LOD1 mainly stores simple block models that represent buildings with flat roofs; LOD2 differentiate the roof structures among dif-

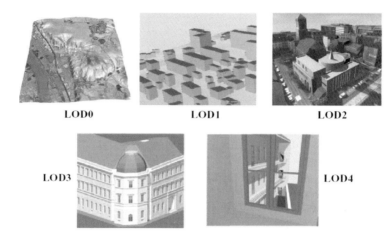

**Fig. 16.21** The five levels of detail (LOD) defined by CityGML, 2006

ferent building. With the same building, different LOD represents different kind of roofs. The different LOD could only be stored in visualization tool, if and only if, it converts into different layer of graphical data.

## 16.6 Conclusions

A number of issues and challenges must be addressed to develop and manage 3D solid objects in database. The problems, challenges and issues could be summarized as:

1. An appropriate 3D datatype that defines 3D primitive for geometry and topology needs to be developed. Different kinds of 3D spatial objects deal with different applications, e.g. TEN deals with terrain modeling and polyhedron addresses with building structures. The 3D datatype should follow the standard specification provided by OGC, and store three-dimensional objects in an DBMS environment. The DBMS structure that stores 3D primitive in spatial column should be able to manage and maintain different kinds of LOD (texture/colour attributes are optional since certain display tools could manage these well). The performance in terms of the size of data storage and management efficiency need to be given attention because DBMS provides the medium for data management and should also be integrated with other aspects like 3D display. Therefore, an efficient DBMS that supports various kinds of 3D primitives is important for 3D spatial modeling.
2. Spatial indexing: It is a mechanism that is usually applied to accelerate the process of queries in the database by keeping some extra in-

formation. Many types of indexing methods have been cited such as the R-tree (Guttman 1984), the K-D-B-tree (Robinson 1981) or the Z-ordering (Orenstein 1986). Although other spatial indices that combine a tree structure and a capacity technique have been proposed (Seeger and Kriegel 1990; Berchtold et al. 2000), novel algorithms and structures that give very high performance for high dimensionalities (e.g. 3-dimension) need to be developed. Again, the performance of 3D indexing will also need to be evaluated. The 3D R-tree indexing available in Oracle Spatial, A-tree (for 3D spatial index by Sakurai 2002) and other 3D indexing should be compared.

3. Functions and operations: There are several geometrical algorithms that deal with spatial and attribute data manipulations for GIS analysis. The importance of this algorithm is directly referred to its application, e.g. calculation of volume for land subsidence, etc. Some DBMSs implement a wide range of functions for database management and spatial analysis. The spatial operations could be divided into several types and need to addressed as well:

- Computational-geometry operations: functions that return a new geometry from two objects intersection, e.g. 3D Intersection, and 3D Union.
- Topological operations: functions that return a Boolean result from 2 object intersection, e.g. extending 3D overlap.
- Metric operations: functions that involve mathematical calculation, e.g. volume calculation of tetrahedron and polyhedron.

4. 3D visualization and interaction: Visualisation is mostly used in the context of display 3D graphics. Realism display of the objects is also an issue and could be categorized into two different approaches: texture/colour and Level Of Detail (LOD). One of the questions like 'how close to the actual real world' one could display and interact with the object. Another issue of 3D visualization is interaction. Ideas about how user-friendly the interactive tool needs to be for individuals to perform tasks in 3D GIS must be adressed. Different principles and applications lead to different approaches of visualization and interaction methods.

We reviewed and described a number of research works pertaining to the 3D solids associated with spatial data modelling and management in DBMS. The discussions cover the 3D datatype, spatial indexing, and functions/operations (from standard specification to implementation; from commercial to research/development). However, many issues must be addressed to improve the current situation of 3D spatial modeling. The most important issue for 3D spatial data modeling is the standardization and specification of GIS. Although some of the specifications (abstract specification) are discussed in this paper, many other standards need to be investigated as well, i.e. 3D operations (geometry and topology) for solid objects. The implementation of 3D operations could be done in DBMS. The spatial operators should involve some procedures that can use, query, create, modify, or delete spatial objects.

Other challenges in the 3D GIS domain include interoperability between different applications, data model, integration between DBMS and visualization, and the link between data modelling and data acquisition.

# References

Abdul-Rahman A, Zlatanova S, Coors V (2006) Lecture Note on geoinformation and cartography – Innovations in 3D Geo Information Systems, Springer-Verlag

Aguilera A (1998) Orthogonal polyhedra: study and application. Ph.D. Thesis, LSI-Universitat Politècnica de Catalunya

Arens CA (2003) Modelling 3D spatial objects in a geo-DBMS using a 3D primitives. Msc thesis, TU Delft, The Netherlands

Arens C, Stoter JE, van Oosterom PJM (2005) Modelling 3D spatial objects in a geo-DBMS using a 3D primitive. In: Computers & Geosciences, 31:165-177

Arya S, Mount DM, Netanyahu NS, Silverman R, Wu AY (1994) An optimal algorithm for approximate nearest neighbor searching. In: Proc. ACM-SIAM Symposium on Discrete Algorithms, pp. 573-582

Beckmann N, Kriegel H, Schneider R, Seeger B (1990) The R* tree: An efficient and robust access method for points and rectangles. In: Proc. ACM SIGMOD Int. Conf. on Management of Data, pp. 322-331

Bentley (2007) available at http://www.bentley.com/

Berchtold S, Keim DA, Kreigel HP (1996) The X-tree: An index structure for high dimensional data. In: Proc. of the Int. Conf. on Very Large Databases

Berchtold S, Keim DA, Kriegel HP, Seidl T (2000) A new technique for nearest neighbor search in high-dimensional space. IEEE Trans. In: Knowledge and Data Engineering, 12(1):45-57

Chen TK, Abdul-Rahman A (2006) 0-D feature in 3D planar polygon testing for 3D spatial analysis. In: Abdul-Rahman A, Zlatanova S, and Coors V (eds), Lecture Note on geoinformation and cartography – innovations in 3D

Geo information systems, Springer-Verlag. pp. 169-183

CityGML available at http://www.citygml.org/

Clark JH (1976) Hierarchical geometric models for visible surface algorithm. In: Communications of the ACM, 19(10), pp. 547-554

ESRI (2007) available at http://www.esri.com/

Ferhatosmanoglu H, Tuncel E, Agrawal D, Abbadi AE (2001) Approximate nearest neighbor searching in multimedia databases. In: Proc. Int. Conf. on Data Engineering, pp. 503-511

Garcia YJ, Leutenegger ST, Lopez MA (1998) A greedy algorithm for bulk loading R-trees. In: Proc. of ACM GIS

Gionis A, Indyk P, Motwani R (1999) Similarity search in high dimensions via hashing. In: Proc. 25th International Conference on Very Large Data Bases (VLDB), pp. 518-529

Guttman A (1984) R-trees: A dynamic index structure for spatial searching. In: Proceedings of ACM SIGMOD, International Conference on Management of Data, Boston, MA, pp. 47-57

Henrich A (1994) A distance scan algorithm for spatial access structures. In: Proc. ACM International Workshop on Advances in Geographic Information Systems, pp. 136-143

Hjaltason GR, Samet H (1995) Ranking in spatial databases. In: Proc. 4th Symposium on Spatial Databases, pp. 83-95

Ledoux H, Gold CM (2004) Modelling oceanographic data with the three-dimensional Voronoi diagram. In: ISPRS 2004-XXth Congress, Istanbul, Turkey,. Vol. 2, pp. 703-708

Kada M, Haala N, Becker S (2006) Improving the realism of existing 3D city model. In: Abdul-Rahman A, Zlatanova S, and Coors V (eds), Lecture Note on geoinformation and cartography – innovations in 3D Geo information systems, Springer-Verlag. pp. 405-415

Kamel I, Faloutsos C (1994) Hilbert R-tree:An improved R-tree using fractals. In: Proc. 20th International Conference on Very Large Databases, pp. 500-509

Katayama N, Satoh S (1997) The SR-tree: an index structure for high-dimensional nearest neighbor queries. In: Proc. ACM SIGMOD International Conference on Management of Data, pp. 369-380

Kolbe T, Groeger G, Czerwinski A (2006) City Geography Markup Language (CityGML). In: OGC, OpenGIS Consortium, Discussion Papers, Version 0.3.0

Les P (1991) On NURBS: a survey. IEEE Computer Graphics and Applications 11(1): 55-71

Lin KI, Jagdish HV, Faloutsos C (1994) The TV-tree: An index structure for high-dimensional data. VLDB Journal, 3:517-542

Lomet DB, Salzberg B (1990) The hB-tree: A multi-attribute indexing method with good guaranteed performance. Proc. A CM Syrup. on Transactions of Database Systems, 15(4):625-658

OGC (1999) Abstract specifications overview. Available at http://www.opengis.org/

OGC (1999a) OpenGIS simple features specification for SQL. Available at http://www.opengis.org/

OGC (2001) The OpenGIS™ Abstract specification, topic 1: feature geometry (ISO 19107 Spatial Schema) Version 5

Oracle Spatial 10g available at http://www.oracle.com/

Orenstein J (1986) Spatial query processing in an object-oriented database system. In: Proceedings of 1986 ACM SIGMOD International Conference on Management of Data, pp. 326-336

Penninga F (2005) 3D topographic data modelling: why rigidity is preferable to pragmatism. In: Spatial Information Theory, Cosit'05, Vol. 3693 of Lecture Notes on Computer Science, Springer. pp 409-425

Penninga F, van Oosterom PJM, Kazar BM (2006) A TEN-based DBMS approach for 3D topographic data modelling. In: Spatial Data Handling 2006

Penninga F, van Oosterom PJM (2007) A compact topological DBMS data structure for 3D topography. In: Fabrikant S, Wachowicz M (eds.), Lecture Notes in Geoinformation and Cartography. ISBN: 978-3-540-72384-4

Pilouk M (1996) Integrated modelling for 3D GIS. PhD Thesis, ITC, The Netherlands

PostGIS (2006) available at http://postgis.refractions.net/

Pu S (2005) Managing freeform curves and surfaces in a spatial DBMS. Msc Thesis, TU Delft

Pu S, Zlatanova S (2006) Integration of GIS and CAD at DBMS level. In: E. Fendel E, Rumor M (eds), Proceedings of UDMS'06 Aalborg, Denmark, TU Delft, pp 9.61-9.71

Rigaux P., Scholl M, Voisard A (2002) Spatial databases - with application to GIS. Morgan Kaufmann Publishers, San Francisco

Robinson J (1981) The K-D-B-Tree: A search structure for large multidimensional dynamic indexes. In: Proceedings of ACM SIGMOD International Conference on Management of Data, pp. 10-18

Sakurai Y, Yoshikawa M, Uemura M, Kojima H (2002) Spatial indexing of high-dimensional data based on relative approximation. The International Journal on Very Large Data Bases, 11(2), pp. 93-108

Seeger B, Kriegel HP (1990) The Buddy tree: an ef[FB01?]cient and robust access method for spatial data base systems. In: Proc. 16th International Conference on Very Large Data Bases (VLDB), pp. 590-601

Vebree E, Zlatanova S (2004) 3D-modeling with respect to boundary representations within geo-DBMS. GISt report No.29, TU Delft

Wang F (1991) Relational-linear quadtree approach for two-dimensional spatial representation and manipulation. IEEE Trans. on Knowledge and Data Engineering, 3(1):118-122

Weber R, Schek HJ, Blott S (1998) A quantitative analysis and performance study for similarity-search methods in high dimensional spaces. In: Proc. 24th International Conference on Very Large Data Bases (VLDB), pp. 194-205

White DA, Jain R (1996) Similarity Indexing with the SS-tree. In: Proc. IEEE 12th International Conference on Data Engineering, pp. 516-523

Zlatanova S (2000) 3D GIS for urban development. PhD thesis, ITC, The Netherlands

# Chapter 17
# Implementation alternatives for an integrated 3D Information Model

Ludvig Emgård[1,2] and Sisi Zlatanova[1]

## Abstract

The 3DIM (3D Integrated Model) is an information model under development which intends to integrate geographic features on the earth surface as well as above and below the earth surface into a common semantic-geometric model. We present and discuss two alternative implementations of the information model for DBMS. In the first alternative semantics are separated from geometry and organized into two table groups while in the second alternative semantic tables incorporates the geometry of the objects.

## 17.1 Introduction

Semantic models describing geographic features are common in many fields. Geo-scientists, geologists, constructors, architects and urban planners have been developing various semantic models to be able to better define objects, their representations and important relationships, which might be of importance for a particular application. However, generic semantic models that include a broad range of geographic features without emphasis on a specific application exist mostly in international or national GIS standards or ontology research. For example, the INSPIRE initiative (INSPIRE 2007) deals with harmonization of topographic features, the North American Data Model (NADM 2004) is focused on geological features while the CONGOO (Pantazis,1997), Towntology project (Caglioni 2006) and CityGML (Gröger et. al. 2006) concentrate on city environments. Although not related to a spe-

---
[1] Delft University of Technology, OTB, section GIS Technology,
Jaffalaan 9, 2628 BX the Netherlands
[2] SWECO Position AB, Sweden
ludvig.emgard@sweco.se, s.zlatanova@tudelft.nl

cific application, these frameworks focus on a set of real-world features (e.g. above, on the surface or under surface). Furthermore, semantic models hardly discuss spatial representations of semantic features. Those dealing with the spatial aspect consider only 2D geometries in 2D space. The most extensive contribution in semantic spatial models focusing on three-dimensional representations of real world is found in the information model of CityGML. The information model takes care of the semantic respectively thematic properties, taxonomies and aggregations of Digital Terrain Model (DTM), sites (including buildings, bridges, tunnels, etc.), vegetation, water bodies, transportation facilities, and city furniture. The semantic part of the model is complemented with geometry corresponding to the Simple Feature Specifications (Herring 2001). Special focus is put on the building features, which are represented in five levels of detail (LOD). The current version of CityGML does not include underground features, however.

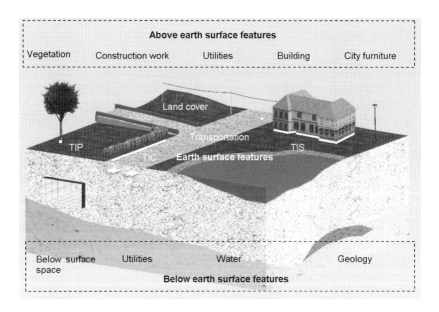

**Fig. 17.1** 3DIM subdivision of real-world features

The section for GIS technology at the Delft University of Technology is currently developing a 3D Information Model (3DIM), which intends to integrate features on surface, above and below the surface (Figure 1). We have presented an initial conceptual model in a previous publication (Emgård & Zlatanova, 2007). This article concentrates on the implementation of 3DIM in a DBMS. In the next section we briefly outline the major concepts of the information model. The third section gives a short overview on possibilities for implementing 3D geometry in a DBMS. We present and compare two alternatives for DBMS implementation of the conceptual model. The last section concludes the paper with a discussion on advantages and disadvantages of the two alternatives and outlines future research.

## 17.2 3D Integrated Model concept

The 3D integrated information model is intended to provide a generic model for 3D environments, which can be used by different domains, but being less specific and avoiding semantics and attributes that are very specific for a certain domain (and thus not of interest for many applications).

3DIM is intended to be used as a data model and contains thematic semantics and mapping to geometry data types for all man-made and natural real-world features on the surface, above and bellow (Figure 1). The model adopts several concepts presented in CityGML but also introduces stricter general rules as follows:

- The features are classified into above surface, integrated in surface and below surface
- The earth surface is fully partitioned. One part of the surface can only be occupied by one feature. Fictional features (Zlatanova 2000, Billen & Zlatanova, 2003) such as thematic land use can also be additionally attached i.e. residential or industrial area or another administrative attribute (not elaborated in the current version of the model).
- The surface accommodates all the intersections (touch) between the features above and below the surface and the surface itself. The idea of *TerrainIntersectionCurves (CityGML)* is therefore extended to include terrain intersection *surfaces*, *curves* and *points*.

The top-level classes are subdivided as follows (Figure 2):

- The Earth surface is represented by a fully partitioned surface consisting of *Transportation*, *Landcover* or *TerrainIntersectionSurface*. The first two classes represent objects on the earth surface. The *TerrainIntersectionSurface* is a special type of class since it represents the intersections of objects above and below the surface with the terrain.
- The above ground features are classified into *Building*, *Vegetation*, *City Furniture*, *AboveSurfaceUtilitiy* and *ConstructionWork*.
- The underground features are classified into *BelowSurfaceSpace*, *Geology*, *Water* and *BelowSurfaceUtilitiy*

The class *Building* is entirely adopted from the CityGML information model while *ConstructionWork*, *Geology*, *BelowSurfaceSpace*, *AboveSurfaceUtility* and *BelowSurfaceUtility* are new developments within this model.

The spatial extent of features may be defined either by a geometry or a topology model. In this paper we concentrate strictly on the geometry model.

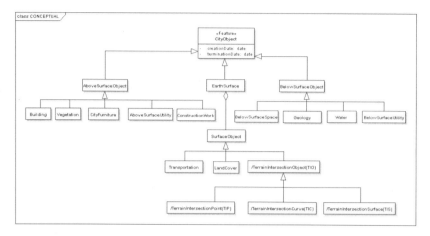

**Fig. 17.2** UML class diagram of top level feature hierarchy in the 3DIM

The possible geometry data types in the model are those available in a DBMS, i.e. point, curve, surface and solid. These are to be represented by simple geometries as described in the Spatial Schema (Herring 2001) and Simple Feature Specifications for SQL (OGC 1999). This means that volumetric objects can be represented only by tetrahedrons and polyhedrons. Freeform curves and surfaces, Constructive Solid Geometry solids, and other complex geometry representations are currently not discussed in the model.

We believe that the subdivision of real-world objects into three groups is defendable. The features bellow and above the surface have rather distinct nature, i.e. geological features are generally continues phenomena, while features above the surface have crisp boundaries. Therefore the modelling approaches differ. Features below the earth surface are preferable modelled in a full partition of space while above ground the features are embedded in the 3D space (i.e. air is not commonly modelled). The features on the earth surface are most important since they allow establishing relationships between all features and considering them in their integrity. Traditionally, the earth surface has also been central in 2D maps, since map features are in most of the cases projections of above- and below- surface features on the earth surface. Very often these projections have been represented on separate 2D maps, i.e. topographic maps (for above and on the surface features) and geologic maps (for below surface features).

The strong subdivision of the objects into above, below and earth surface objects also brings disadvantages. Using the earth surface as a separator may increase the amount of geometric features describing the surface model. In addition, the fact that 3DIM integrates features from many different domains makes it much richer of geometry than a traditional map created for one purpose and thus more complex.

## 17.3 LOD (Level of Detail)

A very important concept in 3DIM is the LOD. In traditional 2D maps scale has been the factor defining the features on the map and for 3D environments the concept of level of detail somehow replaces the concept of scale. The 3DIM includes five levels of detail (LOD0-4), which allow geometry of features to be represented with different accuracy and detail. Features represented in LOD0 use representations that correspond to earth surface features. This mean that above and underground objects are given with their corresponding terrain intersection point, curves and surfaces. Practically, LOD0 is a 2.5 D map and can be integrated with available 2D maps if the z-value is not considered.

Features represented in LOD1 are created by simple geometric shapes (e.g. box models for buildings) consisting of surfaces, some of which are earth surfaces (see Figure 3). The polygons that constitute the boundary of the features are not semantically classified. For example, the polygons constituting a building are not classified into wall-polygons and roof-polygons. LOD2 is more detailed as it includes textures for geometric features and allows semantic classification. LOD3 contains more detailed surface geometries and texture is compulsory for each feature polygon. LOD3 contains the highest resolution that is available for the outside representation of features. LOD4 adds a different type of resolution, i.e. it handles indoor environments of two classes *Building* and *BelowSurfaceSpace*.

The concept of LOD in 3DIM has been adopted from CityGML and the newly introduced features are incorporated in the concept. The *ConstructionWork* and *BelowSurfaceSpace* features are similar to *Building* and therefore use the same definition of LOD. LOD1 describes a simple extruded representation, LOD2 a textured representation where individual polygons are classified, LOD3 a detailed geometric representation and LOD4 indoor environments.

The *AboveSurfaceUtility* and *BelowSurfaceUtility* features are represented by curves and points in LOD0-4. In higher LODs (2-4) the points are replaced for visualization by symbolic surface representations and lines are replaced by the necessary shape. In addition, larger utility objects are that can not be replaced by symbols are stored using surfaces in LOD1-4 The *Geology* feature is represented by borehole TIP points describing observations on the surface in LOD0 as well as a defined borehole datatype to express the observations of the borehole below the surface. In higher LODs the *Geology* features are represented by surfaces and solids.

The point and curve representation in LOD0 is in 3DIM also applied for *Vegetation* and *CityFurniture* features using the terrain intersection concept. The earth surface features *Transportation* and *LandCover* are presented by surface geometries only in all LODs.

## 17.4 Rules

3DIM also contains a set of rules as defined below:

1. A semantic feature must have a geometric representation. The geometry of a feature must be defined before (or simultaneously) with the semantic feature it describes.
2. A semantic feature can have only one geometry representation with respect to a LOD.
3. Texture images, color coding and symbols that are used for visualization of features must be created before (or simultaneously) they are referenced by a feature.
4. The earth surface parts *TIS*, *LandCover* and *Transportation* must together form a fully partitioned surface.
5. If a terrain intersection point, curve or surface is represented, a corresponding geometry must exist for the same feature.
6. A surface geometry or a combination of earth surface geometry and surface geometry must exist for a feature that is defined as a solid
7. Surfaces describing the exterior boundary of a building in LOD1, LOD2 or LOD3 and corresponding earth surface must be specified for each semantic building feature. If an earth surface is defined in one LOD the corresponding surface object for the LOD must also be specified.
8. The relation *TerrainIntersectionSurface* between the *Geology* feature and the earth surface may only exist when the *Geology* feature is to be seen in the open, for instance a mountain outcrop or a beach

More details about the conceptual model can be found in Emgård and Zlatanova (2007).

## 17.5 3D management of geometry in DBMS

Since the mid-nineties, several solutions for DBMS storage of 3D city models have been described in the literature i.e. (Kofler, 1998, Köninger & Bartel 1998, Stoter & Zlatanova 2003, Coors 2003). Köninger and Bartel (1998) presented a DBMS schema for 3D city models based on boundary representation in three levels of details using vertex index arrays to represent faces also including image mappings to texture images stored as BLOBs. Other examples can be found in Coors (2003) and Stoter & Zlatanova 2003 where buildings are stored as sets of faces. Within geology, 3D features are represented as 2.5D surfaces and 3D volumetric primitives for storage of stratum boundary surfaces, folded strata, ore bodies etc. in an object oriented DBMS (Breuning & Zlatanova 2006).

However, natively supported 3D data types within DBMS were not developed until recently. Prototype representations of polyhedron are reported by

Arens et. al. (2005) and of tetrahedron by Penninga et. al. (2006). Commercial support of 3D data types and operations are also expected to be available shortly (Oracle 2007).

At the moment, data types are restricted to point, curves and polygons with 3D coordinates. With some exceptions (e.g. PostGIS) the operations on these data types are 2D dimensional (Zlatanova & Stoter 2006).

Generally, within 3D database research, emphasis has been mostly given to topology and geometry and less work is completed on mapping between thematic semantics and 3D geometry. Some examples of DBMS implementation of a semantic model with 3D geometry are the 3D-geodatenbank Berlin (Plümer et. al. 2007), based on CityGML and the GeoBase21 (Haist & Coors 2005).

## 17.6 Implementations

Since most DBMSs follow the Simple feature specification for SQL (OGC 1999), the initial implementation alternatives we present are restricted to usage of simple features: point, curves and polygons. Given the simple feature types, surfaces and solids are created from a collection of polygons or a multi-polygon data type. Solids are in addition expressed by solid data type, assuming commercial DBMS will be able to maintain solids soon. TINs are expressed by multi-polygons containing only triangular polygons. Texture is either mapped to each polygon in a collection of polygons or draped over a surface consisting of a multi-polygon geometry.

The following thematic semantic features are selected for implementation in the database model: In general, each semantic class as given in Figure 2 is

| Above surface features | Earth surface features | Below surface features |
|---|---|---|
| Building | Transportation | Geology |
| Construction Work | LandCover | BelowSurfaceSpace |
| Vegetation | | Utility |
| CityFurniture | | Water |

implemented as a table and the attributes correspond to a column in a table. The superclass *TerrainIntersectionObject* is not implemented as a table, since it is not a feature class. *AboveSurfaceUtility* and *BelowSurfaceUtility* are merged in one table *Utility* due to the similar properties. An attribute shows whether the *Utility* is located above or below ground. If a utility feature is intersecting the earth surface, the feature is split into two geometries and a *TerrainIntersectionPoint* appears on the earth surface.

Given the chosen semantic entities, two alternatives of implementation are described. In the first alternative the common geometric representations of all entities are identified and created as separate tables. The tables containing the thematic semantic objects are linked to the geometric tables. In the second alternative the thematic semantic tables integrate the geometries. Geometric tables are not shared except data types describing symbols and textures, which features have in common.

Both suggested implementations of the 3D Information Model include an extended amount of semantic features. It should be noticed that some solutions are adopted from the Berlin model because (as mentioned above) some features are adopted from the CityGML.

The two implementation approaches presented here are intended for Oracle Spatial and its object-relational data model. Therefore data types as given by SDO_GEOMETRY and ORDSYS.ORDIMAGE are used in the descriptions.

### *17.6.1 Implementation alternative I*

The first implementation alternative strictly separates semantics from geometry into two table groups. The common geometries are organised in four relational tables (point, curve, surface, earth surface) corresponding to simple geometry data types (point, curve, surface), one compound table giving maintenance of solids and supplementary tables for maintaining textures (Figure 3). The semantic entities are modelled as separate tables each referring to a semantic class and linked to tables containing the geometry.

The geometry tables are divided into point, curve, polygon, multi-polygon, solid and texture tables where the SURFACE GEOMETRY table represents polygons, the EARTH SURFACE table multi-polygons and the COMPUND GEOMETRY table solids. The SURFACE GEOMETRY tables contain polygons with texture or colour that is defined on one or both sides of the polygon while the EARTH SURFACE GEOMETRY table contains multi-polygons that are textured with a draped image or a repeated draped image. In this way a simple colour mapping, a low resolution ortho photo or a high resolution façade texture can be used depending on the semantic feature class. Since two textures can be referenced by a SURFACE GEOMETRY polygon, two relations exist between the SURFACE GEOMETRY table and the TEXTURE IMAGE table. A solid representation of the feature is defined in the COMPUND GEOMETRY table that is referring to a collection of SURFACE GEOMETRY polygons and EARTH SURFACE GEOMETRY multi-polygons. A surface from the SURFACE GEOMETRY table can appear only in one solid. For example, in the case of a building with a common surface the surface is stored twice in the SURFACE GEOMETRY table. In contrast, a surface from EARTH SURFACE GEOMETRY appears only once in the table. A surface from the EARTH SURFACE GEOMERY table can be a part of a feature above or below the surface. When composing the COMPOUND GEOMETRY table (see below), the orientation of the surface has to

# 17 Implementation alternatives for an integrated 3D Information Model

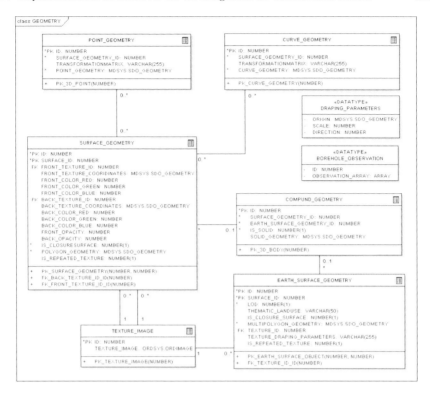

**Fig. 17.3** Geometric table group containing all used geometric feature tables in implementation alternative I

be adapted to create a valid solid. This is to describe that for above features it has to be flipped.

The symbols (e.g. trees, bus stops, streetlights and utility elements) consist of a collection of intersecting or non-intersecting polygons coloured or textured to be placed at different locations in the scene. These locations are maintained in the point and the curve geometry tables. The SURFACE GEOMETRY table is used to store the surfaces of the symbols as well. The relation between a feature and a symbol is many-to-many and it is established by the IDs of the feature and the symbol. For example one point refers to a SURFACE GEOMETRY ID, which may consist of many polygons having different IDs. In this manner, symbols can be referenced by several points and the points can use symbols using several polygons. Symbols can be placed along a curve described in the CURVE GEOMETRY table if the distance between the symbols is specified in the semantic tables. Such cases are possible for *Vegetation* and *CityFurniture* semantic classes.

The COMPUND GEOMETRY table is included to be able to define and validate solids. This actually creates a relation between the SURFACE GE-

OMETRY and the EARTH SURFACE GEOMETRY. It should be noticed that the primary storage tables for the geometries of the features are still the SURFACE GEOMETRY and EARTH SURFACE GEOMETRY tables. The geometry stored in the COMPUND GEOMETRY table is a copy of the geometry but in a *solid* data type, which allows to perform validation and/or other 3D spatial operations (e.g. volume).

Image textures are stored using the ORDSYS.ORDIMAGE data type referred from the TEXTURE IMAGE table. Parameters for texture placement are included in each polygon of the surface geometry table and for each multi-polygon in the earth surface object.

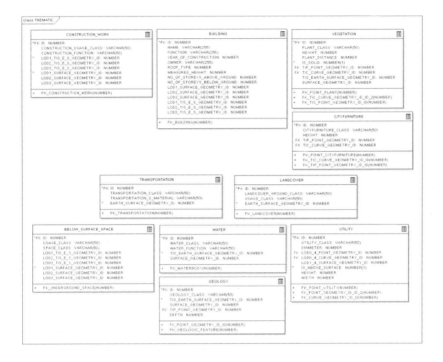

**Fig. 17.4** Semantic table group containing above surface, earth surface and below surface feature tables in implementation alternative I

Each semantic table include relations to one or more of the geometry tables (Figure 4). BUILDING, CONSTRUCTION WORK, BELOW SURFACE SPACE and WATER tables are all related to both SURFACE GEOMETRY and EARTH SURFACE GEOMETRY tables. As mentioned above, features that belong to these classes can be represented by surfaces and solids. TRANSPORTATION and LAND COVER tables are typically surface types of features and can be related only to the EARTH SURFACE GEOMETRY table. The VEGETATION table is related to all geometric tables, since the

*Vegetation* can be represented by point, curve, surface and solid. The CITY FURNTIURE and UTILITY tables are related to point and curve geometry tables, as they can be points and curves. The UTILITY table is also related to the SURFACE GEOMETRY table since a utility feature can be a large object under ground connected to several other utility features and the geometry of the feature is too complex and individual to be replaced by a symbolic geometric feature.

A *Geology* feature that is not intersecting with the earth surface is stored as a solid referring only to the SURFACE GEOMETRY table. Faults and stratums are maintained as surfaces. Borehole geometries are stored using the POINT GEOMETRY table to store the TIP of the Borehole. In addition a data type borehole is created for storage of the observations in the borehole.

BUILDING, CONSTRUCTION WORK and BELOW SURFACE SPACE tables can have different surface models depending on the level of detail. Therefore each LOD is related to a set of polygons in the SURFACE GEOMETRY table. To fulfil the more extended concept of LOD2-3 for buildings our implementation must be complemented with the tables: THEMATIC SURFACE GEOMETRY, THEMATIC SURFACE and BUILDING OPENING as described in the database schemas of 3D-Geodatenbank Berlin (Plümer et. al. 2007).

## *Implementation of rules*

The rules for the conceptual model are implemented as foreign keys and user specifications, which may be further implemented as methods and triggers. Rules 1 and 3 are implemented as foreign keys between some of the defined features i.e. between semantic tables and the POINT GEOMETRY and CURVE GEOMETRY tables. With the foreign keys defined, the given TIP POINT GEOMETRY ID of a semantic point feature i.e. a streetlight must also be defined as an ID in the POINT GEOMETRY table. This fulfils rule 1 defining that the geometry of the feature has to be created before the semantic information can be added. Similar constraints are also applied for the tables containing *TerrainIntersectionObject*. A foreign key is also added to control that a texture that is referenced exists in the TEXTURE IMAGE table (rule 3).

Since the *TerrainIntersectionSurface* identifier ID for the semantic tables are referring to one or several rows in the EARTH SURFACE GEOMETRY table, a foreign key can not be created for EARTH SURFACE GEOMETRY or for SURFACE GEOMETRY (rule1). This constraint is therefore implemented as a user rule instead of a database constraint. Rules 2 and 4-8 are not implemented as constraints in the database.

## 17.6.2 Implementation alternative II

The second implementation alternative is based on complete semantic subdivision where the geometry columns are integrated in the semantic tables.

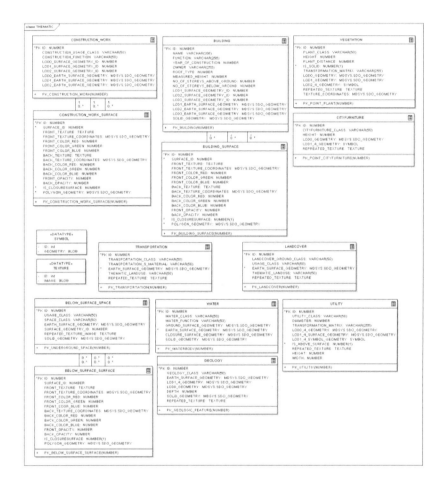

**Fig. 17.5** Table schema representing all tables in implementation alternative II

In this implementation, the geometry column may contain different geometries. Specific tables are defined for geometry only when explicitly required. For example, CONSTRUCTION WORK, BELOW SURFACE SPACE and BUILDING tables require individual textures for each surface in higher LOD. Therefore, the features can not be implemented using the multi-polygon data type. Instead, three corresponding surface tables where created to handle the texturing where the CONSTRUCTION WORK SURFACE table, the

17 Implementation alternatives for an integrated 3D Information Model    325

BUILDING SURFACE table and BELOW SURFACE SPACE table follow the same concept as the surface geometry table in implementation Alternative I storing one polygon in each row. The tables also have three relations to the geometry table, one for each LOD. In the other cases geometries are stored as point, curve or multi-polygon features as an attribute in the table. The earth surface features do not require individual texturing for each polygon and are therefore modelled as attributes within the feature they are part of. Symbols and textures are defined in new data types storing images and geometry as BLOBs.

## *Implementation of rules*

No geometric constraints are implemented in alternative II since the geometry is mostly organised as an attribute to a particular semantic feature. The only exceptions are the BUILDING, CONSTRUCTION WORKS and BELOW SURFACE SPACE tables, which refer to separate tables for the geometry surfaces. This approach is used to allow texture mapping per each individual polygon. An alternative option will be to create a special data type, which would encompass geometry and texture. This alternative, however, complicates the spatial index.

The rules 1- 3 can be fulfilled by developing functions and/or triggers. While the triggers can check the validity during data import (Louwsma et al 2006), functions control the rules after data are loaded. Similarly to the first alternative, rules 4-8 are not implemented.

## *17.6.3 Comparison of implementations*

The two proposed implementations can be strictly compared only after testing with different data sets. Here only some initial expectations are discussed. In general, from a user's point of view, Alternative II has advantages due to the clear and simpler structure (where all features are integrated in the semantic tables). However, Alternative I provides more robust database management, since some of the consistency check can be performed by the DBMS.

Loading data into the database is more straightforward for alternative II, since at least two tables have to be filled for each feature in alternative I. In Alternative I the geometric feature have to be created before the semantic feature while in Alternative II they can be loaded simultaneously with the insertion of the semantics. Therefore we expect better performance (faster import) in the second alternative.

Alternative II may cause more redundancy in geometric storage. A surface cannot be shared by two features in some cases as in Alternative I. For example, a geologic body that is touching another body shares several polygons. Alternative I allow the user to make reference to the same polygons (surfaces) from the two different geological features. This concept is however not allowed for buildings. In Alternative II, surfaces will be stored per feature and therefore twice. However, Alternative II allows for more elaborated use of data types (solid, multi-polygon) at lower LOD when, no texture mapping is applied.

Operations like edit and update would be faster in one or the other Alternatives with respect to the attributes to be changed. If the changes are related to features, the second alternative will be faster. Changes in geometry will benefit from alternative I. It should be noticed that also the implementation of Rule 4 (full partitioning of earth surface) will be rather complicated for the Alternative II.

A query based on semantics of a single feature is less complicated in Alternative II, while always a join of tables is required in Alternative II. In addition less geometry is traversed since all geometries of one class are in the same table. For example a query of all utilities defined as points is less complex in Alternative II. On the other hand a pure geometric selection is less complex in Alternative I since the geometry can be found in a single table depending on geometry type (point, curve and surface). For example, a query of all features represented by surfaces within a specified area will be much faster in Alternative I. Similarly, spatial queries investigating relationships between objects in many cases would be simpler in Alternative I. For example, the query 'find all the neighbouring features of Building 77' can be performed on only the geometry tables. Alternative II would always require a traverse of all the semantic tables.

In Alternative I symbolic representations are stored using individual textured polygons while in Alternative II the symbols are expressed by BLOBs. To assemble the symbols from individual polygons in general has a negative effect with respect to performance compared to symbols stored in BLOBs. This yet has to be tested to be proved true.

## 17.7 Conclusion and future work

We have presented two alternative implementations of the 3D Integrated Model and have discussed advantages and disadvantages. The first approach is more beneficial for geometric queries while the other one is more promising concerning semantic queries. What implementation to choose is much dependent on the application or the purpose. Generally, most of the cases would require both semantic and geometric attributes. This means the query schema (order of querying) should be well-thought to achieve needed performance.

As mentioned above, the model is under development. Some of the classes in the model have to be further elaborated with respect to LOD, for example

*Geology*, *BelowSurfaceSpace* and *ConstructionWork*. The *Utility* attributes have to be extended with domain specific attributes.

The two alternatives will be tested in a case study of the Campus area of Delft University of Technology.

# References

Arens C, Stoter J, van Oosterom P (2005) Modelling 3D spatial objects in a geo-DBMS using a 3D primitive. Computers & Geosciences 31 (2005) 165–177.

Billen R, Zlatanova S (2003) Conceptual issues in 3D Urban GIS, In: GIM International, Vol. 17, No. 1, January 2003, pp.33-35

Breuning M, Zlatanova S (2006) 3D Geo-DBMS, In: 3D large scale data integration: challenges and opportunities, CRC Press, Taylor&Francis Group, pp. 87-135

Caglioni M (2006) Ontologies of Urban Models, Technical report n°4, Short Term Scientific Mission Report, Urban Ontologies for an improved communication in urban civil engineering projects, Towntology Project 5p, http://www.towntology.net/Documents/STSM-Caglioni.pdf [last accessed 2007-08-27]

Coors V (2003) 3D-GIS in networking environments, Computers, Environment and Urban Systems, Volume 27, Number 4, July 2003, pp. 345-357(13)

Emgård L, Zlatanova S (2007) Design of an integrated 3D information model, in: Coors, Rumor, Fendel&Zlatanova UDMS Annual 2007, Taylor & Francis, 26th proceedings of UDMS, 10-12 October 2007, Stuttgart, (in press)

Gröger G, Kolbe T, Czerwinski A (2006) OpenGIS CityGML Implementation Specification. http://www.citygml.org/docs/CityGML_Specification_0.3.0_OGC_06-057.pdf [last accessed 2007-08-27]

Haist J, Coors V (2005) The W3DS-Interface of CityServer3D In: Proceedings of International Workshop on Next Generation 3D City Models 2005, Bonn pp.63-67

Herring J (2001): The OpenGIS Abstract Specification, Topic 1: Feature Geometry (ISO 19107 Spatial Schema), Version 5. OGC Document Number 01-101. http://www.opengeospatial.org/standards/as [last accessed 2007-

08-27]

INSPIRE, The European Parliament (2007) Directive of the European Parliament and of the Council establishing an Infrastructure for Spatial Information in the European Community (INSPIRE) (PE-CONS 3685/2006, 2004/0175 (COD) C6-0445/2006) http://www.europarl.europa.eu/oeil/FindByProcnum.do?lang=2&procnum=COD/2000/0175 [last accessed 2007-08-27]

Kofler M (1998) R-trees for the visualisation of large 3D GIS databases, PhD thesis, TU, Graz, Austria.

Köninger A, Bartel S (1998) 3D-GIS for Urban Purposes. GeoInformatica 2:1,pp 79-103 (1998)

Louwsma J, Zlatanova S, van Lammeren R, van Oosterom P (2006) Specifying and implementing constraints in GIS – with examples from a geo-virtual reality system, *GeoInformatica* 10 (2006): pp 531-550

NADM (2004) North American Geologic Map Data Model Steering Committee Conceptual Model 1.0—A conceptual model for geologic map information: U.S. Geological Survey Open-File Report 2004-1334, 58 p., accessed online at URL http://pubs.usgs.gov/of/2004/1334. Also published as Geo-logical Survey of Canada Open File 4737, 1 CD-ROM.

OGC 1999 OpenGIS Simple Features Specification For SQL, Revision 1.1 http://www.opengeospatial.org/standards/sfs [last accessed 2007-08-31]

Oracle (2007) Oracle Spatial 11g Planned features presentation material

Pantazis D (1997) CON.G.O.O. : A conceptual formalism for geographic database design. In Geographic Information Research, Bridging the Atlantic (London: Taylor & Francis), pp. 348-367.

Penninga, F., van Osteroom, P. & Kazar, B. 2006. A Tetrahedronized Irregular Network Based DBMS approach for 3D Topographic Data Modeling. *Progress in Spatial Data Handling*

Plümer L, Kolbe T, Gröger G, Schmittwilken, J & Stroh V (2007) 3D-Geodatenbank Berlin, Dokumentation V1.0. http://www.3dcitydb.org/index.php?id=259 [last accessed 2007-08-27]

Stoter J, Zlatanova S (2003)Visualisation and editing of 3D objects organised in a DBMS, J., In: Proceedings of the EuroSDR Com V. Workshop on Visualisation and Rendering, 22-24 January 2003, Enschede, The Netherlands, 16p

Zlatanova S, (2000) 3D GIS for urban development, PhD thesis, ISBN 90-6164-178-0, ICG, TUGraz, Austria, ITC publication 69, ISBN 90-6164-178-0

Zlatanova S, Stoter J (2006) The role of DBMS in the new generation GIS architecture, in: Rana&Sharma (eds.) Frontiers of Geographic Information Technology, Springer-Verlag, Berlin Heidelberg ISBN-1- 3-540-25685-7, pp. 155-180

# Chapter 18
# Serving CityGML via Web Feature Services in the OGC Web Services - Phase 4 Testbed

Eddie Curtis

## 18.1 Background

The OGC Web Services – Phase 4 (OWS-4) test-bed is an initiative under the OpenGeospatial Consortium's (OGC) interoperability programme in which 72 organisations collaborated to extend and demonstrate interoperability of internet based geospatial services. The main activities of the test-bed took place between June and December of 2006. These activities included the deployment and integration of a number of software and data components culminating in two demonstrations of the capabilities developed during the test-bed, as well as numerous reports detailing issues encountered and recommendations.

The test-bed used an emergency response scenario to exercise the capabilities of a variety of OGC conformant components. The scenario required numerous components and services to interoperate in order to provide emergency planners with information to coordinate the response to the hypothetical incident. The activity was divided into a number of threads each of which addressed particular set of technical issues affecting the scenario such as security, workflow and sensor web enablement. This paper will consider some of the findings from the thread which covering the integration of Computer Aided Design (CAD) systems, GIS (Geographical Information Systems) and Building Information Models (BIM), known as the CAD/GIS/BIM thread. This thread was concerned with issues related to exchanging information about buildings and 3D geometry between GIS analysts and building designers.

---

Snowflake Software, United Kingdom
eddie.curtis@snowflakesoftware.co.uk

## 18.2 CAD/GIS/BIM Integration

Whilst both GIS and CAD deal with spatial information the approaches used by each are shaped by different requirements and working practices. The GIS world emphasises the analysis of measured information i.e. data collected through survey, satellite imagery etc. whereas the CAD world emphasises the design and construction processes. However, there is clearly an overlap of concerns between geographical information users and the Architecture Engineering and Construction (AEC) industry, which makes extensive use of CAD.

There are a number of scenarios in which the interoperation of BIM and geographical information are of benefit.

In the site planning process for AEC it is useful to consider contextual information about the site in order to understand how a proposed building will interact with its environment. This could include aerial photography, terrain models, and nearby buildings and infrastructure. The ability of CAD systems to discover and import this information is therefore beneficial to the design process. Enabling CAD systems to act as clients to OGC web services is a means to achieve this.

Integration of BIM models for different sites requires the introduction of geographical concepts into the BIM models. Unless the separate BIM models are referenced to geographical coordinate systems it is not possible to relate the positions of objects in the separate models to each other. This type of integration is necessary to handle common infrastructure shared by buildings and can identify potential conflicts between construction projects.

It is also beneficial to bring BIM information into the GI world. Location based services could be extended from the street to building interiors. For example, fire-fighters could be provided with information about infrastructure such as electricity and water supplies both inside and outside a burning building. Geographical analysis and broad scale visualisation could also make use of BIM information allowing, for example, emergency planners to identify buildings prone to particular risks or suitable for conversion to emergency use. Since BIM models often contain highly detailed information it is not practical to create and maintain a single database containing building information for a whole city. On-the-fly integration of information held in a distributed, heterogeneous set of databases through OGC web services represents a more practical approach to making building information available to GIS applications.

## 18.3 CityGML

City Geography Markup Language (CityGML)[2] is an information model and GML application schema for the exchange of 3D city and landscape models. Originally developed by the members of the Special Interest Group 3D (SIG 3D) of the Geodata Infrastructure North-Rhine Westphalia (GDI NRW) initiative in Germany, CityGML has now been adopted as an OGC discussion paper with a view to it becoming an OGC best practices paper.

A key characteristic of CityGML is that it combines the ability to contain complex, geo-referenced 3D vector data along with the semantics associated with the data. In contrast to other 3D vector formats, CityGML is contains a rich, general purpose information model in addition to geometry and graphics content. The CityCML information model includes:

- Digital Terrain Models as a combinaton of triangulated irregular networks (TINs), regular rasters, break and skeleton lines, mass points
- Sites (currently buildings and bridges)
- Vegetation (areas, volumes, and solitary objects with vegetation classification)
- Water bodies (volumes and surfaces)
- Transportation facilities (both graph structures and 3D surface data)
- City furniture
- Generic City objects and attributes

For specific domain areas CityGML also provides an extension mechanism to allow the model to be enriched with additional properties and feature types. Targeted application areas explicitly include urban and landscape planning; architectural design; tourist and leisure activities; 3D cadastres; environmental simulations; mobile telecommunications; disaster management; homeland security; vehicle and pedestrian navigation; training simulators; and mobile robotics.

CityGML provides a model at multiple levels of generalisation. These can be used individually within a model or multiple levels of representation can be modelled together. The levels are:

- LOD 0 – Regional, landscape
- LOD 1 – City, region
- LOD 2 – City districts, projects
- LOD 3 – Architectural models (exterior), landmarks
- LOD 4 – Architectural models (building interiors)

Since CityGML deals with buildings and constructions there is clearly some overlap with BIM. However, the two information models are different in scope. CityGML stops far short of the level of detail supported by BIM. A BIM can contain detail down to the level of component parts within individual fixtures such as doors and windows. CityGML LOD4 provides a level of detail suitable for a "walkable" model of a building for simulation or space analysis purposes. However, CityGML provides for modelling of the building context including roads, street furniture, terrain, vegetation etc. BIM models are concerned only with the building. The two models are complementary with

CityGML providing a geographical view of buildings in their context, and BIM providing a detailed view of buildings and their construction.

## 18.4 The OWS-4 Scenario

The OWS-4 test-bed required a variety of systems and data to interoperate to solve a test scenario. In this scenario a 'dirty bomb' has detonated in a port. After identifying and analysing the problem the emergency planners decide that an emergency field hospital will be required and begin looking for a suitable building to convert to use as a hospital. There are a number of criteria which the building must meet including access to an airstrip, and space requirements for an operating theatre.

Identification of a suitable building requires a number of OGC services to provide information about the site such as aerial photography of the site, terrain models etc. In order to assess the space requirements of the building a 3D model of the building is needed. This is supplied from a Web Feature Server (WFS)[4] providing a CityGML model of the site. A hangar building at a nearby airport is identified as being suitable for the field hospital.

The hangar must be modified for use as a hospital. Here the focus shifts from GIS analysis to CAD design. An operator uses a CAD client to access the BIM model for the hangar via the internet. The operator modifies the hangar model to convert it for use as a hospital. The modified hangar model is made available as both a BIM and CityGML model, thus enabling the GIS analysts to see the modified hangar as CityGML as they did with the original hangar model.

## 18.5 CityGML WFS Challenges

Serving CityGML via a WFS presents a number of technical challenges arising from the characteristics of the CityGML model.

The CityGML model is untypical of GML application schemas in the level of complexity of the data model. CityGML makes extensive use of complex data types for properties and nesting of features within feature collections. Consequently CityGML data can contain very deeply nested data structures.

The geometry types supported in relational databases are often more limited than the range of geometry types used in CityGML. For example, TINs and 3D solid geometries are not supported in Oracle 10g. This presents an obstacle to WFS implemented on top of a relational database.

At high levels of detail data volumes can become large, even for a small geographical area.

## 18.6 Snowflake CityGML WFS

CityGML datasets for the test-bed scenario were created by the Forschungszentrum Karlsruhe institute. The Forschungszentrum Karlsruhe team developed a software tool for the conversion of building information models encoded using Industry Foundation Classes (IFC)[1] building models to CityGML. This tool carries out a number of mapping operations between IFC and CityGML models. The tool maps IFC classes to CityGML classes e.g. an IfcSpace becomes a CityGML interiorRoom. The tool also converts geometry from the local coordinate systems used within the IFC model to geographic coordinate systems.

The Snowflake CityGML WFS was created by deploying Snowflake's GO Publisher. GO Publisher is a data translation engine which translates from relational databases to XML. In order to stand up the GO Publisher WFS the CityGML data produced by Forschungszentrum Karlsruhe was loaded into an Oracle data model using a GML bulk loading tool called GO Loader. This tool has a similar translation capability to GO Publisher but translates from GML to relational models.

Oracle's SDO geometry types were used to store the geometry and conventional relational structures were used to hold the non-spatial properties. Relationships such as the containment relationships between buildings and rooms were represented as table joins in the database. GO Publisher's graphical user interface was then used to configure a translation from the relational model to the CityGML schema. GO Publisher tools were used to bundle this translation with the GO Publisher software into a Web ARchive (WAR) file for deployment within an application server. The CityGML WFS was deployed by uploading this WAR file into the Tomcat application server.

Once deployed within the application server the Snowflake CityGML WFS was able to process requests to the WFS operations getCapabilities, describeFeatureType and getFeature. On receiving a getFeature request GO Publisher translates the WFS filter in the request into an SQL query using the translation configured prior to deployment. The resulting SQL query contains all conditions from the WFS filter including both spatial and non-spatial operations. The SQL query is run against the Oracle database containing the city model. GO Publisher then translates the resulting SQL records into GML, again using the translation configured prior to deployment. The resultant GML is then streamed back to the client. The data is returned in a compressed stream if appropriate.

By carrying out the two directional translation (WFS filter to SQL followed by relational data to GML) GO Publisher is able to make use of the scalability, performance and robustness of the underlying Oracle database. The application server can create multiple instances of the WFS in order to deal with concurrent requests for data. These technologies therefore provide a highly scalable platform for the WFS.

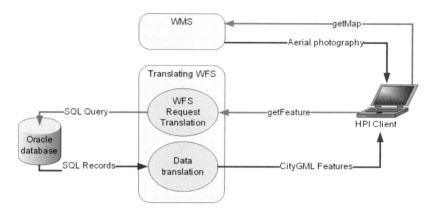

**Fig. 18.1** Data requests and response from the client application

## 18.7 Findings

The testbed proved the feasibility of serving CityGML through the WFS interface. A client application developed by the Hasso-Plattner Institute successfully connected to the WFS, retrieved data and displayed it. The demonstration also showed the utility of connecting to OGC web services as the client was able to connect to several different services provided independently by different organisations around the world and to integrate the data into a single view. For example, the client took a terrain model from the CityGML WFS and draped aerial photography from a Web Coverage Server across the terrain. This allowed the analyst to build up a picture of the situation at the airport by drawing on a variety of independent sources of information.

The CityGML schema is much more complex than those usually deployed in WFS. The unusual level of complexity did not cause problems in the interaction between the client and the Snowflake CityGML WFS. The underlying information model of CityGML was known to the client, so the client was able to form meaningful request and correctly interpret the response.

A specific example of the data complexity is that CityGML buildings are made up of component parts which also contain component parts. A building may be made up of wall surfaces, each of which can contain windows and doors. All of these are objects in their own right with their own identity, properties and geometries. When the client requested the hangar building, the server returned the hangar and also returned some of the component parts of the hangar (walls, doors etc.) even though these were not explicitly requested.

For different parts of the scenario different combinations of objects were required. In particular the spaces (rooms) within the hangar were required for analysis of the building but later on, after editing of the spaces for use as a hospital, the spaces were supplied from a different server whilst the walls and windows continued to be supplied from the Snowflake CityGML WFS. This was handled by setting up a number of alternative WFSs which served different variations of the content. These included different combinations of optional properties of the CityGML model suitable for the different circumstances. Because of the translation capabilities of the Snowflake GO Publisher product this could be done without duplication of the data in the underlying database. Several different translations from the underlying database were set up with some translations omitting some of the optional CityGML properties. The client was thus able to get different views into the single underlying city model by connecting to different WFSs.

The CityGML model contains 5 levels of detail (LOD). In the testbed scenario the WFS client made a series of request for data including some buildings at LOD2 (building shapes with roof shapes) and the hangar at LOD4 (an architectural model including interior rooms, walls and doors). This mixing of LOD allowed the client to build up a view with increased level of detail for the buildings of interest and lower levels of details for the buildings which were requested for context. This created a composite view with LOD in different areas of the model customised, on-the-fly, to the task at hand. The CityGML LOD concept proved useful because it allows the level of detail available in a model to be specified easily in both the data and the description of a data set. This scenario showed that by providing CityGML data through a WFS the LOD concept adds further value by allowing the client to select data at a mixture of levels of detail appropriate to the scenario.

A number of 3D geometry types were served by the WFS in addition to the solid geometries of the buildings. A detailed solid geometry for a helicopter object was served. This geometry is a particularly large geometry in GML consisting of many hundreds of surface patches. A terrain model in the form of a TIN (Triangulated Irregular Network) was also served from the WFS. Both of these geometry objects required the WFS to return large GML files in response to the WFS request. Initially this cause performance problems since the files took approximately 9 minutes to return via the network available for the demonstration, thus preventing the WFS from being used in an interactive manner. This was overcome by using compressed streams to return the data. Both gzip and zip streams were implemented on the Snowflake CityGML WFS and the Hasso-Plattner client was enhanced to read these streams. The zip algorithm compressed the GML output to approximately 5% of its uncompressed size (a total data volume of 24.5 MB was compressed to 1.3 MB). This reduced the time for requesting, receiving and displaying the CityGML data from approximately 9 minutes to around 20 seconds, allowing the service to be demonstrated live from a remote server during the final demonstration of the OWS4 test-bed.

Filters were used within the WFS requests to select features spatially. Filters containing 2D bounding-box geometries were used to select features with 3D solid geometries. The spatial test was carried out by testing the

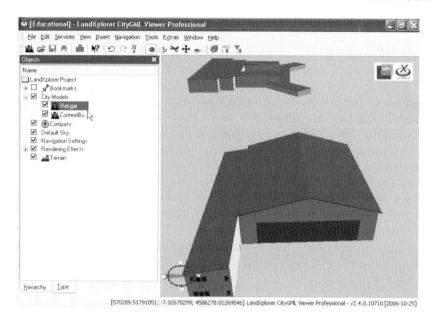

**Fig. 18.2** The LoD4 hangar building in the foreground has architectural detail such as windows and doors whilst LoD2 building in the background does not

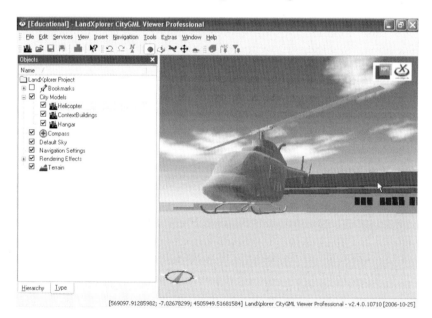

**Fig. 18.3** This screenshot shows illustrates the large number of surface patches and vertices in the helicopter geometry

interaction of the bounding-box with the 2D horizontal projection of the 3D geometries. This proved to be an effective interpretation of the filter even though the third dimension was ignored for purposes of the query. This is because the objects of the city model, although 3D, are distributed around the ground surface. The horizontal distribution of the objects is therefore great in relation to the vertical distribution. Consequently the degree of selection offered by 2D queries is high and corresponds to the use-cases of the test bed.

The limitations of the Oracle 10g geometry types was overcome by storing geometry in existing data structures but changing the interpretation of those structures on generating GML geometries. GML solid geometries where generated by interpreting the polygons within an Oracle multi-polygon as faces in the boundary of a GML solid. This approach to geometry translation is discussed in more detail in the paper 'Extending 2D Interoperability Frameworks to 3D' [3].

## 18.8 Conclusions

Despite the complexity of the CityGML model the WFS interface proved to be suitable for providing web based access to the city model.

Initial performance problems relating to the verbose nature of XML and GML were solved using compression and the WFS interface was shown to be a practical solution for large and complex models including large geometries.

The ability to select and filter data via the web allowed the client to build up a model specific to the problem in hand combining high levels of detail for buildings of particular interest and less detail for contextual objects. The scalable nature of the Snowflake CityGML WFS makes it feasible for a client to draw on very large city models by delegating the task of selecting and filtering the data to the remote server.

## References

1. International Alliance for Interoperability (2006) Industry Foundation Classes,
   http://www.iai-international.org/Model/R2x3\_final/index.htm
2. Gröger G, Kolbe TH, Czerwinski A (2006) City Geography Markup Language, OGC document 06-057r1
3. Mueller H, Curtis E (2005) Extending 2D Interoperability Frameworks to 3D, Paper presented to the International Workshop on Next Generation 3D City Models, Bonn, Germany, 2005
4. OGC (2005) Web Feature Service Implementation Specification, OGC document 04-094

# Bibliography

OGC (2004) Filter Encoding Implementation Specification, OGC document 04-095

OGC (2007) OGC Web Services Architecture for CAD GIS and BIM, OGC Interoperability Program Report

OGC (2007) OpenGIS Geography Markup Language (GML) Implementation Specification, OGC document 03-105r1

# Chapter 19
# Towards 3D environmental impact studies: example of noise

Jantien Stoter[1], Henk de Kluijver[2], and Vinaykumar Kurakula[3]

**Abstract**

Current environmental impact studies supporting environmental policies are mostly based on a 2D approach. The - mostly 3D - output of software that calculates the specific phenomenon (e.g. air or noise pollution) is processed and visualised in 2D and combined with 2D topographical and other data, such as population distribution, to quantify the effects.

The research described in the paper aimed at improving visualisation and quantification of impact of continuous spatial phenomena on the environment by applying a 3D approach. Noise is taken as example. Based on the specific demand, an approach is presented to generate a 3D noise map as basis for noise impact studies. The proposed concept is proofed by applying it to a sample noise impact study. From experiences with the sample it can be concluded that the 3D noise map offers insight into the 3D noise situation where 2D noise maps have limitations. In addition more accurate assessment of noise impact is possible in particular when different floors of a building close to the noise source and/or behind noise barriers are considered, which is specifically relevant in urban areas. The proposed methodology can be applied to other continuous spatial phenomena so that it meets the more general problem of how to represent 3D aspects of environmental impact studies.

---

[1]ITC Enschede, Department of Geo Information Processing,
Hengelosestraat 99, 7500 AA Enschede, the Netherlands,
[2] dBvision,
Vondellaan 104, 3521 GH, Utrecht, the Netherlands,
[3] L.I.G 'B' 543, A.S.Rao nagar,
E.C.I.L (post), Hyderabad, India, pin: 500062
[1]stoter@itc.nl, [2]henk.dekluijver@dBvision.nl, [3]kurakulavinay@rediffmail.com

## 19.1 Introduction

Many continuous spatial phenomena with a negative impact on the environment, such as soil pollution, air pollution and noise, have a 3D component. Although 2D GIS has been widely used to study the impact of these phenomena on the environment, it can be expected that a 3D approach can offer fundamental improvements. The research presented in this paper studied how environmental impact studies can be extended towards 3D by the combination of a continuous surface representing field-based data from which 3D contours can be generated with 3D object-based data (mostly manmade objects such as buildings).

Noise is the spatial phenomenon used as case study in this multi-disciplinary research executed by the company dBvision (providing knowledge on noise and its impact studies) and ITC (providing expertise on geo-information processing). The aim of the presented research is to show how environmental impact studies can be improved using basic 3D GIS functionalities rather than how 3D GIS functionalities can be improved in general. In that respect this research is a typical 'design' research as it is recognized in the discipline of information systems. According to Association for Information Systems (2007) design research should a) show that there is a demand for a design; b) review existing and propose a new design; c) show proof of concept by applying it to sample data; d) show that it can represent concepts that are impossible with existing designs; and, e) show that the improved design can meet more general problems. All these aspects will be addressed in this paper.

In section 2 the case of noise is presented defining the motivation for this research: why is noise a problem; how are noise impact studies currently carried out; what are the problems of current 2D approach in noise assessment and management. Section 3 presents a methodology for a 3D approach in noise impact studies by the integration of a surface representation of noise levels at a surface following the height of the terrain (including buildings) with a 3D city model. The results of the 3D noise map, including applying it to a noise impact study, are presented in section 4. The paper ends with conclusions in section 5.

The solution proposed in this paper provides improvements in noise impact visualisation and in accuracy of noise assessment specifically in urban areas. In rural areas, where considerable noise fall with height up a building is less relevant a 2D representation can still do. The cities Paris and Hong Kong already produced 3D noise maps (see Butler, 2004; respectively Wing and Kwong, 2006). Also the MITHRA-tool (CSTB, 2007) provides a 3D presentation of noise levels. These 3D noise maps look promising. The additions of the research presented in this paper is the generation of 3D contours that show a more detailed presentation of the noise situation, as well as a flexible method for visualisation: the visualisation is just another representation of the information. Conversions to virtual reality environments are not necessary to visualise the 3D information. As a consequence the noise surface as produced in this research can be used directly to quantify noise impact in 3D as will be seen in this paper.

Main aim of noise impact studies, as object of this research, is to visualise and quantify the overall noise impact. Therefore increasing accuracy of calculated noise levels on specific locations is not the objective of this research. However improving accuracy of calculated noise levels is an important scientific issue. It is also relevant in other noise applications, for example to comply with noise limits by insulating houses. Since much money is involved in the insulation of houses it is important to precisely indicate which houses are exposed to high noise levels. When the accuracy of calculated noise levels is addressed, future research can focus on improved calculation methods, on improved interpolation methods and on how to make use of developments in the area of 2.5D and 3D interpolation (see for example Boissonat and Flötotto, 2002; Ledoux and Christopher, 2004; FIELDS, 2006, and GRASS, 2007).

## 19.2 Describing the case of noise

This section presents the case of noise. The noise problem including policies to reduce noise problems are introduced in section 2.1. Section 2.2 describes current practice of noise impact studies. The demand for a 3D approach for noise assessment and management is described in section 2.3.

### *19.2.1 Noise problem*

Noise pollution in large urban areas, mainly caused by industry and road and railway traffic, is considered as a serious environmental problem (Silvia et al., 2003). For the management of these noise problems many governments require that the environmental impact of noise produced by planned constructions such as infrastructure and industries is assessed before construction starts. If negative effects are expected, measures need to be taken. These measures may comprise a change of the plan, construction of noise barriers, use of quiet road surfaces and insulation of houses.

Besides the prevention of future noise problems, steps are being taken to reduce present noise effects. In order to have a common European mitigation program to control noise levels the European Union has formulated a directive on noise pollution (European Union, 2002). The directive prescribes a common approach for all member countries to prevent and reduce the harmful effect of noise. A major component of this approach is a common method to produce strategic noise maps for all major cities, roads, railways, airports and industrial sites. The strategic noise map presents noise levels caused by existing noise producers. The EU-directive requires the noise map to represent noise levels at a height of four meters from the surface. The EU-directive further requires publishing the noise maps to the public and updating the maps every five years. Based on these maps, plans need to be made to reduce the impact of noise, also every five years. The EU-directive does not contain common noise limits. These can be determined by each member state.

## 19.2.2 Noise impact studies

Noise impact studies consist of two stages: 1) the calculation of noise levels and 2) the combination of other spatial and non spatial data with the calculated noise levels to produce insight into the impact of noise.

**Calculation of noise**

In noise impact studies, noises levels are determined with computer simulations models rather than with noise measurements. There are several reasons for this. First of all field measurements are time consuming since the noise levels concern the yearly averaged values and can only be done under the appropriate weather conditions. In practice it is difficult to execute an adequate number of measurements in order to produce reasonable noise maps. Furthermore it is impossible to determine future noise levels by measurements whereas noise simulation models can deal with future situations. In addition it is shown that models can predict noise levels within an acceptable level of uncertainty for most situations. Therefore noise calculation methods, which have been validated and calibrated extensively with field measurements, are widely accepted to provide reliable information on noise levels. In computer models that implement the calculation methods, noise levels are calculated on 'virtual microphones' (observation points). A virtual microphone, specified with a x,y,z coordinate, is a point that reports what the noise level would be at a certain location under given circumstances. In the computer models noise levels are computed on 3D data points based on:

1. Information on the noise source (roads in this case): traffic intensity, maximum speed, road surface type, average emission of different vehicle types.
2. Information on aspects that influence noise propagation such as noise obstruction by objects (like buildings, noise barriers) and noise absorption (like open areas with grass or bare soil). This information also covers heights of buildings and of other topography.
3. 3D distance and direction of the data points with respect to the location of the noise source.

## Determining the impact of noise

GIS functionalities are commonly used to assess the impact of noise by producing strategic noise maps. Noise maps can be made with the combination of interpolated surface of noise levels and spatial data covering the area of the study. An example of a noise map is shown in figure 1 (Kluijver and Stoter, 2003). This figure shows several noise contours that represent same noise levels along either side of road and railway.

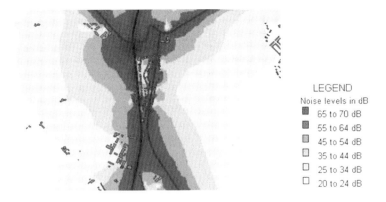

**Fig. 19.1** 2D noise map (Kluijver and Stoter, 2003)

The noise maps are used as input for the assessment of noise impact on the environment, for example determining the area which is affected by severe noise; determining the number of noise sensitive buildings or the area of natural parks where a certain noise level is exceeded; determining the number of citizens who are annoyed by noise etc. Quantifying the impacts of noise facilitates the comparison of several designs in order to choose the design with the smallest noise impact on the environment.

### 19.2.3 Problems of current 2D approach

Most of the noise calculation software calculates noise with the three dimensional data, i.e. heights of buildings, of noise barriers and of other topography are taken into account. Although output of noise calculation software (observation points with calculated noise levels) is described in 3D, most current noise maps are in 2D representing noise levels at one selected height (for example at four meter as required by the European directive). Disadvantage of this 2D mapping method is the lack of insight into the three dimensional character of noise. In many situations noise levels at one particular selected height (for example at four meters) do not provide complete information for assessing noise impact at higher floors of a building.

2D noise maps are used to quantify impact of noise, e.g. overlaid with a 2D building map. This can cause considerable differences between calculated impacts and impacts that (will) occur in reality. The difference is especially large when a building of interest is located close to the noise source or when a noise barrier is present. People living on lower floors of an apartment building benefit more from a noise barrier than people living on higher floors. Therefore number of annoyed people might be overestimated when based on 2D analyses. Summarising, 2D noise maps and 2D analyses are insufficient to discriminate between noise impacts at different heights which is specifically relevant in urban areas. Although current noise simulation models predict noise levels in 3D, noise maps generated in 2D cannot be used directly to study the 3D impact of noise. To use the 3D information in 3D noise impact studies, firstly the output of noise software needs to be processed in 3D.

## 19.3 Methodology

This section presents a 3D approach for noise impact studies. It starts with a description of the study area (section 3.1). Section 3.2 presents the calculation method used in this research. The process of selecting the locations of observation points, which is extremely important, is described in section 3.3. Section 3.4 describes how the 3D noise map is generated and section 3.5 describes how the 3D noise map is applied in a noise impact study.

### *19.3.1 Study area*

The study area is a small part of the city centre of Delft, the Netherlands. Delft is a city of around 95,000 people in the densely populated South Holland province of the Netherlands. The study area is approximately 30,000 m$^2$ and contains about 185 residential buildings with an average height of 15 meters. A 3D city model covering the study area, containing all details of buildings, was provided by Vosselman et al. (2005). The city model, shown in figure 2, is constructed based on an interactive segmentation of the parcel boundaries using several tools for splitting the polygons along height jumps edges. The roads, canals and trees were also reconstructed from the combination of parcel boundaries and laser altimetry data.

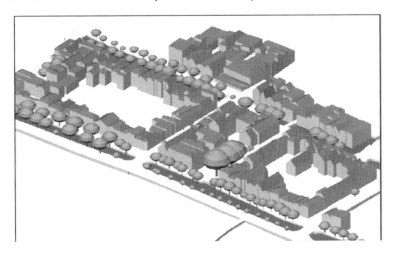

**Fig. 19.2** 3D city model of study area

## 19.3.2 Noise calculation method

As in other countries, also in the Netherlands calculation methods have been standardised and are commonly accepted as appropriate for noise impact studies after having been validated and calibrated with extensive measurements in 1970s and 1980s (VROM, 1999). From the available Dutch methods the Standard Calculation Method 1 (SCM1) was selected for this research. SCM1 (VROM, 2004) was established by the Ministry of Housing, Spatial Planning and the Environment, according to advise of noise experts and after extensive testing, for assessing relatively simple noise situations, such as determining noise hot spots, quantifying overall effects and visualising noise levels. SCM1 was chosen since it takes the obstruction of noise by objects such as buildings into account but it is still relatively simple to use, also for non-noise experts. At the same time it meets the requirements for this research (to see how noise studies can be improved by a 3D approach). Other more sophisticated noise calculation methods could also have been used. In this study these methods are not relevant, since the focus is on a method to improve the visualisation and quantification of noise impacts using 3D approach and not to improve the accuracy of the calculated noise level on one specific location.

Inputs for the noise computer model implementing SCM1 are noise sources (location and characteristics), noise propagation factors and observation points. This input information was generated using the 3D city model. Fictitious data were used for traffic intensities. It must be noted that noise levels on 3D observation points are calculated in SCM1 by considering 3D distance and direction of the observation points to the source. Consequently SCM1, as other noise methods, implements a 3D approach for noise calculation.

## 19.3.3 Locating 3D observation points

Key issue was to optimally locate the observation points that were used in a second step to produce a continuous noise surface with 2D interpolation (see section 3.4). There are several conditions that prescribe the best location. One condition was that the observation points should be located on the height surface of the terrain since the interpolated noise surface will be draped over the 3D city model in a later stage. Another condition was related to the spatial distribution. In this case in 2D since a 2D spatial interpolation method was used. The decision about the spatial distribution of points for noise simulation is not straightforward. Point density should be sufficiently high to reach adequate accuracy of interpolation results. On the other hand too many points should be avoided in order to considerably reduce computation time of the noise software.

Characteristics of noise propagation can be taken into account in order to optimally distribute points. Noise reduces continuously and logarithmically with distance in absence of obstacles. Furthermore noise reduces discontinuously at obstacles, such as buildings. A previous study showed that point density should be adjusted to these characteristics in order to minimise the error introduced with interpolation (Kluijver and Stoter, 2003). This implies higher density (1 m spacing in the test area) of observation points near the noise sources and buildings and lower density further away from noise sources and buildings (2 m spacing).

Most optimally points should be located at facades of buildings, i.e. with same x,y coordinates but with varying z coordinate. However since 2D interpolation, as applied in this research, can only be used if points are located on different x,y coordinates, points with similar x,y coordinates were simulated by giving them an offset of 0.1 m leaning towards the buildings (see figure 3 (a)). The maximum offset cumulates to about 1m (compare top and bottom of building in figure 3 (a)).

Summarising, there are three types of point densities in the generated observation points data set when only considering x,y coordinates: 1 meter between points near roads and buildings; 2 meter between points further away from roads and buildings (where noise variance is low) and 0.1 meter between points at facades of building (to facilitate 2D interpolation of the observation points). In vertical direction (considering z coordinate) all points are located at 2 meter distance from each other.

The total number of points generated (in ArcScene) was around 16,800, see figure 3 (b). The resulting point density is rather high for noise computer models, although appropriate for the densely built study area. Calculation

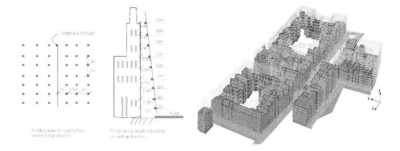

**Fig. 19.3** (a) spacing of points in horizontal and vertical direction on facades of building, (b) observation points to be used as basis for interpolated noise surface

time was acceptable because of the relatively small size of the study area. Further optimisation of the density of observation points was therefore not necessary but would be necessary in case of a larger area.

### 19.3.4 Generating a noise surface by interpolating noise levels

The noise surface was built by interpolating noise levels at known 3D observation points, only taking x,y coordinates of points into account, i.e. using 2D interpolation. It was considered to use 2.5D and even 3D spatial interpolation, however currently available techniques have limitations, e.g. only a limited number of input data was allowed (Ledoux and Christopher, 2004); it was not able to produce a noise surface from the interpolated 3D solid model (FIELDS, 2006); the solid model algorithm is implemented to interpolate attribute values from depth intervals of strata such as soil, rock, or ground water and not from individual points (GOCAD, EVS, Rockworks); or a closed surface was formed which was not appropriate for the city model in our study (Boissonat and Flötotto, 2002, implemented in CGAL).

The 3D analyst tool of ArcScene was used to generate the noise surface with 2D interpolation. There is no standard spatial interpolation method that can deal with the logarithmically reduction of noise levels with distance. However there were some prerequisites that motivated the selection of Triangular Irregular Network (TIN) for the interpolation. If noise levels on facades are calculated with noise levels above the road or above buildings errors are introduced due to the high variance in noise level on facades of building. This high variance in 2D (i.e. noise changes quickly with x,y distance) is a result of the effect that noise levels are calculated in the noise computer model based on 3D distance. When projecting these observation points in 2D, sudden change in noise levels occur on a relatively short distance. TIN only takes the closest three observation points into account (distance measured in 2D) when calculating noise level at any unknown point, avoiding that noise values above roads and buildings contribute to interpolated values on facades. Therefore TIN was selected as interpolation method. To proof the assumption that other spatial interpolation methods are less suitable also experiments were done with Inverse Distance Weighting (IDW), Natural Neighbourhood and Kriging. For an explanation of the principles and advantages and disadvantages of each interpolation method, see Watson (1992).

### 19.3.5 3D noise impact study

After the interpolation the noise surface is draped over the 3D city model to generate the 3D noise map. This is done in ArcScene. The noise surface is made transparent so that the buildings can be seen through the surface. Also contours are generated which are extended towards 3D by draping them over the city height model. Using this methodology the 3D noise map is easy to construct and suited for quantitative analyses such as for finding noise hot spots, calculating area that are effected by high noise levels, and estimating population annoyed by noise. To improve the reality look virtual reality functionalities could have been applied such as textures (see also Wing and Kwong, 2006). One should however realise that the more realistic the visualisation looks the more accurate decision makers expect it to be. This expected accuracy does not always coincide with the intended accuracy of the noise representation. The 3D noise map is used as input to quantify noise impact in 3D using basic spatial analysis tools in ArcScene.

## 19.4 Results

In this section the results of the 3D noise map are presented. Section 4.1 presents the results of the noise calculation. The 3D noise map is assessed in section 4.2 and the improvements of the 3D noise map compared to the 2D approach are presented in section 4.3 by applying it to some aspects of a noise impact study.

### 19.4.1 Accuracy of noise calculation

The offset of 0.1 m for points on facades introduces an error, which was analysed to see if the error is acceptable. At a distance of 5 m from the centre of the road the error is $\pm 0.7$ dB(A) which is only minor compared to the calculated difference of ±10 dB(A) between noise levels at the top and bottom of the building (see figure 4).

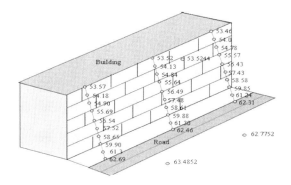

**Fig. 19.4** Observation points near buildings with computed noise levels

Furthermore this difference is not audible for human beings. For general noise impact studies, where the selected SCM1 method is designed for, this inaccuracy is acceptable. However when complying with noise limits a minor difference could be relevant and other more accurate calculation methods should be applied. As stated before an accurate calculation method is not the aim of this research.

## 19.4.2 Results of the 3D noise map

To assess the accuracy of the interpolated noise surface, cross validation was applied (Davis, 1987). Observation points were removed before interpolation and interpolated values on these points were compared to values calculated by the noise software. This yielded a mean error of 0.3 dB(A) and was therefore also acceptable. This implies that 2D interpolation can be used for building 3D noise map.

The result of the 3D noise map integrating the noise surface using the TIN interpolation method and the 3D city model is shown in figure 5. From figure 5 it can be seen that the 3D noise map is able to properly process and visualise the 3D output of noise calculation software. The 3D representation offers insight into the impact of noise at any particular height on the terrain surface and on facades of buildings: high noise levels occur on road surfaces and low noise levels occur on top and backside of buildings.

2D noise contours (interval of 1 dB(A)) were generated and projected on the city model to extend them towards 3D. The IDW noise surface was generated with power 2, search radius 2 m, and cell size 0.1 m. The cell size of

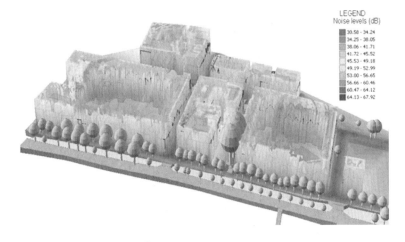

**Fig. 19.5** 3D noise map obtained with TIN interpolation

0.1 m was chosen to cover the high point density on facades of buildings when only considering x,y coordinates. Contours from TIN interpolation method and IDW interpolation method at facades of buildings (location with highest variance in noise levels) are shown in figure 6 (a) respectively 6 (b). Noise observation points are shown as well.

As expected IDW interpolation produces irregular contours. This does not reflect the real noise behavior since noise levels reduce similarly with decreasing distance if no other variables, such as noise barriers and absorption properties, are met or changed. On the contrary TIN contours do show straight contours, as can be seen in figure 6 (a). Other interpolation methods (Natural Neighbourhood and Kriging) yielded similar results as IDW.

These irregular IDW-contours were found on locations were noise reduces very fast with distance (when only considering x,y) represented by high point density, as is the case on facades. IDW (as the other alternative methods) is based on one search radius for the whole area, by which values on roads and above buildings are used for calculating values at facades. This causes the faulty effects on locations where noise reduces very quickly with distance, as shown in figure 6 (b) and 7 (b).

TIN takes only three observation points into account when calculating a value at an unknown location. It is appropriate for situations with high noise variance represented by high point density because it generates more triangles with relative small areas at these locations (see figure 7 (a)). As a consequence noise levels on facades are calculated based on observation points on facades only and it is avoided that observation points on roads and above buildings contribute to interpolated values on facades. This explains why TIN is the most optimal method for generating the 3D contours for the 3D noise map, using 2D interpolation applied to 3D observation points.

# 19 Towards 3D environmental impact studies: example of noise

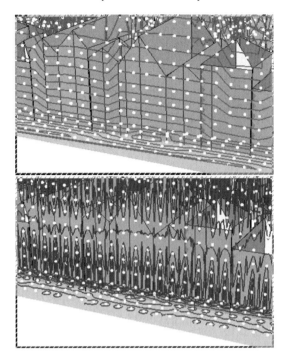

**Fig. 19.6** Noise contours of two interpolation methods projected on 3D city model (top: contours generated from TIN interpolation method, bottom: contours generated from IDW interpolation method)

## 19.4.3 Results of applying 3D noise map to noise impact study

The results of the 3D noise map with respect to improved 3D functionalities were tested by applying it to different aspects of noise impact studies. The aspects that are addressed here are:

- Assessing reduction of noise levels by noise barriers
- Estimation of population annoyed by high noise levels

### Assessing reduction of noise levels by noise barriers

A 3D noise map was produced with the methodology described in section 3, using information on seven fictious noise barriers in order to assess the 3D impact of several characteristics of noise barriers. Figure 8 shows the impact of the different noise barriers varying in height, width and distance from the

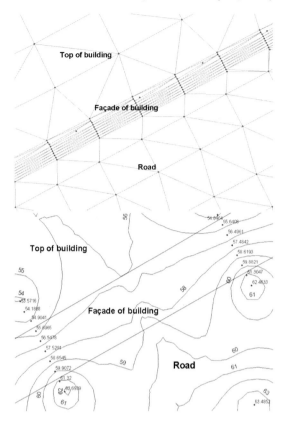

**Fig. 19.7** (a): TIN can deal very well the spatial irregularly distribution of observation points on facades of buildings, (b) results of noise contours based on IDW (viewpoint is from above in both examples)

road. Details of the different barriers are shown in the bottom left of figure 8.

The first three barriers (a), (b), (c) are of height 3 m and located at a distance of 3 m, 6 m, respectively 9 m from the road. As can be seen in figure 8, the effect of the barrier reduces when the distance of the barrier to the road increases. Furthermore it shows that there is no effect of the barriers on higher floors.

The next three barriers (d), (e), (f) are of different heights (2 m, 3 m, and respectively 4 m) and located at equal distance of 5 m from the road. Figure 8 shows that noise reduction due to the noise barriers increases when the height of the barrier increases. Still no effect of the noise barrier is found at higher floors. Barrier (g) is located where there is no building. Barrier (g) shows therefore the effect on the ground surface.

19 Towards 3D environmental impact studies: example of noise

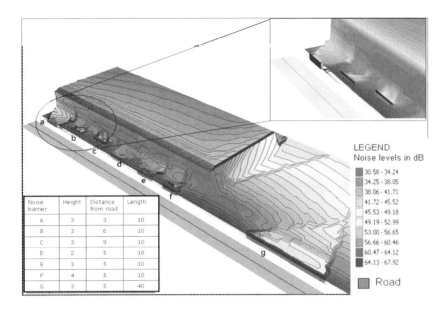

**Fig. 19.8** Effect of noise barriers represented in 3D noise map

Table 1 shows the noise impact on the facade of the building just behind barrier (a) with height 3 m and located 3 m from the road. From this table it can be seen how the noise barrier reduces noise levels at different heights. This case study shows that a noise barrier should be high enough and sufficiently close to the road to have a reducing effect for all floors. A 2D map representing noise levels at only one height (e.g. 4 m) cannot provide this information. In case of 2D map, noise levels on lower floors could be overestimated and on higher floors underestimated.

| Height above the ground surface (m) | Without noise barrier (dB(A)) | With noise barrier dB(A) | Effect dB(A) |
|---|---|---|---|
| 2 | 59 | 38 | -21 |
| 4 | 58 | 41 | -17 |
| 6 | 56 | 44 | -12 |
| 8 | 55 | 46 | -9 |
| 10 | 53 | 48 | -5 |
| 12 | 51 | 51 | -0 |

**Table 19.1** Noise levels at different heights on facade of building with and without noise barrier (a) as indicated in figure 8

**Estimation of population annoyed by high noise levels**

For estimating population annoyed by noise, the considered threshold of annoyance is 55 dB(A). This noise level is considered as hazardous by . Table 2 shows the comparison of annoyed population estimation using 3D noise map (taking floors of buildings into account) and using a 2D approach at a height of 4m. For the 3D noise map calculation, population numbers were assigned to 3D points and for the 2D approach, population numbers were assigned to 2D points. These points are selected in such a way that they coincide with centres of living units. Based on the number of population points covered by surfaces with high noise levels in 3D case and by areas with high noise levels in 2D case, population annoyed by noise was estimated.

The results in table 2 show that annoyed population calculated using the 3D noise map is considerably less than using the 2D noise map. This is because in case of 2D assessment all floors are considered to be effected by the same noise level, even though the noise levels are calculated for one specific height (i.e. 4 m). In reality there are several floors above 4 m and only one floor below 4 m. Floors above 4 m are effected by lower noise levels than at 4 m (when there is no noise barrier) since noise levels decrease with distance from the road and therefore with height. Consequently the 2D assessment results in an overestimation of the number of annoyed people. From this case study it can be concluded that the 3D noise map analysis provides a much more accurate estimation of annoyed population than the 2D noise map analysis.

| 2D noise map | 11000 |
|---|---|
| 3D noise map | 7200 |

**Table 19.2** Population annoyed by noise levels > 55 dB(A)

## 19.5 Conclusions

In this paper a research was presented that shows how impact studies of continuous spatial phenomena, such as air pollution, soil pollution and noise, can be improved by applying a 3D approach to the output of software that predicts the spatial phenomenon on 3D observation points. In the study noise is used as example.

In section 2 the demand for a 3D approach of noise impact studies was described based on a review of existing approaches. In section 3 a new approach was proposed in order to appropriately address the 3D aspect of noise when visualising and assessing impact of noise. The proposed concept was proven by applying it to a sample noise impact study in section 4. From these experiments it can be concluded that the 3D noise map offers insight into the 3D noise situation where 2D noise maps have limitations. Current noise simulation software already has a 3D approach in predicting noise levels. The 3D noise map provides the possibility to actually process and visualise this 3D information. As a result more accurate assessment of noise impact is possible in particular when different floors of a building close to noise sources or noise barriers are considered, which is specifically relevant in urban environment. Since a 3D noise map is easy to 'understand' they are also beneficial for communication purposes with the public in city planning processes.

The research presented in this paper showed that a 3D noise map can be generated by integration of a 3D city model with a noise surface. For producing the noise surface, TIN interpolation was applied to 3D observation points. The noise surface and generated noise contours were draped over the 3D city height model to obtain a 3D noise map. From the results it can be concluded that this approach serves its purposes also with respect to the accuracy required by the specific application, which is visualising and quantifying overall noise impact in 3D where 3D effects are relevant.

Since this research is specifically aiming at improvement of the overall picture, improving the accuracy of calculated noise levels for specific locations is not the main concern. When this accuracy is concerned, e.g. when complying to noise limits, future research can focus on how to make use of developments in the area of 2.5D and 3D interpolation. In that case however the whole process of a noise application - from data collection and prediction to applying it for a specific purpose - should be taken into account. Accuracy is influenced at each operation such as during generation of observation points, spacing of points, noise calculation, spatial interpolation and analysis. Ambitions for further improving accuracy are obviously supported by the authors but not without emphasising the need for error assessment and presentation of the uncertainties in all phases of the process, also with respect to the purposes the study has to serve. It is inappropriate to put effort in obtaining accurate noise levels if noise levels are only used for visual presentation and/or if noise levels are combined with less accurate spatial information in order to quantify overall noise impacts.

The methodology presented in this paper can be applied to other continuous spatial phenomena as well so that it meets the more general problem of how to represent 3D aspects of environmental impact studies.

## Acknowledgements

We are grateful to Professor George Vosselman since he provided us with the 3D city model and to MSc. Monika Kuffer since she gave very useful comments for this paper.

# References

Association for Information Systems, 2007, http://www.isworld.org/researchdesign/drisISworld.htm

Boissonnat, J. and J. Flötotto, A local coordinate system on a surface. Proceedings 7th ACM Symposium on Solid Modeling and Applications, 2002, p. 116-126.

Bruel, P.V. and V. Kjaer, 2002, Environmental Noise, Denmark. http://www.bksv.com/pdf/Environmental\%20Noise\%20Booklet.pdf, Access date 25-09-06.

Butler, D., 2004, Noise management: Sound and vision. Nature, 427(6974): 480-482.

CSTB, 2007, http://www.cadcorp.com/press_releases/2006_pr9.htm

Davis, B.M., 1987, Uses and abuses of cross-validation in geostatistics. Math. Geol.(19, pages 241-248).

FIELDS, 2006, U.S.EPA, FIELDS Rapid Assessment Tools. http://www.epa.gov/region5fields/htm/software.htm, Access Date: 30-12-06

GRASS, 2007, http://grass.itc.it/applications/index.php, Access Date: 15-9-07

Kluijver de Henk and Stoter, J., 2003, Noise mapping and GIS: optimising quality and efficiency of noise effect studies. Computers, Environment and Urban Systems, 27(1): 85-102. http://www.sciencedirect.com/science/article/B6V9K-44GHTN5-3/2/f75bca60cefff030ea2e379d5be56c4b

Kurakula, V., 2007, A GIS-Based Approach for 3D Noise Modelling Using 3D City Models, MSc thesis, ITC, Enschede, The Netherlands, GEM thesis number: 2005-04

Ledoux, H., and M. Christopher, 2004, An efficient natural neighbour interpolation algorithm for geoscientific modelling, in: Developments in Spatial Data Handling - 11th International Symposium on Spatial Data Handling, Springer, editor: Fisher, Peter F., pages 97-108

Silvia R, Ricardo H, Luis C J, 2003, Evaluation and prediction of noise pollution levels in urban areas of Cdiz (Spain). Acoustical Society of America Journal: Volume 114, pp-2439-2439.

Vosselman, G., Kessels, P. and Gorte, B., 2005, The utilisation of airborne laser scanning for mapping. International Journal of Applied Earth Observation and Geoinformation, 6(3-4): 177-186. http://www.sciencedirect.com/science/article/B6X2F-4F2VS7P-1/2/bf5c1ceeb35b2919a497a7fea2529864

VROM, 1999, Handleiding Meten en Rekenen Industrielawaai (calculation and measurement methods industry noise). Ministry of Housing, Spatial Development and the Environment. (In Dutch)

VROM, 2004, Regeling omgevingslawaai, bijlage 3 Karteringsvoorschriften weg- en railverkeerslawaai (calculation method for producing strategic noise maps). Ministry of Housing, Spatial Development and the Environment. (In Dutch)

Watson, D. F. Contouring: A guide to the analysis and display of spatial data. Pergamon Press (1992).

Wing, K., Kwong, 2006, Visualisation of Complex Noise Environment by Virtual Reality Technologies, Environment Protection Department (EPD), Hong Kong. http://www.science.gov.hk/paper/EPD\_CWLaw.pdf, Access Date: 30-01-07.

# Chapter 20
# The Kinetic 3D Voronoi Diagram: A Tool for Simulating Environmental Processes

Hugo Ledoux

## Abstract

Simulations of environmental processes are usually modelled by partial differential equations that are approximated with numerical methods, based on regular grids. An attractive alternative for simulating a fluid flow is the Free-Lagrange Method (FLM). In this paper, I discuss the use of the FLM—based on the Voronoi diagram (VD)—for the modelling of fluid flow in three dimensions (e.g. the movement of underground water or of pollution plumes in the ocean). Such a technique requires the *kinetic* three-dimensional VD, which is a VD for which the points are allowed to move freely in space. I present a new algorithm for the movement of points in a three-dimensional VD, and show that it can be relatively easy to implement as it is the extension of a simple two-dimensional algorithm.

## 20.1 Introduction

The integration of simulation models and geographical information systems (GISs) is a major source of problems because GISs have not been designed for handling time, and even less for handling processes which involves continual movements [50]. Full integration is almost unheard of, and the two are usually simply 'linked', i.e. a GIS is used as a pre-processing tool (e.g. to prepare a dataset or convert formats) and as a post-processing tool (e.g. visualisation and further spatial analysis), althoug the simulation itself is done using a specialised tool. Many argue for 'full integration', as both tools would ultimately gain [19, 43]. The simulation of some environmental processes, e.g. the track-

---

Delft University of Technology (OTB—section GIS Technology)
Jaffalaan 9, 2628 BX Delft, the Netherlands
h.ledoux@tudelft.nl

ing of pollution plumes in the ocean or dispersion models in meteorology, is even more problematic because these phenomena are three-dimensional by nature, and three-dimensional GISs are still their infancy (see Zlatanova et al. [52] and Raper [44] for discussions). Simply to be integrated into a GIS, the results of environmental simulations must often be 'sliced' into several 2D datasets [8, 41].

Many disciplines require simulations of real-world processes, and the methods they use obviously differ. Simulations in geoscience or in engineering have usually been based on partial differential equations (PDEs) that describe the behaviour of some fluid (e.g. the fluid flow around an aircraft, or the movement of underground water) [10]. PDEs are solved, or rather approximated, by mainly two numerical methods: the *finite difference method* (FDM), which was developed for regular tessellations; and the *finite element method* (FEM) [48], which is possible on any tessellations, regular or irregular. The solution of a PDE is obtained by first approximating the behaviour of the process studied for each element of the tessellation, and the final solution is obtained by accumulating all the results. The FDM performed on grids, as used for instance by systems for weather forecasting, is well-known, efficient and mostly accurate. However, the use of grids can sometimes lead to unreliable results [5], and some other technical problems also arise (for instance the curvature of the Earth is problematic for large datasets).

For the modelling of fluid flow, an alternative approach to using a fixed/rigid tessellation is the *Free-Lagrange method* (FLM) [20]. With this method, the flow is approximated by a set of discrete points (called particles) that are allowed to move freely and interact, with each particle having a mass and a velocity. A tessellation of the space is still required, although this will be modified as the particles are moving. As briefly explained in Sect. 20.2, the Voronoi diagram 'naturally' tessellates the space based on a set of points, and has therefore been used (see Erlebacher [16] for instance). Although the FLM can theoretically better represent a physical process [20], it is handicapped by the many difficulties encountered when implementing it. As Mostafavi and Gold [40] note, the adjacency relationships of the cells of the tessellation must be kept consistent at all times, and there must be a way to model time, because fixed time steps can comprise the adjacency of the tessellation. Indeed, earlier implementations of the FLM were very slow as the adjacency relationships between cells had to be rebuilt at each step of the process. With Mostafavi and Gold's solution, the *kinetic* VD, all the topological events are managed locally, and the time steps that were previously used (which could lead to overshoots and unwanted collisions) can be avoided as topological events are used. They further show the advantages of the kinetic VD with the simulation of global tides on the Earth (thus using the VD on a sphere).

In this paper, I extend the work presented in Mostafavi and Gold [40] to three dimensions, and present a novel algorithm for keeping a 3D VD up-to-date as the points defining it are moving over time. In other words, an algorithm for the kinetic VD in three-dimensional space is presented. Such

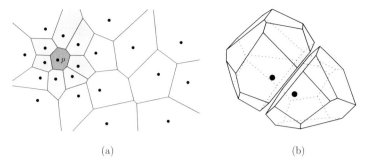

**Fig. 20.1** (**a**) VD of a set of points in the plane. The point $p$ (whose Voronoi cell is dark grey) has seven neighbouring cells (light grey). (**b**) Two Voronoi cells adjacent to each other in $\mathbb{R}^3$ (they share the grey face).

an algorithm is interesting for two reasons: (i) it permits us to perform simulation with the VD (which potentially yields more accurate results); and (ii) it is a step in the direction of integrating simulation models and GIS (that is, if the VD is used as an alternative spatial model to the usual point-line-polygon model, as in the work of Gold [25], Gahegan and Lee [21] and Chen et al. [11]). Because the paper is fairly technical, important concepts related to the VD are first introduced in Sect. 20.2, and then the algorithm itself is presented in Sect. 20.3, along with a literature review of other potential methods. Sections 20.4 and 20.5 discuss the potential applications of the algorithm presented, and also briefly discuss its implementation. Notice that most of the concepts and methods discussed are firstly introduced by describing their two-dimensional counterparts (because readers are often more familiar with these), and that most figures are for the 2D case, as they are much simpler to understand.

## 20.2 Voronoi Diagram & Related Issues

Let $S$ be a set of $n$ points in an $n$-dimensional Euclidean space $\mathbb{R}^n$. The Voronoi cell of a point $p \in S$, defined $\mathscr{V}_p$, is the set of points $x \in \mathbb{R}^n$ that are closer to $p$ than to any other point in $S$. The union of the Voronoi cells of all generating points $p \in S$ form the Voronoi diagram of $S$, defined VD($S$). Fig. 20.1 shows two- and three-dimensional examples. The VD is arguably one of the most important geometric/spatial structures in sciences because it is very simple, and yet is so powerful that it helps in solving many theoretical problems, and also helps in many real-world applications; Aurenhammer [6] and Okabe et al. [42] offer exhaustive surveys.

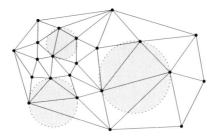

**Fig. 20.2** DT in 2D of the same set of points as in Fig. 20.1(a).

*The Delaunay triangulation.*

The VD is closely related to another structure: the Delaunay triangulation (DT). The DT is popular in 2D in many application domains because it has many useful properties, among others are the fact that the triangles created are as equilateral as possible, and that it can be modified locally. In 3D, the Delaunay *tetrahedralization* is defined by the partitioning of the space into tetrahedra—where the vertices of the triangles are the points in $S$ (generating each Voronoi cell)—that satisfy the *empty circumsphere* test (a sphere is empty is no points are in its interior, but points can lie directly on the sphere). Fig. 20.2 illustrates the idea in 2D.

*Duality.*

The VD and the DT are dual structures, which means that the knowledge of one implies the knowledge of the other one. In other words, if one has only one structure, he can always extract the other one. The concept of duality is important for the construction, the manipulation and the storage of the VD, because all the operations can be performed on its dual, and when needed, the VD extracted. The algorithm in Sect. 20.3 uses this idea.

*General position.*

An important concept when discussing the VD and the DT is that of the position of the points in a set of points. A set $S$ of points is said to be in *general position* when the distribution of its points does not create any ambiguity in the structures derived from the points (e.g. the VD or the DT). For the VD and/or the DT in $\mathbb{R}^d$, the degeneracies, or special cases, occur when $d+1$ points lie on the same hyperplane and/or when $d+2$ points lie on the same ball. For example, in three dimensions, when five or more points in $S$ are *cospherical* there is an ambiguity in the definition of DT($S$): since all the points lie on a sphere, all the tetrahedralizations of the points respect the Delaunay criterion. When four or more points are coplanar in 3D, DT($S$)

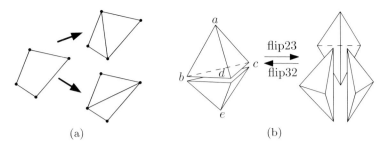

**Fig. 20.3** Flips in (**a**) two dimensions, and (**b**) three dimensions.

and VD($S$) are unique, but problems with the computation of the structures can arise (see for instance Sugihara and Inagaki [49] and Field [18]).

## 20.2.1 Construction of the VD/DT

The construction of a VD, or a DT, is a well-known problem in computational geometry and different efficient algorithms, based on different computational geometry paradigms, are available (see for instance Edelsbrunner and Shah [15], Watson [51] or Cignoni et al. [12]).

*Predicates.*

An important consideration is that for all the construction algorithms, essentially only the following two basic geometric tests (called *predicates*) are required: ORIENT determines if a given point is over, under or lies on a plane defined by three points; and INSPHERE determines if a given point is inside, outside or lies on a sphere defined by four points. Both tests can be reduced to the computation of the determinant of a matrix, see Guibas and Stolfi [30] for more details.

## 20.2.2 Flips

A flip is a local topological operation that modifies the configuration of some adjacent tetrahedra [34, 15]. The concept of flip is valid in any dimensions for triangulations. In 2D, many algorithms to construct and modify a triangulation use the *flip22*, as illustrated in Fig. 3(a). It permits us to transform a triangulation of four points into the only other one possible. In 3D, the same idea can be applied to a set $S = \{a,b,c,d,e\}$ of five points in general position. According to Lawson [34], there are exactly two ways to tetrahedralize such

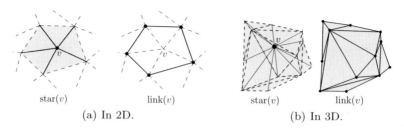

**Fig. 20.4** The star and the link of a vertex $v$.

a polyhedron: either with two or three tetrahedra. As illustrated in Fig. 3(b), in the first case, the two tetrahedra share a triangular face *bcd*, and in the latter case the three tetrahedra all have a common edge *ae*. A *flip23* is the operation that transforms one tetrahedralization of two tetrahedra into another one with three tetrahedra; and a *flip32* is simply the inverse operation. Notice that the numbers refer to the number of triangles/tetrahedra before and after the flip.

It is worth noticing that three-dimensional flips do not always apply to adjacent tetrahedra [32]. For example, in Fig. 3(b), a flip23 is possible on the two adjacent tetrahedra *abcd* and *bcde* if and only if the line *ae* passes through the triangular face *bcd* (which also means that the union of *abcd* and *bcde* is a convex polyhedron). If not, then a flip32 is possible if and only if there exists in the tetrahedralization a third tetrahedron adjacent to both *abcd* and *bcde*.

### 20.2.3 Star, Link and Ears

Three concepts related to triangulations are introduced here, and they will be used in Sect. 20.3 where the proposed algorithm is described.

*Star.*

Let $v$ be a vertex in a $d$-dimensional triangulation. Referring to Fig. 20.4, the star of $v$, denoted star($v$), consists of all the simplices that contain $v$; it forms a star-shaped polytope. For example, in two dimensions, all the triangles and edges incident to $v$ form star($v$), but notice that the edges and vertices disjoint from $v$—but still part of the triangles incident to $v$—are not contained in star($v$). Also, observe that the vertex $v$ itself is part of star($v$), and that a simplex can be part of a star($v$), but not some of its facets.

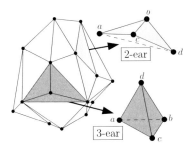

**Fig. 20.5** Perspective view of the outside of a polyhedron. Two adjacent triangular faces (e.g. in light grey) form a 2-ear, and three triangular faces incident to the same vertex (e.g. in dark grey) form a 3-ear.

*Link.*

The set of simplices incident to the simplices forming star($v$), but 'left out' by star($v$), form the link of $v$, denoted link($v$), which is a ($d-1$) triangulation. For example, if $v$ is a vertex in a tetrahedralization, then link($v$) is a two-dimensional triangulation formed by the vertices, edges and triangular faces that are contained by the tetrahedra of star($v$), but are disjoint from $v$.

*Ear.*

Let $P$ be a simplicial polyhedron, i.e. made up of triangular faces. An ear of $P$ is conceptually a potential, or imaginary, tetrahedron that could be used to tetrahedralize $P$. As shown in Fig. 20.5, such a tetrahedron—that does not exist yet—can be constructed by the four vertices spanning either two adjacent faces, or three faces all sharing a vertex (the vertex has a degree of 3). The former ear is denoted a 2-ear, and the latter a 3-ear. A 3-ear is actually formed by three 2-ears overlapping each other. In practice, a 2-ear can be identified by an edge on $P$ because only two faces are incident to it.

A polyhedron $P$ will have many ears, but note that not every ear is a potential tetrahedron to tetrahedralize $P$, as some adjacent faces form a tetrahedron lying outside $P$. Referring again to Fig. 20.5, a 2-ear *abcd* is said to be *valid* if and only if the line segment *ad* is inside $P$; and a 3-ear *abcd* is *valid* if and only if the triangular face *abc* is inside $P$.

## 20.3 Moving Points in a VD/DT

It should first be said that when a point in a VD/DT is continually moving over time and if one is interested in every intermediate state of the VD/DT, it makes no sense to continually insert, delete and reinsert it again somewhere else, as this is a computationally expensive operation. A more efficient option is to literally *move* the point and update the topological relationships of the VD/DT when needed. In other words, instead of using 'discrete updates', 'continuous updates' to the VD/DT are made. Discrete updates are nevertheless an adequate solution for many applications where points move a lot and where the intermediate states are not important (just the start and end states are of interest). For example, De Fabritiis and Coveney [13] use a combination of discrete and continuous updates (depending on the situation) for the simulation of fluids. Similarly, for the simulation of physical processes (where molecules are moving only by very small distances), Guibas and Russel [29] found that continuous updates permit them to update the VD/DT in approximately half to three quarters the time it takes to recompute the entire structure.

The algorithms to maintain a VD/DT of a set $S$ of points up-to-date as one or more points in $S$ are moving are based on the following observation.

**Observation 1** *Let $\mathcal{T}$ be the DT(S) of a set $S$ of points in $\mathbb{R}^d$ in general position. If one point $p$ is moved by a sufficiently small amount so that $S$ stays in general position at all times, then the combinatorial structures of DT(S) (and of VD(S)) will not change (see Fig. 20.6).*

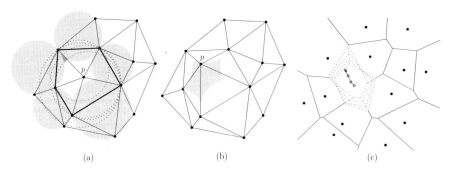

**Fig. 20.6** (a) The combinatorial structure of the DT will not change as long as the vertex $p$ is moved within the white polygon. (b) $p$ has moved but $S$ is still in general position. The combinatorial structure of DT(S) has not changed. Only the location of $p$ has changed, and so have the edges incident to it. (c) Example of the consequences of moving $p$ on the VD.

Notice that $S$ will remain in general position until $p$ is cospherical (lies on the same ball in $\mathbb{R}^d$) with $d+1$ other points in $S$. At the critical moments when the loss of general position arises, the topological structures of VD(S) and DT(S) will be modified; the critical moments are called *topological events*. Observe in Fig. 20.6(c) that the VD where one point is moving looks different as the point is moving, but that the adjacency relationship of the Voronoi cell of the moving point remain the same.

The result of Observation 1 is that in order to move one or more points in $S$, one has to detect when the topological events will arise, and modify VD($S$) and/or DT($S$) consequently.

## 20.3.1 Related Work in Two Dimensions

Observation 1 was used by Roos [45] to analyse the complexity of the movement of points in a two-dimensional VD, and proposed an algorithm to update the VD. Although he discusses the movement of points in a VD, the algorithm he developed is based on the dual (because it is simpler and because the VD can be trivially extracted). When a topological event arises, the update to the DT is made with a flip. He considers that all (or most) of the points in $S$ are moving according to a linear trajectory and that they have a constant velocity. His algorithm starts by computing DT($S$), then all the potential topological events for all the quadrilaterals (every pair of adjacent triangles in DT($S$) is tested) are computed and put in a priority queue (e.g. a balanced search tree), sorted according to the time they will arise. The time is computed by finding the zeros of the INCIRCLE (two-dimensional counterpart of INSPHERE, as briefly discussed in Sect. 20.2.1) developed into a polynomial; in other words, the aim is to find when the four points forming a quadrilateral will become cocircular (if ever). After that, the first topological event is popped from the queue, DT($S$) modified with a flip22, and the queue is updated because the flip has changed locally some triangles. The algorithm continues until there are no topological events left in the queue. The algorithm is efficient as only $\mathcal{O}(\log n)$ is needed for each topological event ($n$ being the number of points in the set).

Similar algorithms have also been proposed, see for instance Bajaj and Bouma [7] and Imai et al. [31]. Moreover, Gold [23] and Gold et al. [26] (with more details in Mostafavi [38] and Mostafavi and Gold [40]) propose a different algorithm and give more implementation details. They focus on the operations necessary to move a single point $p$, and then explain how to have many points move. To detect topological events, only the triangles inside star($p$) and the ones incident to the edges of link($p$) need to be tested, and the changes in the DT are also made with flip22 operations.

## 20.3.2 Related Work in Three Dimensions

The work of Roos [45] has been generalised to three- and higher-dimensional space by Albers et al. [1, 3, 2]. Their work is mostly theoretical as they aim to find upper bounds on the number of topological events when the points are moving according to some trajectories. They state that only the two-dimensional case has been implemented, and they demonstrate that in three dimensions the flip23 and flip32 can be used to update the DT. Gavrilova and Rokne [22] discuss the movement of $d$-dimensional balls (not only points, but balls with defined radii) while the *additively weighted* Voronoi diagram (or Apollonius diagram) is maintained up-to-date with flips; the operations are performed on the dual of the Apollonius diagram. Their algorithm is exactly the same as the one used by Albers et al., but they show how the INSPHERE test must be modified to consider the radius of each ball.

The major impediment to the implementation of the algorithm used by Albers et al. in three dimensions is that, as Gavrilova and Rokne [22] observe, calculating the zeros of the function INSPHERE cannot be done analytically, as is the case for the INCIRCLE function. Indeed, the polynomial for the three-dimensional case has a high degree (8th degree) and iterative numerical solutions must be sought. That results in a much slower implementation, and it could also complicate the update of the DT when the set of points contains degeneracies. On the other hand, Guibas et al. [28] recently proposed a generic framework for handling moving objects. The methods they use for the kinetic 3D VD/DT is theoretically the same as in Roos [45], although they use different methods for finding the zeros of polynomials (INSPHERE) using fixed precision and exact arithmetic, and they claim that 3D VDs/DTs can be updated relatively fast in most cases.

It appears that the computational geometry community is more interested in studying the complexity of the problem than implementing it. To my knowledge, the only reports of implementations are that of Guibas and Russel [29] and Guibas et al. [28], whose algorithm has recently been added to CGAL[1], and some reports in related disciplines where there is a real need. For instance, Ferrez [17] and Schaller and Meyer-Hermann [46] did practical implementations of the algorithm for respectively the simulation of granular materials and cell tissues. Ferrez's algorithm is for spheres in Laguerre space (power distance is used), and thus the regular tetrahedralization is built; this is almost the same as the DT, for only the INSPHERE test has to be modified slightly. Both seem to have missed out several theoretical issues, e.g. they do not consider Observation 1, and use 'time steps' to move a point. In other words, a point is simply moved to a certain location without first verifying if topological events will arise. Flips are performed after each move to restore the Delaunay criterion, and their only constraints is that the combinatorial structures must stay valid between two steps, i.e. a point is not allowed to penetrate another tetrahedron. While this may work for some cases, defining a time step that works for all cases is impossible, and they do not consider the fact that unlike in two dimensions, it is sometimes impossible to flip adjacent tetrahedra. Their solution could therefore not work for every cases, and more importantly, their tetrahedralization does not respect the Delaunay criterion at all times, which could be problematic for some applications.

---

[1] The Computational Geometry Algorithms Library (www.cgal.org).

## 20.3.3 A Flip-based Algorithm

Here I discuss a new algorithm used to move points in $\mathbb{R}^3$ and update the VD/DT when topological events arise. Since the implementation of Albers et al.'s algorithm is intricate, I describe a generalisation to three dimensions of Gold and Mostafavi's method [23, 26, 38, 40]. Unlike Albers et al.'s method where all the pairs of simplices must be tested, the algorithm I present permits the movement of one or only a few points in the set $S$ by using only local information, i.e. if $p$ is moved, only the geometry of the neighbouring tetrahedra of $p$ will be used, and tetrahedra not in the neighbourhood of $p$ or the trajectory do not need to be tested. Moreover, it is not necessary to find the zeros of the function INSPHERE because the topological events are detected by testing the intersections between the circumsphere of neighbouring tetrahedra and the trajectory of $p$.

The different types of tetrahedra that must be considered are first discussed, then the algorithm to move a single point is presented, and finally the movement of several points in $S$ is discussed. The concepts described are direct generalisations of the algorithm of Gold and Mostafavi, and I describe the intricacies that one more dimension brings. The algorithm is based on the same operations that are necessary for constructing a VD, the flips, and is thus conceptually very simple and easy to implement.

### 20.3.3.1 Types of tetrahedra

Three types of tetrahedra, with respect to the moving point $p$, must be defined (see Fig. 20.7(a) for an example in the plane):

Real tetrahedra: are the tetrahedra $\tau_i$ that are incident to the faces of link($p$), but outside star($p$).

Imaginary tetrahedra: are the ears $\sigma_i$ of star($p$), as defined in Section 20.2.3. They are imaginary because they do not exist yet, but some would exist if $p$ was removed or moved somewhere else. Remember that 2-ears and 3-ears can exist.

Behind tetrahedra: are real tetrahedra that are 'behind' $p$ and its trajectory. In theory, they are not mandatory, but in practice they permit us to test fewer tetrahedra (for the intersection with the trajectory), and not to retest tetrahedra that have been previously tested. The criterion for a real tetrahedron $\tau$ to be a behind tetrahedron is if the orthogonal projection of the centre of its circumsphere, denoted sphere($\tau$), onto the trajectory falls behind $p$, see Fig. 20.7(a).

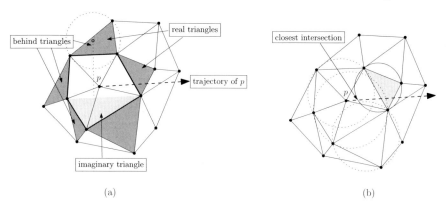

**Fig. 20.7** (a) The different types of triangles needed to move the vertex $p$ along the trajectory. Real triangles are the dark shaded ones, and one example of an imaginary triangle is light shaded. Notice that a behind triangle is always also a real triangle. (b) $p$ must be moved to the closest intersection of a circumcircle along the trajectory. The triangle having the closest intersection is the shaded triangle (a real triangle).

#### 20.3.3.2 The algorithm

The general idea of the algorithm is that, given the moving point $p$ and its final destination $x$, we must move $p$ step-by-step to the closest topological event, perform a flip, and then do these two operations again until $p$ reaches the location $x$. As shown in Fig. 20.7(b), the closest topological event is the location along the trajectory (the line segment $\overline{px}$) where the intersection between $\overline{px}$ and the circumspheres of the real tetrahedra of $p$ and the imaginary tetrahedra of $p$ (only the valid ears are tested) is the closest. Observe that there are two possible cases (this is illustrated in Fig. 20.8):

(1)    $p$ is 'moving in' the circumsphere of a real tetrahedron;
(2)    $p$ is 'moving out' of the circumsphere of an imaginary tetrahedron.

The new algorithm I present, MOVEONEPOINT (Fig. 20.3.3.2), is for moving a single point in a set $S$ (while all the other ones are fixed). It is assumed that $S$ is in general position. MOVEONEPOINT start by initialising a list, denoted $B$, containing all the behind tetrahedra of $p$. $B$ is built by checking if the orthogonal projection of every centre of sphere($\tau_i$) falls 'before' the trajectory $\overline{px}$; if it is then $\tau_i$ is added to $B$. Although it is not a necessity to store lists for the real and imaginary tetrahedra, it may be a good idea to built them at the beginning and simply update them as flips are performed. The lists are not necessary because at each step of the process the real and imaginary tetrahedra can be efficiently retrieved on the fly by simply identifying all the tetrahedra forming star($p$); the data structure used to store the DT should permit that, but these are not discussed here due to space constraints.

In order to get the tetrahedron (real or imaginary) whose circumsphere has the closest intersection, a simple test that computes the intersection between

20 Kinetic 3D Voronoi Diagram for Simulation

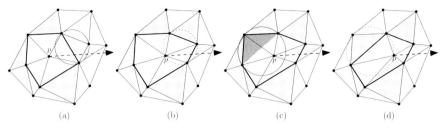

**Fig. 20.8** Two cases are possible when $p$ is moved along a trajectory. At all times, the real triangles are the light shaded ones. (**a**) The closest intersection is with a real triangle. (**b**) $p$ is moved to the circumcircle, a flip22 is performed and the real triangles are updated. (**c**) The next closest intersection is with an imaginary triangle (dark shaded triangle). (**d**) $p$ is moved to the circumcircle, a flip22 is performed and the real triangles are updated. Notice also that the behind triangles are updated, while they were not when $p$ moved in a real triangle.

**Input:** DT($S$) $\mathcal{T}$; the point $p$ to move; final destination $x$
**Output:** $\mathcal{T}$ is modified: $p$ is at location $x$
 1: initialise $B$ {let $B$ be a simple dynamic list}
 2: **while** $p$ is not at location $x$ **do**
 3:    $\tau_l \leftarrow$ get tetra (real or imaginary) having closest intersection with trajectory
 4:    move $p$ to intersection
 5:    **if** $\tau_l$ is a real tetrahedron **then**
 6:      $\tau \leftarrow$ get tetrahedron inside star($p$) adjacent to $\tau_l$
 7:      **if** $\tau \cup \tau_l$ is convex **then**
 8:        flip23($\tau, \tau_l$)
 9:      **else**
10:        flip32($\tau, \tau_1, \tau_2$)
11:      **end if**
12:    **else** {$\tau_l$ is an imaginary tetrahedron}
13:      **if** $\tau_l$ is a 2-ear **then**
14:        remove from $B$ the 2 tetrahedra outside star($p$) incident to $\tau_l$
15:        flip23($\tau_l$)
16:        add $\tau_l$ to $B$ {the ear becomes a behind tetrahedron}
17:      **else** {$\tau_l$ is a 3-ear}
18:        remove from $B$ the 3 tetrahedra outside star($p$) incident to $\tau_l$
19:        flip32($\tau_l$)
20:        add $\tau_l$ to $B$ {the ear becomes a behind tetrahedron}
21:      **end if**
22:    **end if**
23: **end while**

**Fig. 20.9** Algorithm MoveOnePoint($\mathcal{T}, p, x$)

a line segment and a sphere is used. As explained in Bourke [9], the idea is to start with the parametric equation of the line segment $\overline{px}$, such that at $t = 0$ we are at location $p$, and at $t = 1$ we are at location $x$. By substituting the equation of $\overline{px}$ with the equation of a sphere, we can get a quadratic equation that has no solution (no intersection), one solution (sphere is tangent to $\overline{px}$), or two solutions (the line intersects the sphere). We also know where along the line segment the intersection(s) occur(s): if $t < 0$ then it is before $p$, and if $t > 0$ it is after $x$. Observe that when we are dealing with imaginary tetrahedra, the intersection with the highest value of $t$ is to be considered as we are moving out of a sphere, while the smallest value of $t$ is considered for real tetrahedra.

When $S$ is in general position, the rest of the algorithm is straightforward. Indeed, Albers et al. [2] show that when $p$ is moved to the closest topological event, then a flip (either flip23 or flip32) is always possible, i.e. there will not be unflippable cases.

An important point is that when $p$ moves out of an imaginary tetrahedron, the list $B$ must be updated. As shown in Fig. 20.8, the behind tetrahedra are modified when an ear is flipped, but not when $p$ moves in the circumsphere of a real tetrahedron. In three dimensions, the ear $\sigma$ flipped becomes a tetrahedron $\tau$ (spanned by the four vertices of $\sigma$) that must be added to $B$. The two or three neighbours of $\tau$ (depending if $\sigma$ was a 2- or 3-ear) outside star($p$) must also be deleted from $B$.

### 20.3.3.3 Moving several points

Let $S$ be a set of points in $\mathbb{R}^3$, where several points are moving over time. Each point is moving according to a linear trajectory and has a velocity $v$; the time taken to reach its next topological event, called $t_t$, is therefore $t_t = d/v$, where $d$ is the distance to the closest intersection between its trajectory and the circumspheres of the real and imaginary tetrahedra. As explained in Mostafavi and Gold [40], to ensure that at a given time $t$ all the moving points are where they should be, a global time needs to be kept (e.g. since the start of the simulation at $t = 0$). If only $t_t$ was considered, then a case where a point having many topological events close to each others would delay the movement of other points.

Three following types of time must therefore be considered: $t$ is the current time (time elapsed since $t = 0$); $t_t$ is the time to the next topological event; and $t_p$ is the predicted time (global time) for a point to reach the next topological event. The three types are linked:

$$t_p = t + t_t. \tag{20.1}$$

To ensure that the points are moved in the correct order, i.e. in such a way that the combinatorial structure of the VD/DT stays valid, a priority queue containing the $t_p^i$ of every moving point $i \in S$ is kept. At each step, the point $i$ whose $t_p^i$ is the earliest is popped from the queue and processed. After the movement of the point $i$, $t_p^i$ must be recalculated: the updated $t^i$ becomes $t_p^i$, and a new $t_p^i$ must be computed with the new $t_t^i$. The types of time are depicted in Fig. 20.10 for the movement of two points $a$ and $b$. Observe that $a$ went through two topological events before reaching its current position,

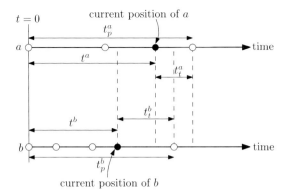

**Fig. 20.10** The three types of time needed for moving several points at the same time. The next movement to be made is for point $b$, since $t_p^b$ is before $t_p^a$. (Figure after Mostafavi and Gold [40])

and that $b$ went through three. The total predicted time for point $b$ is before that of point $a$, so $b$ will be moved before $a$ (although $t_t^a < t_t^b$).

After a point $i$ has been moved, the $t_p^j$ of all the points $j$ that were 'influenced' by the movement must update their $t_t^j$ (and thus their $t_p^j$). The movement of $i$ modifies the shapes of all the tetrahedra incident to $i$, and since these can be the real tetrahedra of other points, the points $j$ that form link($i$), plus the points forming the link of these, must be updated.

## 20.4 Applications

The FLM based on the VD could obviously be used for simulation purposes in three dimensions, provided that we can formalise the physical forces applying to every location in space. Because the movement of points in a VD is rather computationally expensive (when all the points are moving simultaneously), the simulation of atmospheric or oceanographic phenomena on a large scale might not be the most suitable examples right now—we want to obtain the weather forecast for tomorrow, today! A representative example is the simulation of underground water, for instance for a city. Questions such as 'where does the ground water come from?', 'how does it travel?', and 'where do water contaminants come from, and where are they going?', can all be answered if we can adequately model the phenomenon. The work presented in this paper has already been used, by Dr Mostafavi at the Université Laval, Québec City, Canada [39], for the development of a prototype GIS modelling underground water. His team is currently working on defining the governing equations to obtain the vector and the velocity of every point in three dimensions, and it is hoped that the kinetic VD will yield results that are more accurate than the ones with methods currently used.

## 20.5 Discussion

It should be noticed that the simulation of environmental processes is one of many of the potential applications of the kinetic three-dimensional Voronoi diagram. As briefly mentioned, it can be very useful in other disciplines, such as in engineering of physics. It is also potentially a tool for the interactive modelling of datasets [4], and one can think of an application where the user can 'play' at will with a dataset by adding, removing or moving objects.

The description of the algorithm MOVEONEPOINT, as presented in this paper, is not totally complete because the degenerate cases were not covered (it assumed that the set of points is in general position). Fixing an algorithm so that it is robust for all inputs/situations is usually a cumbersome and difficult task, especially for 3D geometric algorithms that are plagued with special cases [47]. The handling of degenerate cases (coplanar/cospherical points, collisions between points, etc.) was excluded due to of space constraints, but details and insights have been published previously by Ledoux [35].

It should also be stated that the algorithm presented offers one solution to the general problem of managing temporal data in a GIS. Indeed, if, as Gold has been advocating for years [24, 27], the VD is used to manage topological relationships between objects in a map, then we obtain a spatial model where insertion, deletion and movement of objects is possible *locally*, without the need to reconstruct from scratch the topological relationships when there is a modification to the map. This means that every operation is reversible. As shown in Gold [25] and Mioc [37], by simply keeping a 'log file' of every operation done it is possible to rebuild each 'topological state' of a VD, at any time. There is no need to keep various 'snapshots' of the data at different times for further analysis: when a representation at a specific time is required, it is reconstructed from the original data and from the log file. The work of Gold [25] and Mioc [37] was made for the VD in 2D, but, as shown in Sect. 20.2, the construction is also possible in 3D, and so is the deletion of vertices [14, 36]. At this moment, only algorithms for the VD of points in 3D have been implemented, but, as is the case for 2D [33], we can envision that in the forseeable future efficient algorithms for 3D VD for a set in which lines and surfaces are present will be available.

## Acknowledgements

I would like to thank Christopher Gold and Maciej Dakowicz for several useful discussions concerning the kinetic Voronoi diagram in two dimensions.

## References

[1] Albers G (1991) *Three-dimensional dynamic Voronoi diagrams* (in German). Ph.D. thesis, Universität Würzburg, Würzburg, Germany.

[2] Albers G, Guibas LJ, Mitchell JSB, and Roos T (1998) Voronoi diagrams of moving points. International Journal of Computational Geometry and Applications, 8:365–380.

[3] Albers G and Roos T (1992) Voronoi diagrams of moving points in higher dimensional spaces. In *Proceedings 3rd Scandinavian Workshop On Algorithm Theory (SWAT'92)*, volume 621 of *Lecture Notes in Computer Science*, pages 399–409. Springler-Verlag, Helsinki, Finland.

[4] Anselin L (1999) Interactive techniques and exploratory spatial data analysis. In PA Longley, MF Goodchild, DJ Maguire, and DW Rhind, editors, *Geographical Information Systems*, pages 253–266. John Wiley & Sons, second edition.

[5] Augenbaum JM (1985) A Lagrangian method for the shallow water equations based on the Voronoi mesh-flows on a rotating sphere. In MJ Fritts, WP Crowley, and HE Trease, editors, *Free-Lagrange method*, volume 238, pages 54–87. Springer-Verlag, Berlin.

[6] Aurenhammer F (1991) Voronoi diagrams: A survey of a fundamental geometric data structure. ACM Computing Surveys, 23(3):345–405.

[7] Bajaj C and Bouma W (1990) Dynamic Voronoi diagrams and Delaunay triangulations. In *Proceedings 2nd Annual Canadian Conference on Computational Geometry*, pages 273–277. Ottawa, Canada.

[8] Bivand R and Lucas A (2000) Integrating models and geographical information systems. In S Openshaw and RJ Abrahardt, editors, *Geocomputation*, pages 331–363. Taylor & Francis, London.

[9] Bourke P (1992) Intersection of a line and a sphere (or circle). `http://astronomy.swin.edu.au/~pbourke/geometry/sphereline/`.

[10] Burrough PA, van Deursen W, and Heuvelink G (1988) Linking spatial process models and GIS: A marriage of convenience or a blossoming partnership? In *Proceedings GIS/LIS '88*, volume 2, pages 598–607. San Antonio, Texas, USA.

[11] Chen J, Zhao R, and Li Z (2004) Voronoi-based $k$-order neighbour relations for spatial analysis. ISPRS Journal of Photogrammetry & Remote Sensing, 59:60–72.

[12] Cignoni P, Montani C, and Scopigno R (1998) DeWall: A fast divide & conquer Delaunay triangulation algorithm in $E^d$. Computer-Aided Design, 30(5):333–341.

[13] De Fabritiis G and Coveney PV (2003) Dynamical geometry for multi-scale dissipative particle dynamics. Computer Physics Communications, 153:209–226.

[14] Devillers O (2002) On deletion in Delaunay triangulations. International Journal of Computational Geometry and Applications, 12(3):193–205.

[15] Edelsbrunner H and Shah NR (1996) Incremental topological flipping works for regular triangulations. Algorithmica, 15:223–241.

[16] Erlebacher G (1985) Finite difference operators on unstructured triangular meshes. In MJ Fritts, WP Crowley, and HE Trease, editors, *Free-Lagrange Method*, volume 238, pages 21–53. Springer-Verlag, Berlin.

[17] Ferrez JA (2001) *Dynamic triangulations for efficient 3D simulation of granular materials*. Ph.D. thesis, Département de Mathématiques, École Polytechnique Fédérale de Lausanne, Switzerland.

[18] Field DA (1986) Implementing Watson's algorithm in three dimensions. In *Proceedings 2nd Annual Symposium on Computational Geometry*, volume 246–259. ACM Press, Yorktown Heights, New York, USA.

[19] Freda K (1993) GIS and environment modeling. In MF Goodchild, BO Parks, and LT Steyaert, editors, *Environmental modeling with GIS*, pages 35–50. Oxford University Press, New York.

[20] Fritts MJ, Crowley WP, and Trease HE (1985) *The Free-Langrange method*, volume 238. Springler-Verlag, Berlin.

[21] Gahegan M and Lee I (2000) Data structures and algorithms to support interactive spatial analysis using dynamic Voronoi diagrams. Computers, Environment and Urban Systems, 24(6):509–537.

[22] Gavrilova ML and Rokne J (2003) Updating the topology of the dynamic Voronoi diagram for spheres in Euclidean $d$-dimensional space. Computer Aided Geometric Design, 20:231–242.

[23] Gold CM (1990) Spatial data structures—the extension from one to two dimensions. In *Mapping and Spatial Modelling for Navigation*, volume 65, pages 11–39. Springer-Verlag, Berlin, Germany.

[24] Gold CM (1991) Problems with handling spatial data—the Voronoi approach. CISM Journal, 45(1):65–80.

[25] Gold CM (1996) An event-driven approach to spatio-temporal mapping. Geomatica, Journal of the Canadian Institute of Geomatics, 50(4):415–424.

[26] Gold CM, Remmele PR, and Roos T (1995) Voronoi diagrams of line segments made easy. In *Proceedings 7th Canadian Conference on Computational Geometry*, pages 223–228. Quebec City, Canada.

[27] Gold CM, Remmele PR, and Roos T (1997) Voronoi methods in GIS. In M van Kreveld, J Nievergelt, T Roos, and P Widmayer, editors, *Algorithmic Foundations of Geographic Information Systems*, volume 1340 of *Lecture Notes in Computer Science*, pages 21–35. Springer-Verlag.

[28] Guibas L, Karaveles M, and Russel D (2004) A Computational Framework for Handling Motion. In *Proceedings 6th Workshop on Algorithm Engineering and Experiments*, pages 129–141. New Orleans, USA.

[29] Guibas L and Russel D (2004) An empirical comparison of techniques for updating Delaunay triangulations. In *Proceedings 20th Annual Symposium on Computational Geometry*, pages 170–179. ACM Press, Brooklyn, New York, USA.

[30] Guibas LJ and Stolfi J (1985) Primitives for the manipulation of general subdivisions and the computation of Voronoi diagrams. ACM Transactions on Graphics, 4:74–123.

[31] Imai K, Sumino S, and Imai H (1989) Geometric fitting of two corresponding sets of points. In *Proceedings 5th Annual Symposium on Computational Geometry*, pages 266–275. ACM Press, Saarbrücken, West Germany.

[32] Joe B (1989) Three-dimensional triangulations from local transformations. SIAM Journal on Scientific and Statistical Computing, 10(4):718–741.

[33] Karavelas MI (2004) A robust and efficient implementation for the segment Voronoi diagram. In *International Symposium on Voronoi Diagrams in Science and Engineering*, pages 51–62. Tokyo, Japan.

[34] Lawson CL (1986) Properties of $n$-dimensional triangulations. Computer Aided Geometric Design, 3:231–246.

[35] Ledoux H (2006) *Modelling three-dimensional fields in geoscience with the Voronoi diagram and its dual*. Ph.D. thesis, School of Computing, University of Glamorgan, Pontypridd, Wales, UK.

[36] Ledoux H, Gold CM, and Baciu G (2005) Flipping to robustly delete a vertex in a Delaunay tetrahedralization. In *Proceedings International Conference on Computational Science and its Applications—ICCSA 2005*, volume 3480 of *Lecture Notes in Computer Science*, pages 737–747. Springer-Verlag, Singapore.

[37] Mioc D (2002) *The Voronoi spatio-temporal data structure*. Ph.D. thesis, Département des Sciences Géomatiques, Université Laval, Québec, Canada.

[38] Mostafavi MA (2001) *Development of a global dynamic data structure*. Ph.D. thesis, Département des Sciences Géomatiques, Université Laval, Québec City, Canada.

[39] Mostafavi MA (2006) Personal communication.

[40] Mostafavi MA and Gold CM (2004) A global kinetic spatial data structure for a marine simulation. International Journal of Geographical Information Science, 18(3):211–228.

[41] Nativi S, Blumenthal MB, Caron J, Domenico B, Habermann T, Hertzmann D, Ho Y, Raskin R, and Weber J (2004) Differences among the data models used by the geographic information systems and atmospheric science communities. In *Proceedings 84th Annual Meeting of the American Meteorological Society*. Seattle, USA.

[42] Okabe A, Boots B, Sugihara K, and Chiu SN (2000) *Spatial tessellations: Concepts and applications of Voronoi diagrams*. John Wiley and Sons, second edition.

[43] Parks BO (1993) The need for integration. In MF Goodchild, BO Parks, and LT Steyaert, editors, *Environmental Modeling with GIS*, pages 31–34. Oxford University Press, New York.

[44] Raper J (2000) *Multidimensional geographic information science*. Taylor & Francis, London.

[45] Roos T (1991) *Dynamic Voronoi diagrams*. Ph.D. thesis, Universität Würzburg, Germany.

[46] Schaller G and Meyer-Hermann M (2004) Kinetic and dynamic Delaunay tetrahedralizations in three dimensions. Computer Physics Communications, 162(1):9–23.

[47] Schirra S (1997) Precision and robustness in geometric computations. In M van Kreveld, J Nievergelt, T Roos, and P Widmayer, editors, *Algorithmic Foundations of Geographic Information Systems*, volume 1340 of *Lecture Notes in Computer Science*, pages 255–287. Springer-Verlag, Berlin.

[48] Strang WG and Fix GJ (1973) *An analysis of the finite element method*. Prentice-Hall, Englewood Cliffs, USA.

[49] Sugihara K and Inagaki H (1995) Why is the 3D Delaunay triangulation difficult to construct? Information Processing Letters, 54:275–280.

[50] Sui DZ and Maggio RC (1999) Integrating GIS with hydrological modeling: Practices, problems, and prospects. Computers, Environment and Urban Systems, 23:33–51.

[51] Watson DF (1981) Computing the $n$-dimensional Delaunay tessellation with application to Voronoi polytopes. Computer Journal, 24(2):167–172.

[52] Zlatanova S, Abdul Rahman A, and Pilouk M (2002) 3D GIS: Current status and perspectives. In *Proceedings Joint Conference on Geo-spatial Theory, Processing and Applications*. Ottawa, Canada.

## Chapter 21
# Techniques for Generalizing Building Geometry of Complex Virtual 3D City Models

Tassilo Glander and Jürgen Döllner

**Abstract**

Comprehensible and effective visualization of complex virtual 3D city models requires an abstraction of city model components to provide different degrees of generalization. This paper discusses generalization techniques that achieve clustering, simplification, aggregation and accentuation of 3D building ensembles. In a preprocessing step, individual building models are clustered into cells defined by and derived from its surrounding infrastructure network such as streets and rivers. If the infrastructure network is organized hierarchically, the granularity of the cells can be varied correspondingly. Three fundamental approaches have been identified, implemented, and analyzed: The first technique uses cell generalization; from a given cell it extrudes a 3D block, whose height is calculated as the weighted average of the contained buildings; as optimization, outliers can be managed separately. The second technique is based on convex-hull generalization, which approximates the contained buildings by creating the convex hull for the building ensemble. The third technique relies on voxelization, which converts the buildings' geometry into a regular 3D raster data representation. Through morphological operations and Gaussian blurring, aggregation and simplification is yielded; polygonal geometry is created through a marching cubes algorithm. The paper closes with conclusions drawn with respect to the characteristics and applicability of the presented generalization techniques for interactive 3D systems based on complex virtual 3D city models.

Hasso-Plattner-Institute at the University of Potsdam
tassilo.glander, doellner@hpi.uni-potsdam.de

## 21.1 Introduction

A virtual 3D city model represents both a technical and conceptual framework to assemble, integrate, present, and use 3D geoinformation as well as for 3D geovisualization [8]. Besides their application in GIS, they provide an effective user interface to complex spatial 3D information in a growing number of IT applications and system such as enterprise systems, navigation, telecommunication, disaster management, simulators, and e-government. In particular, virtual 3D city models facilitate the integration of heterogeneous 2D and 3D geo data, and their interactive visualization offers comprehensible and efficient communication, exploration, and analysis of complex geoinformation. In addition, 3D city models form part of administrative geoinformation databases, services, and infrastructures [7].

A common problem for implementation and usability of virtual 3D city model systems arises from their complexity with respect to the number of individual components, their computer graphics representations, and the rendering resources [2,35]. A typical 3D city model consists of several hundreds of thousand objects, including models of buildings, vegetation, and infrastructure elements. To achieve efficient rendering and interactive frame rates, level-of-detail geometry representations are required that control the polygon count and texture resources [3]. Furthermore, the 3D city model needs to be represented at different generalization levels, for example, to enable context-&-detail views, to enhance comprehensibility of depictions, and to support hierarchical navigation and browsing.

Our investigation addresses a fundamental problem of today's visualization of fine granular, complex 3D city models: Their visualization frequently shows 'noise' and 'flickering' that appears in areas that are far away from the view point because 3D objects are mapped to few pixels or even only to a fraction of a pixel. Similarly, pedestrian or car driver perspectives tend to produce a 'sea of houses' beyond a certain distance, making it impossible for the user to identify these areas. Furthermore, abstract information, such as hierarchy information such as whole districts and quarters of a city, is not explicitly visible if the full model resolution is used for depiction. All these phenomena result because *information density* of the 3D city model is not adjusted.

As a key technique to control information density, generalization both helps to reduce the graphics complexity as well as the cognitive complexity of 3D city models. Similar to maps, features of city models should be visualized at different scales to accomplish different spatial tasks. In addition, '[...] this abstraction and concentration also helps to discern between relevant and irrelevant information: only those objects have to be presented, that are important for the current task – irrelevant information can be suppressed.'[29] Traditionally, generalization of map objects requires experienced cartographers, who are using the human ability to abstract. For 2D vector data as well as for 3D objects, downscaling does not suffice because readability and comprehensibility have to be preserved [13]. These requirements are more important than exactly scaled geometry or its exactly preserved appearance. For this reason, generalization does not only simplify objects but also deforms, drops, aggregate, classifies, or unifies objects and their features. That is, generalization techniques apply operations such as simplification, aggregation, classification, and displacement.

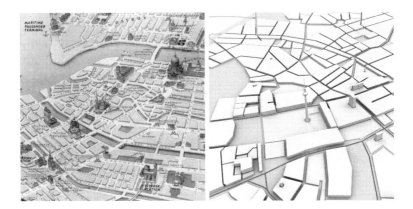

**Fig. 21.1** Example of artistic 3D city map of St. Petersburg (from http://www.escapetravel.spb.ru) (left) and an automatically generalized 3D city model of Berlin, generated by the presented cell-based method (right)

This paper concentrates on generalizing 3D building ensembles as one main category of city model objects. The presented techniques first cluster 3D building models into ensembles according to a given hierarchical street and infrastructure network (e.g., car navigation data) and then automatically generalize the ensembles in the underlying cells defined by the network (Fig. 1). Three different approaches, cell-based, convex-hull-based, and voxel-based generalization algorithms, are outlined and evaluated.

Our work emphasizes the generalization of clustered 3D building models in contrast to various 2D footprint generalization algorithms as well as various 3D building generalization techniques that simplify and abstract single 3D building models. Furthermore, our core goal is to automatically derive generalized building ensembles at various levels of granularity, optimized for using these generalized models in interactive 3D city model visualization in contrast to generalized 2D map production.

## 21.2 Related Work

Meng and Forberg [24] give an overview of state-of-the-art and challenges of 3D building generalization, describing the current scale space of 3D buildings as 'a linear continuum, along which an arbitrary number of milestones can be said to exist referred to as Levels of Detail (LoD)'. The LoD provide different representations of the buildings with different degrees of generalizations. However, there exist different classifications of the LoD, so no standards compared to the scales in cartography have been established yet. Elementary 2D map generalization approaches are described, e.g., by [13][14][30].

Among the first techniques for 3D building generalization, the application of morphological operations on 3D geometry was suggested by Mayer [22][23]: a curvature space simplification has been developed, which detects local curvature and shifts the adjacent polygons accordingly. Both methods apply to specific geometry structures but are costly in terms of processing time. In [11] generalization is based on moving near parallel faces of building geometry to a common plane and merging them if possible. Unfortunately, the algorithm requires orthogonal buildings to work. An automated algorithm for generalizing 3D building geometry is described in [26].

The feature removal algorithm of Ribelles et al. [27] was applied on buildings by Thiemann [33] to create a constructive solid geometry (CSG) representation of the given building geometry. It uses the planes of the building's faces to subdivide the geometry into a body and features, which can be integrated or left out of the generalized representation, depending on the degree of generalization. Similarly, Kada [17] uses a few approximating planes to remodel the building with simpler geometry. Another technique for LoD creation [26] flattens roofs and merges adjacent polyhedra, followed by collapsing facades while respecting visually important walls.

While these 3D generalization approaches focus on single buildings, aggregation of multiple buildings currently is usually referred to as the next important step. Stüber [31] presents a framework for generalization of building models while preserving visual correctness. This approach simplifies buildings based on feature removal and aggregates buildings depending on their visibility. However, automatic aggregation appears to be limited to simple configurations.

Motivated by classical cartography, Anders [1] applies generalization techniques on 2D projections of linear building groups. For each of the main axes' projections, the shapes are aggregated and simplified using a specific generalization technique [28]. The simplified shapes are extruded and intersected to form the generalized building group. The approach achieves aggregation and simplification, however it is limited to linear building groups. Sester [29] suggests a 3D visualization providing simplification, aggregation, displacement and enlargement by extruding the processed ground plans to a certain height. In addition, the height can be used to further emphasize special buildings in for pedestrian navigation.

CityGML [5], a proposed interchange format for virtual 3D city models currently discussed by the Open Geospatial Consortium (OGC), differentiates five consecutive levels of detail (LOD-0 to LOD-4) [18], where objects become more detailed with increasing LOD regarding both geometry and thematic differentiation. CityGML models can contain multiple representations for each object in different LOD simultaneously but does not address the way these LODs are created or transferred. In [6] a continuous level-of-detail concept for individual building models has been introduced, but no automated derivation of generalized building ensembles is considered.

Real-time 3D rendering relies on efficient treatment of polygonal 3D data sets, and it provides a variety of LoD techniques, which can be classified into static and dynamic techniques in general. Static techniques provide discrete LoD representations (e.g., [9][19][12]), whereas dynamic techniques transform polygonal surfaces partially according to the current viewing situation (e.g.,[16]). Common to all techniques, the original high-resolution 3D object is simplified such that its appearance is preserved, but it is not generalized nor do the techniques consider specific semantics or characteristics of the type of the 3D object to be simplified.

## 21.3 Generalization Techniques for 3D Buildings

In this section, three generalization techniques are presented that are primarily based on simplifying and aggregating 3D building geometry. Inspired by classical city plans and bird's eye views (Fig. 1), these techniques achieve generalized 3D building ensembles that facilitate context views of geovirtual 3D environments.

As a control parameter, we introduce the term *degree of generalization* (DoG) in contrast to level of detail (LOD). While LOD usually refers to simplified representations of a single object, DoG describes the level of abstraction, which allow us, for example, to represent a group of neighboring buildings by a single block. In addition, to achieve a simplified and comprehensible visualization, we do not use façade textures that would amplify the visual complexity.

The input data include 3D building models and 2D vector-based, hierarchical street and infrastructure networks (e.g., streets, rivers). In the following examples, part of the Berlin 3D city model (Fig. 2) is used for illustration together with a navigation street data set. The streets are attributed by weights, which differentiate four street types. The weights are used by the techniques to define the streets' width.

### 21.3.1 Simple Cell Generalization

This technique aggregates all buildings within one cell defined through the enclosing network system by extruding the cell's boundary to a certain height creating a prismatic block. To leave enough space for the streets, they are

**Fig. 21.2** Snapshot from the visualization of part of the 3D city model of Berlin containing approx. 60,000 3D building models and approx. 6,500 streets without generalized building models)

buffered with a characteristic width and cut out of the block ground plan before the extrusion stage using Boolean operations. The degree of generalization (DoG) can be controlled by the hierarchy level of the streets considered for parceling.

### 21.3.1.1 Parceling

Each street is defined by a consecutive points and a weight. The Computational Geometry Algorithms Library (CGAL [4]) is used to intersect the curves defined by the streets and to calculate the cell geometry [10][34]. After the parceling, *CGAL* supports aggregated point location queries and returns the hit cell for each point. This is done to cluster the buildings represented by their individual centroids according to shared cells. The results of this stage are cells (Fig. 3) and a mapping from each cell to a set of contained buildings, the *building ensemble*.

### 21.3.1.2 Calculating the Cell Height

The cell height is calculated by the weighted average height of the buildings of a cell. The weight of a building can be defined by its footprint area.

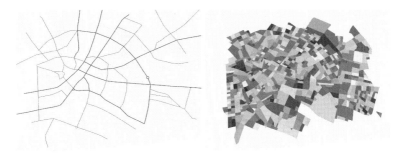

**Fig. 21.3** Street network (left) and calculated cells (right)

Let $height(b_i)$ be the height of building $b_i$ and $area(b_i)$ its area, then the weighted average height $\bar{h}$ of a set of $n$ buildings is

$$\bar{h} = \frac{\sum_{i=1}^{n} height(b_i) \cdot area(b_i)}{\sum_{i=1}^{n} area(b_i)}$$

For low-density cells the calculated value obviously does not reflect the real situation. Instead of dividing by the sum of the buildings' area, the cell area should be used. However, in medium to high-density cells this leads to very small blocks. Therefore, if the ratio of the building area sum to the cell area gets too small, either no block should be created, or the original buildings have to be preserved.

### 21.3.1.3 Subtracting the Network Elements

The cells have to be shrunken to leave enough space for the network elements such as the streets. To accomplish this while supporting different street weights, the streets are buffered before to yield polygons (Fig. 4). Then the adjacent street polygons are subtracted from the cell polygon using 2D Boolean operations.

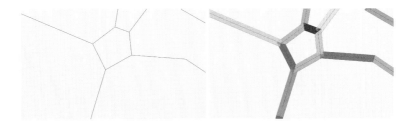

**Fig. 21.4** Vector-based network elements (left) and buffered variant (right)

### 21.3.1.4 Handling of Outlier Buildings

To improve the appearance of generalized building ensembles, outliers can be excluded from the aggregation and placed separately into the generalized version. By outliers we refer to

- **landmark buildings**, i.e. buildings that explicitly have been assigned a higher weight (e.g., landmark buildings, public buildings);
- **outlier buildings**, i.e. buildings that stand out locally as they are considerably higher than their neighborhood.

Outlier building can be determined by comparing the building height with the calculated weighted average height of the cell. To respect the characteristics of the local height distribution, the variance, respectively the standard deviation, is calculated:

$$var(h_1,\ldots,h_n) = \sigma^2 = \frac{1}{n}\sum_{i=1}^{n}(h_i - \overline{h})^2 \text{ with } h_i = height(b_i)$$

The standard deviation $\sigma^2$ can be used more intuitively as it has the same scale and units as the height. Thus, a building is considered as an outlier, if its height is larger than the average height $\overline{h}$ plus $k$-times the standard deviation:

$$is\_outlier(b) = \begin{cases} true & height(b) \overline{h} + k \cdot \sigma \\ false & else \end{cases}$$

With $k = 2$, a reasonable identification of outliers can be done for large scales. A smaller selection in smaller scales can be achieved with bigger values for $k$.

### 21.3.1.5 Results

The final generalized geometry of a cell is given by extruding the cell polygon to its calculated average height. Fig. 5 shows the generalized geometry for different network hierarchy levels.

**Fig. 21.5** Generalized cells for three degrees of generalization (no outlier handling)

The results of the simple cell generalization come close to depictions found in many 2D topographic maps. The cutout of the streets contributes most to the familiar map-look. The height of the cells observed from appropriate viewing angles gives a hint toward the real-world situation and enables relative assessment. As main advantages, this technique requires little processing time and the geometric complexity of the generalized models is low as well. The abstraction of the complex models might pose a way for content providers to offer an overview version of a 3D city model. As a disadvantage, the bare cell blocks do not preserve the appearance. In the case of top views, different heights are barely noticeable even with appropriate shading. With the additional integration of outliers and landmark buildings (Fig. 6), the effectiveness can be improved, as orientation is by far easier especially from low perspectives.

**Fig. 21.6** Generalized cells with outlier handling (left), which are particularly important for the recognition of a city's panorama (right)

## 21.3.2 Convex-Hull-Based Generalization

This generalization technique achieves closer representations of the original buildings of an ensemble using the 3D *convex hull* as an approximation of contained buildings. Since for maps an exact representation is not the primary concern, the convex-hull operation is applied as a mean to simplify and to aggregate.

Compared to the simple cell generalization, buildings are merged to a geometry that reflects the original height distribution in a more sensitive way, as the highest building inside a block creates a peak in the hull.

### 21.3.2.1 Implementation

For each cell the geometry of the contained buildings is extracted and the points are fed into the convex hull calculation. For convex hull computation, a number of libraries exist, including CGAL and qHull [25]. The result is returned by a set of polygons representing the hull, which is finally used as the generalized cell geometry.

### 21.3.2.2 Results

The results –for typical large city– show an 'organic look' (Fig. 7) as the convexity generally induces smoother structures. This is against the principle of visual correctness and especially does not show the typical characteristics of buildings orthogonal, parallel and sharp-edged structures. But, comprehensibility is the main task of maps and clearly the hull hides many details while preserving in a way the height characteristics. A more severe problem is the convexity when a cell is concave. In this case, the generalized block overlaps its cell possibly leading to intersections with other geometry and damages the appearance of the network elements. Additionally, the landmark visualization presented before cannot be integrated in a straightforward way.

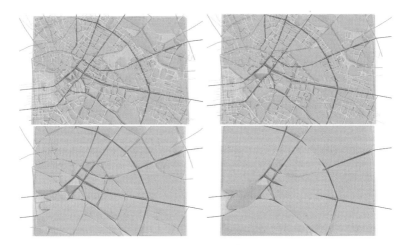

**Fig. 21.7** Examples of the convex hull technique, shown at 4 different levels of generalization

As a solution to these problems and to provide for a finer selection of the degree of generalization, clusters based on neighborhood could be created and used as input for the convex hull calculation instead of the mere cells. This would allow us to use the parameter of the minimal cluster distance to define the DoG, which is currently limited to the levels of the network's hierarchy.

### 21.3.3 Voxel-Based Generalization

This generalization technique applies raster data filter operations to 3D raster data gained from the buildings' geometry [15]. To do so, the geometry is sampled within a 3D grid. Then, morphological opening and Gaussian blurring is performed. To convert the 3D grid to geometry, we apply the marching cubes algorithm to the grid. In addition, the raster data is processed with morphological opening operations to perform an aggregation.

As stated above, morphological operations have been applied to vector data representations before [22][23]. However, we also wanted to experiment with other raster data filter operations like Gaussian blurring.

#### 21.3.3.1 Voxelization

The first step transforms the geometry from vector space to 3D grid space. This is done through a regular grid that is laid over the geometry and samples it at equidistant points. The grid is set up as follows: The resolution can be given as the ratio $res = \frac{realworldunits}{gridunits}$. Thus, the dimensions of the grid **dim** can be calculated by taking the geometry's extent represented through the bounding box's diagonal $d_{bbox}$, and dividing it by the resolution:

$$dim = \begin{pmatrix} dim_x \\ dim_y \\ dim_z \end{pmatrix} = \left[ \frac{1}{res} \cdot d_{bbox} \right] + 2 \cdot \begin{pmatrix} pad \\ pad \\ pad \end{pmatrix} \text{ with } pad \in N$$

To support raster data operations that expand the extent of the original geometry, a padding *pad* specified in grid units is inserted on each axis' ends. For example a dilation applied to the raster data may enlarge connected structures. The grid's origin is set as the lower left front point of the geometry's bounding box minus the padding. For each grid cell, the object space is sampled to be either inside (1) or outside (0) a solid (building). This leads to a three dimensional binary image.

For a better sampling quality, a box filter can be used. This was done experimentally in this work: For each voxel, not one point is sampled in the original geometry, but 8 points of a box centered at the point. The average of these points is then assigned to the grid cell, which leads to a gray scale image. Currently, the filtering is implemented in a simple way and, therefore, takes 8 times the processing time. The benefits can be seen in Fig 8, where the appearance is smoothed and less artifacts occur if a filtered sampling is performed.

As a simplification of the current implementation, only prismatic building shapes are assumed and thus a point considered inside the building, if it lies within the ground plan polygon and within the building's height. After all grid cells have been set, the grid is a voxel representation of the original

**Fig. 21.8** Voxelization: original 3D building ensemble (left), result of voxelization wit a resolution of 2m (middle), and result of filtered voxelization (right))

geometry. However, the rasterization naturally implies a degradation of the original data, which leads to alignment artifacts when transformed back to a polygonal model. For the images, the original building geometry in the example (Fig. 8) has been sampled with a resolution of 2 meters of the real world geometry reflected one grid unit. Even this coarse resolution leads to a grid size of $156 \times 56 \times 170 = 1,485,120$ grid points.

#### 21.3.3.2 Raster Data Operations

Many filter operations have been developed for use on raster data images. Typically, filters to remove noise and to smooth images rely on morphological operations. In [11][23], morphological operations have been used as vector space operations to aggregate and simplify building geometry. Since after voxelization, the building geometry has been transformed to 3D raster data representation, these operations can now be applied directly.

The elementary operations can be defined as follows: The raster data is the 3D input image, where for each element (voxel) a structuring element $B$ is applied. The structuring element in this case is a cube of $3 \times 3 \times 3$ units. It is moved over the input image. For each voxel, the structuring element is compared with the input image (source grid) and the output voxel is determined as follows:

- **Erosion**: A grid cell in the target grid is set to 1, if *all* voxels of the structuring element $B$ can also be found in the source grid. Otherwise it is set to 0. This is done for the whole grid.
- **Dilation**: A grid cell in the target grid is set to 1, if *one* voxel of the structuring element $B$ can be found in the source grid. Otherwise it is set to 0. This is done for the whole grid.

Erosion and dilation are elementary operations, which can be used to realize morphological opening and closing. Opening is achieved by applying dilation followed by erosion to the image; it leads to an aggregation of near structures. Closing is achieved by applying erosion followed by dilation of the image; it is useful to eliminate small elements.

However, the effect on the raster data is determined by the grid's resolution, which sets the size of one grid unit. Thus, opening and closing are always parameterized with multiples of one grid unit. In addition, the current implementation only supports binary images, i.e., no filtered voxelization can be applied before. Fig. 9 shows a series of morphological operations applied to our test models.

Apart from morphological operations, there are also other filters such as tent, cubic, or quartic filters as well as the Gaussian blur filter. Its usage on the rasterized building geometry has two benefits: First, a Gaussian blur as a low-pass filter further eliminates small features of the geometry that are still in the raster data. Second, the alignment structures introduced by the sampling are dampened. Generally, sharp edges and the surface are smoothed. Fig. 10 shows how the smoothed geometry looks after performing morphological opening.

To rely on a robust implementation of the raster data operations, the *nrrd* library from the *teem* project [32] was chosen. The *nrrd* library is accessed through the command-line tool *unu*, which works on a simple file format. *NRRD* (for 'nearly raw raster data'). It supports a wide range of operations of which resampling is used for this work. With the aim to create smoother surfaces with less alignment artifacts, the Gaussian blur is applied for the reasons mentioned.

### 21.3.3.3 Marching Cubes

In the final step, the processed raster data is transferred to polygonal representation. For this, we extract isosurfaces, i.e. surfaces with a common isovalue everywhere on the surface, producing a surface from the samples of a scalar field defined as a mapping and a given isovalue, setting the threshold to separate between inside and outside [21]. To accomplish this, a freely available marching cubes implementation from [20] is used to create the geometric model. Note that all images presented in this section including the intermediate steps (Fig. 8, 9, 10) are created after applying the Marching Cubes to get a renderable polygonal representation.

### 21.3.3.4 Results

The workflow of this technique currently prohibits a completely automatic handling since the parameters (e.g., grid resolution, morphological operations step-width, Gaussian blur factor) have been set manually in our test model.

The result is characterized by its 'bubble look', lacking a typical city model look. Through the blurring, soft shapes are introduced, while without filtering or blurring the alignment steps are clearly visible. To control the visualization, the resolution of the grid can be adapted, and the offset used in the opening operation can be varied. Finally, also the Gaussian blurring could be used

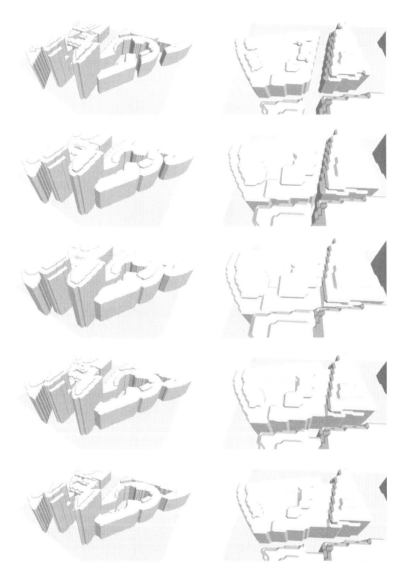

**Fig. 21.9** Application of morphological operations. Starting from the model in row 1, rows 2 and 3 show subsequent dilation operations, and rows 4 and 5 show subsequent erosion operations

**Fig. 21.10** Application of voxelization, opening and Gaussian blurring (left to right)

to tune the DoG, however it might be only used to smooth the alignment structures introduced by the sampling.

## 21.4 Comparison

It is difficult to define appropriate and objective quality measures, as abstraction and generalization are ambiguous in their result, being understood differently by different individuals. In addition, yet few conventions exist for digital 3D city maps. While it is possible to define quantitative measures for different generalization techniques, the current results are rather preliminary and serve as the basis for further research. Usability tests should help to provide these more quantitative measures in the future.

### 21.4.1 Qualitative Measures

In the following, we concentrate on qualitative measures of the presented generalization techniques to evaluate their potential:

- **Similarity to Maps:** How do the results show similarities to 2D map presentations?
- **Similarity to Reality:** How do the results show similarities to (photo-realistic) depictions of non-generalized 3D city model depictions?
- **Aggregation Capabilities:** How do the techniques support building aggregation?
- **Landmark Handling:** How do the techniques handle landmark and outlier buildings?
- **Controllability:** How can the application control the degree of generalization?
- **Dynamic Adaptation:** How can the application change the technique's parameters in response to dynamically changing viewing parameters?

## 21.4.2 Simple-Cell Generalization

This technique creates 3D city model depictions similar to traditional generalized bird's eye view maps if looked from above. The block structure made from cells of intersected streets is typical and can be understood easily. In terms of realism, the model still permits –while being an abstraction– to recognize the extents of a block. The added height gives a further hint about the original building ensemble. The aggregation is done rather naively, depends very loosely on the concrete building geometry and thus is insensitive against potential complexity. Landmarks can be effectively emphasized because their contours can be simply cut out of the cell and the landmark placed into the gap.

The DoG can be controlled by the hierarchy level selected for the network and the width used for network elements. A continuous transfer between different DoG representations has not been investigated so far; a continuous blending with a hysteresis during interactive zooming might be one solution.

## 21.4.3 Convex-Hull-Based Generalization

The convex-hull approach has not been used so far in maps or map-like visualizations. The 'organic' shapes are in contrast to photorealistic depictions. Constrained by the network, the convex hull still reveals and emphasizes the original block boundaries. The height of the largest building contributes much to the occurrence. However, concave cells are handled wrong by the convex hull; in a future version, we plan to replace the convex hull by alpha shapes [9], which can adapt the resulting hull more closely to the building geometry. In addition, network segments running into blocks are covered by the hulls. Handling landmarks and outliers is difficult because they would have to be cut out of the block using 3D Boolean operations. Very tall landmark buildings could just be placed within the hull, but the result would create intersecting geometry. Nevertheless, 3D convex hull creation relies on a mature algorithm working on simple points. For this reason, the technique is robust and insensitive against geometric problems.

The parameter to control the convex hulls is the size of the cells defined by the network. To provide more flexibility as well as to solve the aggregation problems, one could run a clustering algorithm initiated with a given minimal distance. The current solution does not suggest an easy mechanism for the continuous change between different DoG representations.

## 21.4.4 Voxel-Based Generalization

The voxel-based technique is limited to small areas due to the huge amount of required grid data. The result is unlike every map or map-like visualization. The bubble shapes remind the original buildings but do not feature the typical sharp edges and orthogonal structures. Still, aggregation and simplification can be achieved and the unusual look underlines the abstraction. The viewers know from the first view that they look at an abstraction of reality, not a photorealistic visualization.

Landmarks can be excluded from the common voxelization at all and inserted later. As an alternative, they could be voxelized separately using a higher resolution and then inserted without further simplification. This would have the advantage that no visual break occurs in the image, but still the building would be emphasized.

To control this technique, there is first the grid resolution. While a higher resolution yields better quality with less alignment artifacts, it also leads to an explosion of the data to be processed. The second parameter is the buffer size, which controls the morphological opening. Finally the Gaussian blur (or also other filters) can be executed with a given variance and cutoff. Though, the blurring can be seen as a post-processing step independent from the generalization but just to smooth the alignment stair-structures.

Voxel-based generalization cannot be used in conjunction with a dynamic DoG in the current implementation, as the processing is not done all natively by the prototype system. The filter operations are done using an external utility and are re-read later from a file. Additionally, the computation time and the memory needed prohibit a change on-the-fly.

The marching cubes algorithm used introduces a number of redundant polygons, which is no problem for small scenes but might be a problem for bigger city plans. Here, another post-processing step and / or a dynamic level of detail adaptation might be necessary.

## 21.5 Conclusions

This paper presented and discussed three different techniques to automatically derive generalized 3D building ensembles for a cell structure defined by hierarchical networks that divide the reference plane of a virtual 3D city model. The generalized models can be applied to improve the comprehensibility and effectiveness for complex, large-area 3D city models.

The comparison revealed that while the cell-based generalization technique leads to convincing results, the voxel- and convex hull-based techniques currently are less feasible.

Therefore, in the future we will investigate how to sharpen the presented methods towards characteristic architectural elements of 3D building ensembles. In addition, we want to expand the methods towards further city model elements such as vegetation and site objects. An important remaining challenge concerns the handling of multiple scales: A continuous mapping of the DoG to a geometric representation would allow us to combine continous scales in one view of the scene. Also, an optical zoom could be accompanied with a smooth semantic zoom. We will work on this problem when moving forward with our research.

## Acknowledgement

This work has been funded by the German Federal Ministry of Education and Research (BMBF) as part of the InnoProfile research group '3D Geoinformation' (www.3dgi.de).

## References

1. K.-H. Anders. Level of Detail Generation of 3D Building Groups by Aggregation and Typification. *Proc. $22^{nd}$ International Cartographic Conference, La Coruña, Spain*, 2005.
2. M. Beck. Real-Time Visualization of Big 3D City Models. *International Archives of the Photogrammetry Sensing and Spatial Information Sciences*, Vol. XXXIV-5/W10, 2003.
3. H. Buchholz, J. Döllner. View-Dependent Rendering of Multiresolution Texture-Atlases. *Proc. IEEE Visualization 2005*, Minneapolis, 2005.
4. CGAL – Computer Geometry Algorithm Library, www.cgal.org
5. CityGML, www.citygml.org
6. J. Döllner, H. Buchholz. Continuous Level-of-Detail Modeling of Buildings in Virtual 3D City Models. *Proc. 13th ACM International Symposium of Geographical Information Systems* (ACMGIS 2005), 173-181, 2005.
7. J. Döllner, T. H.Kolbe, F. Liecke, T. Sgouros, K. Teichmann. The Virtual 3D City Model of Berlin - Managing, Integrating and Communicating Complex Urban Information. *Proc. 25th International Symposium on Urban Data Management UDMS 2006*, Aalborg, Denmark, 2006.
8. J. Dykes, A. MacEachren, M.-J. Kraak. *Exploring Geovisualization*. Elsevier Amsterdam, Chapter 14, 295-312, 2005.
9. H. Edelsbrunner, E. Mücke. Three-Dimensional Alpha Shapes. *ACM Transactions on Graphics*, 13, 43-72, 1994.
10. E. Fogel, R. Wein, B. Zukerman, D. Halperin. 2D Regularized Boolean Set-Operations. In C. E. Board (Ed.) *CGAL-3.2 User and Reference Manual.* 2006.

11. A. Forberg, H. Mayer. Generalization of 3D Building Data Based on Scale-Spaces. *The International Archives of the Photogrammetry, Remote Sensing and Spatial Information Sciences*, (34)4, 225-230, 2002.
12. E. Gobbetti, F. Marton. Far Voxels - A Multiresolution Framework for Interactive Rendering of Huge Complex 3D Models on Commodity Graphics Platforms. *ACM Transactions on Graphics*, 24(3):878-885, 2005.
13. G. Hake, D. Grünreich, L. Meng. *Kartographie*. Walter de Gruyter, Berlin, New York, 8. Ed., 2002.
14. L. Harrie. *An Optimization Approach to Cartographic Generalization*. PhD thesis, Department of Technology and Society, Lund Institute of Technology, Lund University, Sweden, 2001.
15. T. He, L. Hong, A. Kaufman, A. Varshney, S. Wang. Voxel Based Object Simplification. *Proc. $6^{th}$ Conference on Visualization*, 296, 1995.
16. H. Hoppe. Progressive Meshes. *Computer Graphics Proceedings, Annual Conference Series*, 1996 (ACM SIGGRAPH '96 Proceedings), 99-108, 1996.
17. M. Kada. 3D Building Generalisation. *Proceedings of 22nd International Cartographic Conference*, La Coruña, Spain, 2005.
18. T.H. Kolbe, G. Gröger, L. Plümer. CityGML – Interoperable Access to 3D City Models. *Proc. $1^{st}$ International Symposium on Geo-Information for Disaster Management*, Springer Verlag, 2005.
19. A. Lakhia. Efficient Interactive Rendering of Detailed Models with Hierarchical Levels of Detail. *Proc. 3D Data Processing, Visualization, and Transmission, 2nd International Symposium on (3DPVT'04)*, 275-282, 2004.
20. T. Lewiner, H. Lopes, A. W. Vieira, G. Tavares. Efficient implementation of marching cubes´ cases with topological guarantees. *Journal of Graphics Tools*, 8(2):1-15, 2003.
21. W. E. Lorensen, H. E. Cline. Marching Cubes: A High Resolution 3D Surface Construction Algorithm. *SIGGRAPH '87: Proc. 14th Aannual Conference on Computer Graphics and Interactive Techniques*, ACM Press, 163-169, 1987.
22. H. Mayer. Three Dimensional Generalization of Buildings Based on Scale-Spaces. *Report, Chair for Photogrammetry and Remote Sensing, Technische Universität München*, 1998.
23. H. Mayer. Scale-Space Events for the Generalization of 3D-Building Data. *International Archives of Photogrammetry and Remote Sensing*, 33:639-646, 1999.
24. L. Meng and A. Forberg. 3D Building Generalization. In W. Mackaness, A. Ruas, and T. Sarjakoski (Eds.) *Challenges in the Portrayal of Geographic Information: Issues of Generalisation and Multi Scale Representation*, 211-32. 2006.
25. qHull, www.quhull.org

26 J.-Y. Rau, L.-C. Chen, F. Tsai, K.-H. Hsiao, W.-C. Hsu. Lod generation for 3d polyhedral building model. In *Advances in Image and Video Technology*, 44-53, Springer Verlag, 2006.
27 J. Ribelles, P. Heckbert, M. Garland, T. Stahovich, V. Srivastava. Finding and Removing Features from Polyhedra. *American Association of Mechanical Engineers (ASME) Design Automation Conference, Pittsburgh PA, September* 2001.
28 M. Sester. Generalization Based on Least Squares Adjustment. *International Archives of Photogrammetry and Remote Sensing*, 33:931-938, 2000.
29 M. Sester. Application Dependent Generalization - the Case of Pedestrian Navigation. *Proc. Joint International Symposium on GeoSpatial Theory, Processing and Applications (ISPRS/Commission IV, SDH2002), Ottawa, Canada, July*, 8-12, 2002.
30 K. Shea, R. McMaster. Cartographic generalization in a digital environment: when and how to generalize. $9^{th}$ *International Symposium on Computer-Assisted Cartography*. 56-67, 1989.
31 R. Stüber. *Generalisierung von Gebäudemodellen unter Wahrung der visuellen Richtigkeit*. PhD thesis, Rheinische Friedrich-Wilhelms-Universität zu Bonn, 2005.
32 teem, teem.sourceforge.net/unrrdu
33 F. Thiemann. Generalization of 3D Building Data. *Proc. Joint International Symposium on GeoSpatial Theory, Processing and Applications (ISPRS/Commission IV, SDH2002), Ottawa, Canada, July*, 34(3), 2002.
34 R. Wein, E. Fogel, B. Zukerman, D. Halperin. 2D Arrangements. In: C. E. Board (Ed.), *CGAL-3.2 User and Reference Manual*. 2006.
35 J. Willmott, L.I. Wright, D.B. Arnold, A.M. Day. Rendering of Large and Complex Urban Environments for Real-Time Heritage Reconstructions. *Proc. Conference on Virtual Reality, Archaeology, and Cultural Heritage*, 111-120, ACM Press, 2001.

# Chapter 22
# Automatic Generation of Residential Areas using Geo-Demographics

Paul Richmond and Daniela Romano

**Abstract**

The neighbourhood aspect of city models is often overlooked in methods of generating detailed city models. This paper identifies two distinct styles of virtual city generation and highlights the weaknesses and strengths of both, before proposing a geo-demographically based solution to automatically generate 3D residential neighbourhood models suitable for use within simulative training. The algorithms main body of work focuses on a classification based system which applies a texture library of captured building instances to extruded and optimised virtual buildings created from 2D GIS data.

## 22.1 Introduction

Virtual environments, games and serious games often require the use of large scale urban models which capture essential attributes of real world locations. For example to build a drug dealers serious game, that is virtual training environment for drugs enforcement officers and community support workers, the key requirements are that the training environment must demonstrate realistically linked residential neighbourhood areas as well as being automatically generated to alleviate the need for complex CAD style of modelling. In addition to this the models created need to be optimised for use within a game engine to make use of its animating facilities and platform, and hence should have a limited number of polygons and be easily ported to existing game engines.

The key focus of this paper is on examining how residential neighbourhood profiling systems (geo-demographics) can be used to model realistic neigh-

University of Sheffield
paul@dcs.shef.ac.uk, d.romano@dcs.shef.ac.uk

bourhoods and automatically produce 3D models for simulative training. Its novel contribution focuses on how geo-demographics can be used within a classification system to realistically create virtual neighbourhood areas. First the existing methods of model production are reviewed, and then the implementation of an extrusion and building based classification algorithm which links the geo-demographical and physical properties of building footprint data to an extendable texture library of captured building examples is described.

## 22.2 Previous Work

Existing work in generating city models can be partitioned into two distinct styles which correlate well with the two concluding approaches highlighted by Shiode [1]. The first centres around creating Geographic Information System (GIS) based descriptive models and the second focuses on more detailed constructive models generated either manually or automatically. The next section examines how these two differing methods have been used and highlights some of the problems and limitations with regards to the proposed system.

### 22.2.1 Descriptive Models using GIS data

Urban Planning models use Geographic Information System (GIS) data to create realistic models of real world locations using high resolution data sets. Typically with highly descriptive methods of city modelling aerial photography is used as a texture draped over an extruded model. In order to extrude the models accurately previous techniques have focused on the use of photogrammetry methods of building outline detection [2] and extrusion focusing on GIS building boundary maps [3,4]. Whilst these techniques have been successful the improved availability and structure of GIS data in new Extensible Markup Language (XML) formats has lead to improved model generation methods allowing extremely large datasets to be visualised even in real-time [5].

With the advantage of new GIS datasets there are also new problems introduced regarding copyright, none more apparent than with the Virtual London project [6, 7], which uses Ordnance Survey (OS) Mastermap data in combination with Infoterra's LIDAR (*Light-Imaging Detection and Ranging*) data to extrude buildings. Unfortunately the government funded project which aimed to 'help Londoners visualise what is happening to their city' [8] is currently unavailable to the public through its intended interface (the Google earth application) due to OS licensing actively protecting their intellectual property. Despite a national incentive launched from the UK newspaper the guardian who has proposed the 'free our data' campaign it has recently been suggested that talks between OS and Google have collapsed and it is unlikely that the virtual London model will be available to the public any time soon.

As an alternative to using GIS building boundaries Google offers a community based alternative through the use of Google Sketch-up for Google Earth. This open modelling approach allows the community of Google Earth users to quickly create 3D content, which is easily integrated over the core imagery and terrain data. Despite overcoming the requirement of using OS data neither the Google earth platform or the before mentioned building extrusion methods are particularly well suited to a first person perspective. The reason for this is a matter of resolution as typical aerial photography data is only available at several pixels per meter [9] making distantly viewed urban planning models visually acceptable but lacking in resolution suitable for detailed texture application. In order to create more high resolution and detailed models of cities (and buildings in particular) the attention of using GIS data has shifted to constructive modelling techniques which use either manual or automatic methods to specify building details.

### 22.2.2 Constructive Models using Manual and Automatic City Modelling

With regards to manual modelling the games industry has recently demonstrated its interest in creating realistic virtual cities through the recent games releases of Grand Theft Auto IV and The Getaway, which feature detailed models of the cities of New York and London respectively. The techniques used in this instance are likened by Peter Edward (senior producer) as 'like a western movie. They don't have wooden slates at the back but they are just the fronts' [10]. Whilst the London model makes the use of photorealistic textures captured during an 18 month period the budget and time required to first manually create a wireframe model from the photographic data and then map the photography is far beyond the reach of most modelling projects.

The use of automatic methods to generate virtual cities elevates the budgetary requirement of manually modelling a city but does however introduce a degree of guesswork in representing cities or buildings structure. Despite rule based systems being able to produce entirely procedural (pseudo) infinite cities [11] the models rely heavily on the rules specifying the grid like road structure and hence lack the realism of capturing an actual cities structure, making its application to training somewhat limited. Parish and Muller [12] describe a more intuitive alternative for procedural generation of cities which uses L-systems (inspired by Prusinkiewicz and Lindenmayer [13]) with a basic set of data defining land and water boundaries to subdivide into realistic roadmaps, land, lots and buildings. Although the results of this approach are relatively successful at defining a cities structure in terms of road layout they are limited in application towards building generation, as highlighted in a later paper by Wonka et al [14] which states that a buildings structure;

*'does not follow a growth like process in the same way that plants and streets do, but instead follow a sequence of partitioning steps.'* (p670)

For this reason the automatic generation of buildings tends to focus on the use of shape and split grammars.

### 22.2.3 Building Generation

Split grammars originally introduced by Stiny [15] are used by Wonka et al. [14] as a grammar operating on shapes. By using the split grammar approach, basic shapes (referred to as building blocks) are then recursively replaced by further shapes which have attributes that describe the material and help to determine further subdivision steps. In addition to the extensive set of split grammar rules the notion of a control grammar is also introduced. The aim of this control grammar is described as a way to specify design decisions spatially in a way that corresponds to architectural principles. More simply this control grammar is used to set attributes in a spatially related way ensuring that split grammar rules are propagated in a way which represents realistic architecture. Despite the visual success of this method its complexity requires collaboration between a designer (with a clear understanding of the workings of a split grammar) and an architect, which has lead to the development of similar but slightly less complex split grammar implementations such as the one from Larive [16] who concentrates on 2.5D building frontages for extruded buildings.

More recent work by Muller et al [17] extends the CityEngine application to apply the idea of a shape grammar to create building scripts capable of capturing intricate detail. This is achieved by combining the previous split grammar approach with a complex mass modelling system allowing more complex primitives to be split without the previous limitations of axis aligned shapes. In addition to this the CGA shape system allows the division and placement of other aspects of the environment allowing the generation of pathways and placement of greenery including trees and shrubs.

Although there is a considerable amount of work focusing on the generation of buildings it is important to consider the full spectrum of city generation. Larive et al. [18] uses the notion of 'Urban Zones' to present a hierarchical division of seven stages of city generation; namely, Urban Zones, Road Networks, Blocks, Lots, Exteriors, Building Plans and Furnished Buildings. Whilst the majority of research is concerned with stages 1 through to 5 the focus is primarily on the building stage and concentrates on creating realistic buildings. Whilst this is important in generating a residential area little consideration is given to the creation of realistic neighbourhoods which in the case of the previously described project is more essential than the exact representation of building models.

## 22.3 Implementation

### 22.3.1 Overview

Whilst previously successful implementations are able to build realistic virtual cities and neighbourhoods either procedurally or through using attributes to control architectural variance there is little attention paid to the importance of architectural discrepancies between certain types of housing estates. In addition to this, methods of texture application are either limited to being highly descriptive, in which a large amount of time must be invested to capture buildings or procedural, which although may not require the same investment of time regarding texture capture certainly requires in depth knowledge of architecture and grammars.

The following section of this paper focuses of the implementation of a system which offers the following contributions;

1. Shows how geo-demographics can be used within a classification system to build realistic virtual neighbourhood areas.
2. Demonstrates how an extendable captured building library and classification system can be used to crudely determine a buildings appearance and hence define visual zoning of houses

By achieving the above it is our intention to offer a proof of concept of how geo-demographics can be used to make assumptions of vast city models. In addition to this our work will demonstrate a method somewhere between the descriptive and constructive methods previously described which allows buildings to be represented realistically in a way which creates observable neighbourhood areas. The subsequent section proceeds as follows; Section 3.2 discusses the general extrusion process of the buildings from the floor plan data including how the extrusion process, roofing, polygon face tessellation and any assumptions made of the building structure. The role of captured building textures or Building Instances (BI's) is then discussed in section 3.3 which describes how these are used to influence a buildings visual appearance. Finally section 3.4 describes the classification system used to determine the suitability of each member of the BI library with regards to each building in the dataset.

## 22.3.2 Building Structure & Extrusion

As there is no guarantee that LIDAR data is available for the areas being extruded a method combining a LIDAR data and floor height estimation is used. This is implemented by pre-processing any ASCII LIDAR data (which could be replaced by alternative ASCII data such as cadastral or observed building heights) to calculate an average building height for each building structure in the data set which is then stored in a lookup table. Where LIDAR data is unavailable some assumptions are made towards the heights of buildings. As a rough guide the square of the buildings ground floor surface area is used to calculate a preliminary height, either the averaged LIDAR height or estimated height is rounded to an integer number of floors by division by the standard building floor height of 2.75m. As the majority of buildings within a residential area constitute two and three floor family homes and flats, most buildings have a primary floor calculation of between two and three floors before any adjustments are made to the extreme exceptions. If the calculated number of floors is however less than one then it is assumed that the structure is not a building and is ignored (it is not uncommon that OS classify out buildings such as barns, greenhouses, temporary buildings and some garages as buildings). Likewise buildings with an estimated floor height greater than fifteen are assumed to be larger than average residential dwellings and are reduced to a warehouse/ council flat style of building with at most three floors and a flat roof.

Where as exiting techniques have used split grammars to split a wall face into smaller face areas, allowing procedural textures, the technique employed in this approach is more commonly used within game level design. In particular this method is used in the half life 2 game using the source engine (http://www.valvesoftware.com) for specifying wall brushes (roofs are however described by model entities). In order to reduce the number of faces for each building, a building is first split into floors and then each wall is split (where possible) into a grid of uniform squares. The remainder of the wall which does not constitute a full squares horizontal length is then split evenly between the walls first and last wall section symmetrically. After the walls have been split into wall sections the texture coordinates are then calculated presuming that a square texture is to be applied. For the smaller split wall edge sections the texture coordinates are calculated allowing tessellation where the section meets a full square. This method, although not requiring the manual specification of a split grammar or likewise does require that a reasonable texture library is available providing a rich background of textures to create suitably complex building facades.

Like previous methods of building extrusion and roof generation [19] we have followed Felkel and Obdrzalek's Straight Skeleton Implementation method to apply a simple hipped roof which has been extended to allow a gable/cross gable roof style which is applied to the majority of buildings.

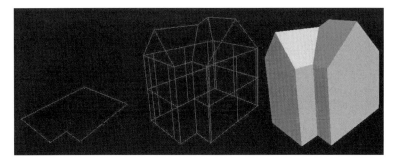

**Fig. 22.1** The wireframe example in demonstrates how a building is extruded, split and applied a gable roof

## 22.3.3 Building Instances

Although there has been significant research into the use of grammatical approaches of texture application to buildings this is not the focus of this paper and as such a simple method has been used to apply textures to buildings. This is achieved by considering a number of captured building textures from the real world which make up what is described in this paper as a Building Instance. Although the choice of which buildings should be captured for the library is somewhat subjective, this method allows the user to capture buildings which stereotype the surrounding area well (see figure 3 for an example of houses from differing ACORN types) and chose buildings where foreground distraction are not present. The subjectivity of BI choices is also considered in section 4.3 which considers the accuracy of the classification for each building in the data set. In addition to capturing a number of textures for any captured building, a number of physical building properties are also recorded for use within the classification system described in section 3.4. For each captured building (or BI) in the texture library, a square texture must be provided for description of a plain brick wall, a windowed area, a doorway and a gable roof section (if the building is flat roofed this is not required). A simple list based description of the building for the ground and above ground floors is then required to provide a base to map each texture to the buildings within the dataset. An example of a captured buildings data is given below where the buildings structure consists of a two floor and four wall structure, where each of the four walls contained room for two descriptive wall elements.

In order to apply the texture to buildings of varying size the description is simply repeated around the building.

**Fig. 22.2** The example building above has a texture description for brick, a doorway, a window and a gable roof section

**Fig. 22.3** Example of two Sheffield houses in close proximity and of a similar structure but with varying ACORN types

## *22.3.4 Neighbourhood Profiling and Geo-Demographics*

The socio-demographic system which has been used in this work is the ACORN (A Classification of Residential Neighborhood) classification system from CACI Limited that contains a hierarchy of classifications spanning down from category, group and type. The categories range from wealthy achievers to hard pressed with the group level giving further definition such as for the wealthy achievers category; wealthy executives, affluent greys and flourishing families. The ACORN type ranges from type 1) 'Affluent mature professionals' in large houses to type 56) 'Multi-ethnic crowded flats'. For each classification there are a number of lifestyle topics and demographic topics which are used to classify UK postcodes. Whilst the lifestyle topics are concerned with interests such as internet cars and shopping the demographic topics consider in addition to other interests dwelling height, size, house hold structure and tenure all of which are particularly useful with regards to classifying the style of a building or neighbourhood. In addition to loading a GML (Geography Markup Language) data file in the World Generator application a table of postcode acorn values must be provided where the postcode is specified as a spot location. In order to calculate the ACORN classification for any building loaded into the system the buildings proximity to each postcode spot location is considered and the appropriate ACORN type code is returned.

In order to determine the most appropriate look of a neighbourhood and hence each building within it, each building read from the GML data is compared to each of the building instances which contain the following additional information which is pre loaded with the textures;

- Name – for reference
- Floors – the number of floor levels above ground level of the building
- Surface area – the surface area of building in square meters at ground level
- External walls - number of external walls (i.e. complexity of the building)
- Roof type – at the moment this is limited to either flat or gable
- ACORN category – the acorn category within the range of 1 to 5
- ACORN group – the acorn group within the range of 1 to 17
- ACORN type – the acorn type within the range of 1 to 56

Buildings and instances are then compared directly by using a distance measure of similarity for each of the above attributes (excluding name). The distance measure is a weighted combination of the summed distance measures between each of the different attributes. This is formalised below by the following expression.

$$D = \frac{\sum \left( \frac{W}{d} \right)}{n}$$

Where
D = Distance measure (0<D<1) between building and BI
W = individual attributes weight value
d = normalised distance (clamped to the range of 0-1) between the building attribute value and BI attribute value
n = number of contributing attributes

By considering each attribute in this way (excluding the roof type, ACORN category and ACORN Type attribute) a value between 0 (infinitely dissimilar) and 1 (exactly similar) is attributed to each attribute. Unlike numeric attributes the roof type attribute which can have two values is either the same (value of 1) or different (value of 0). By considering more than just the ACORN classification building instances with architectural variances but sharing the same ACORN type can therefore co-exist and be attributed to buildings within the model providing realistic variances within the neighbourhoods themselves.

## 22.4 Results & Conclusion

In order to test the building classification method above and its ability to create realistic neighbourhoods a limited number of three building instances were created and applied to a small GML data set of Basingstoke which consisted of four varying ACORN classifications.

### 22.4.1 Dataset & Performance

The GML data set which was used consists of 23037 unique topographic elements of these 6005 are topographic areas (others mainly constitute topographic lines and point data which are usually a repetition of fully defined polygon areas of unclosed polygons which are not used). Within the 6005 topographic areas there are eighteen unique featured codes each with a different combination of the hierarchical elements; theme, descriptive group and descriptive term (a specification of the OS GML specification is available on their website http://www.ordnancesurvey.co.uk), of these eighteen the key feature code (10021) which is used for building generation has the theme 'Building' and descriptive group 'Buildings' (there are 2354 in total). All other features within the GML set are treated according to an XML file which can set the following properties of the hierarchical theme, descriptive group and descriptive term elements.

- materialColour - The surfaces rgb material colour in format 'r,g,b' where r,g,b are java ints
- extrusion - The height in km of a vertical extrusion of the layer above ground level (java double format)
- texture - The texture name to be used to texture objects in the layer
- cleanUpPointTolerance - The tolerance used to clean up points which do not specify any additional detail (i.e. cause an angular difference between the previous and next point of the tolerance value).

This allows the buildings to be set within an environment which displays real world boundaries such as road, pavement and housing boundaries as well as defining areas of the natural environment such as treed areas and scrubland.

In order to generate the 3D model the system operates in two stages first all data is read from the GML file using a SAX parser and stored as a simple object containing the objects properties and polygon points. After all objects are read the surfaces are then extruded, textured, cleaned up and in the case of buildings, constructed using the described method. Running on a Pentium Xeon 2.33 Mhz with 2 Gb of RAM the reading stage takes roughly 3.4 seconds and the extrusion stage takes 2.1 seconds with the production of 195375 faces. Currently all objects are in the visual system are stored in memory for the purposes of displaying them, however it is possible to generate the buildings serially and output them directly to .obj format using a command line interface.

## 22.4.2 Discussion

As demonstrated by figure 4 and figure 5 (which display only the building objects) the effect of using a geo-demographic classification system creates neighbourhoods which reflect the geo-graphic area and assigns buildings a building instance which closely reflects the buildings structure. Despite using only three building instances in this example the complexity of the final model regarding distribution of the building instances is indicative of a realistic neighbourhood. The extendibility of the system also makes the integration of variances in a particular area a matter of simply creating a number of new building instances which reflect new building styles.

In addition to the creation of realistic neighbourhoods, the building/object extrusion and texture application methods create a realistic model of a residential area which despite the lack of integration of decorative objects such as trees, lampposts, etc. provides an ideal platform for a realistic training environment. The use of real GIS data to create the model not only means that the road layouts and housing placements are rational but also offers the possibility of using further GIS data to power the game logic and intelligence models within a simulation.

In order to make a direct comparison of both the classification algorithm and the texturing system a direct comparison is made in figure 7. Although there are some discrepancies, the orientation of the automatically generated terrace roofing being one, the virtual representation demonstrates the correct application of building textures. Despite the texture restriction of only tessellating square textures the low cost texturing method allows the virtual buildings to provide an accurate representation or their real world counterparts.

**Fig. 22.4** ACORN Types

**Fig. 22.5** Building Classification

## 22.4.3 Conclusion and Further Work

The generation of virtual cities seems to have a clear division between descriptive models generated using aerial photography combined with geographic data such as LIDAR and models built using constructive methods such as modelling and shape grammars. Whilst constructive methods offer the only realistically detailed solution of model generation suitable for first person games, the models produced begin to differ from the real world as a function of the time spent either physically modelling, constructing descriptive grammars [14] or manually applying ground captured textures.

**Fig. 22.6** Textured Buildings

The solution offered by this work proposes a method which meets the two objectives set out in section 3.1 by allowing the generation of virtual residential areas and by proving an extendable texturing system. Whilst the building creation methods do not differ greatly from existing techniques the integration of geo-demographics and a classification system allows assumptions to be made of building appearances whilst still maintaining realistic virtual neighbourhood areas. The main weakness of the method described in this paper is the subjective choice of the buildings captured (BI's) within the texture library as a simple extension the system has been implemented to display a buildings colour as a function of its classification accuracy (figure 8). This therefore provides an excellent indication of where extensions to the BI library should be made to improve the model.

With respect to spatial extendibility, it is fair to assume that without the addition of geo-spatial attributes to the classification system it is unlikely that buildings from one area in the UK could be realistically represented by those from another (although the boundaries of neighbourhood areas may still be realistically defined). On a larger scale it is therefore suggested that geo-spatial attributes are included a part of future work. In addition to this it would also be interesting to consider using a more advanced method of texture application for each BI. As a grammar system would be required to achieve this there would be a significantly longer cost of time in preparation of the BI library however it is likely that the building would achieve a more realistic and less granular appearance (a trade off that was made for generation speed and reduced polygon counts in the current implementation).

**Fig. 22.7** Image of Sheffield road compared to Virtual Representation (above with sky post processed)

## References

1. N. Shoide, 3D urban models: Recent developments in the digital modelling of urban environments in three-dimensions, Centre for Advanced Spatial Analysis, University College London, GeoJournal Volume 52, Number 3, pp 263-269, 2000

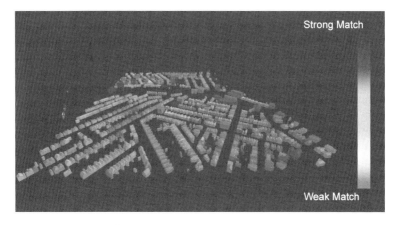

**Fig. 22.8** Classification strength of buildings in Sheffield example

2  I. Suveg, G. Vosselman, 3D reconstruction of Building Models, Int. Archives of Photogrammetry and Remote Sensing, vol XXXIII, part B2, pp. 538-545, 2000
2. R. Tse, M. Dakowicz, C.Gold, D. Kidner, Automatic Building Extrusion from a TIN model Using LiDAR and Ordnance Survey Landline Data, University of Glamorgan, Treforest, Mid Glamorgan
3. R. Laycock, A. Day, Automatically generating large urban environments based on the footprint data of buildings. In IProceedings of the Eighth ACM Symposium on Solid Modeling and Applications, ACM Press, New York, NY, pp. 346-351, 2003
4. A. Steed, S. Spinello, B. Croxford, Richard Milton, Data Visualization within Urban Models, Theory and Practice of Computer Graphics 2004 , pp. 9-16, 2004
5. A.Steed and E.Frecon, "Building and Supporting a Large Scale Collaborative Virtual Environment", Proceedings of 6th UKVRSIG, University of Salford, pp 59-69, 1999
6. A.Steed, E.Frecon, D. Pemberton and G. Smith, "The London Travel Demonstrator", Proceedings of the ACM Symposium on Virtual Reality Software and Technology, pp 50-57, ACM Press, ISBN 1-58113-141-0, 1999
7. Online Article, http://technology.guardian.co.uk/weekly/story/0,,1981821,00.html
8. Wikipedia Link, http://en.wikipedia.org/wiki/Google_Earth#Resolution_and_accuracy
9. Blog article on "The Getaway" PS2 Game, http://digitalurban.blogspot.com/
10. S. Greuter, J. Parker, N. Stewart, G. Leach, Real-time procedural generation of `pseudo infinite' cities, In Proceedings of the 1st international

Conference on Computer Graphics and interactive Techniques in Australasia and South East Asia, 2003
11. Y. Parish, P. Müller, Procedural modeling of cities, In Proceedings of the 28th Annual Conference on Computer Graphics and interactive Techniques SIGGRAPH '01, pp 301-308, 2001
12. P. Prusinkiewicz and A. Lindenmayer, The algorithmic Beauty of Plants, 1991
13. P. Wonka, M. Wimmer, F. Sillion, W. Ribarsky, Instant Architecture, ACM Transactions on Graphics, Volume 22, no. 4, pp. 669-677, 2003
14. G. Stiny, Pictorial and formal aspects of shape and shape grammars: on computer generation of aesthetic objects, Birkhauser, 1975
15. M.Larive and V. Gaildrat, Wall grammar for building generation, Proceedings of the 4th international conference on Computer graphics and interactive techniques in Australasia and Southeast Asia, pp429-437, 2006
16. P. Muller, P. Wonka, S. Haegler, A. Ulmer, L. Gool, Procedural Modeling of Buildings, in Proceedings of ACM SIGGRAPH 2006 / ACM Transactions on Graphics (TOG), ACM Press, Vol. 25, No. 3, pp 614-623, 2006
17. M. Larive, Y. Dupuy, V. Gaildrat, Automatic Generation of Urban Zones, WSCG'2005, Plzen, Czech Republic, 2005
18. R. Laycock and A. Day, Automatically generating roof models from building footprints, The 11-th International Conference in Central Europe on Computer Graphics, Visualization and Computer Vision, 2003

# Part III
# Position papers

# Chapter 23
## Working Group I – Requirements and Applications – Position Paper:
# Requirements for 3D in Geographic Information Systems Applications

Andrew U. Frank

Geoinformation systems (GIS) contain information about objects in geographic space; the focus on geographic space [1] determines the scale of spatial objects and processes of interest at a spatial resolution of approximatively 0.1 m to 40.000 km and to changes occurring once a minute to once a million years. Geographic information is a diverse field which includes many special applications, each of which has special requirements, with special kinds of geometry and particular geometric operations.

The wide variety of requirements of individual geo-applications motivates my first (not new) requirement [2]:

Requirement 1: Construct a fully general 3D (volume) geometry management system based on a clean mathematical foundation (e.g., algebraic topology [3], specifically cw complexes).

Any 3D geometry must be represented with no special cases excluded. Many current packages are optimized for one application (e.g., 3D city models) and restrict the geometry; for example, only volumes with horizontal or vertical boundaries may be accepted. Specialized geometry software, optimized for particular applications, creates difficulties later when data from multiple applications must be integrated to construct a comprehensive view. Restrictions to particular geometries must be possible and the formulation of the corresponding consistency constraints simple (e.g., partitions of 2D, graphs in 3D, 2D surfaces embedded in 3D).

The wide range of spatial objects in a GIS is conceptually structured by level of detail; anybody can experience how one can zoom in on the world untill one sees only one's own front yard (e.g., in Google Earth)! We often conceive this as a hierarchy; however it is better to use a (mathematical)

---

Department of Geoinformation and Cartography,
Technical University Vienna
Gusshausstrasse 27-29/E127,1 , A-1040 Vienna, Austria
frank@geoinfo.tuwien.ac.at

lattice structure [4, 5]. Political subdivisions are typically hierarchies (continent - country - province - county - town), as are watersheds (for example, the watershed of the Fugnitz is part of the watershed of the Pulkau, which flows into the Thaya, which goes to the Danube); if political subdivisions and watersheds are combined, the watershed of the Fugnitz is in the county of Horn, but the watershed of the Pulkau covers parts of several counties and is contained in the province of Lower Austria, whereas the watersheds of the Thaya and the Danube overlap several countries. Hence, a combined representation of the 'part-of' relation of the hierarchical structures of political and watershed subdivisions requires a lattice structure to handle the partial overlaps.

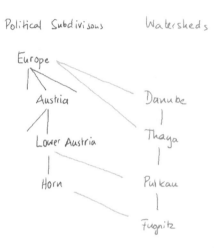

Requirement 2: Support for level of detail: a full or partial containment relation between geometric objects must be maintainable. Applications should be able to view and manage one or a few levels of detail without consideration of other levels. Consistency constraints that connect between the levels are important.

GIS Applications show an approximated current state of what exists. The trend is toward including the temporal aspect and focusing on processes that occur in time and change the world[6, 7]. Processes, not states, are the focus of geography as a science [8, 9, 10]! This requires, first, that updates do not overwrite past states, but that time series of previous states are maintained. Tools to visualize and exploit such timeseries statistically and with data mining operations are needed. This requires, second, separately representable processes and the simulation of future states. Management of time series must be completed with representations of processes that can be calibrated with time series of observed past states and used to simulate future states, e.g., to predict unusual events in order to avoid them, preventing catastrophic results.

Requirement 3: Extend the fully general 3D geometry management with level of details to deal with time and processes[11]. The conceptualization of time should be very general and include continuous and discontinuous changes; it must support a lattice of partial containment relations and different temporal granularities.

These three seemingly simple requirements are, judging from past experiences, very difficult to fulfill. I therefore list here points on which I am willing to compromise:

- It is not required for the designed structures to be efficient or highly efficient (first, computer speed increases steadily; second, optimization of a working solution is often automatable).
- The representation does not need be compact, given the low prices of storage media; however, I fear that high redundancy introduces inconsistencies and increases program complexity [12].

I expect some of the current application areas to extend to 3D+T but also new applications enabled by support for 3D or time. The following examples can be used as tests for proposed approaches to see if these approaches are general enough to support all of them:

- Geology: models of the processes of deposit, folding, and erosion[13];
- Traffic management: cars moving along a street graph. Note the frequency of cars entering and leaving street segments and compare with the frequency of changes in the street graph [14]!
- Cadastral systems [15, 16]: Current systems manage a partition of 2D space that is changing in time. Requirements for 3D are emerging, and it is probably a 3D (volume) topology [17, 18];
- Flood protection: a system is needed to model water flow over a 2D surface embedded in 3D; note that water flow disappears from the surface and reappears somewhere else;
- Organization of pictures taken with a digital camera equipped with GPS having references to location in space and time;
- City planning: Visualize how the city grew and changed in the past and simulate the future;
- Disaster mitigation: models to predict the extension of a substance (e.g., oil or a hazardous gas) over a surface or in a volume under the influence of external forces (gravity, wind, water, flow).

# References

[1] Montello, D.: Scale and multiple psychologies of space. In Frank, A., Campari, I., eds.: Spatial Information Theory: A Theoretical Basis for GIS. Volume 716. Springer Verlag, Heidelberg-Berlin (1993) 312–321
[2] Frank, A.U., Kuhn, W.: Cell graph: A provable correct method for the storage of geometry. In Marble, D., ed.: Second International Symposium on Spatial Data Handling, Seattle, WA (1986) 411–436
[3] Alexandroff, P.: Elementary Concepts of Topology. Dover Publications, New York, USA (1961)
[4] Gill, A.: Applied Algebra for the Computer Sciences. Prentice-Hall, Englewood Cliffs, NJ (1976)
[5] Mac Lane, S., Birkhoff, G.: Algebra Third Edition. 3 edn. AMS Chelsea Publishing, Providence, Rhode Island (1991)
[6] Langran, G., ed.: Time in Geographic Information Systems. Technical Issues in GIS. Taylor and Francis (1992)
[7] Newell, R.G., Theriault, D., Easterfield, M.: Temporal gis - modeling the evolution of spatial data in time. Computers and Geosciences **18**(4) (1992)
[8] Varenius, B.: Geographia Generalis. Elsevier, Amsterdam (1650)
[9] Goodchild, M.F., Egenhofer, M.J., Kemp, K.K., Mark, D.M., Sheppard, E.: Introduction to the varenius project. International Journal of Geographical Information Science **13**(8) (1999) 731–745
[10] Mark, D., Freksa, C., Hirtle, S., Lloyd, R., Tversky, B.: Cognitive models of geographical space. IJGIS **13**(8) (1999) 747–774
[11] Pigot, S., Hazelton, B.: The fundamentals of a topological model for a four-dimensional gis. In Bresnahan, P., Corwin, E., Cowen, D., eds.: Proceedings of the 5th International Symposium on Spatial Data Handling. Volume 2., Charleston, IGU Commission of GIS (1992) 580–591
[12] Gröger, G., Plümer, L.: Exploiting 2d concepts to achieve consistency in 3d gis applications. In: GIS, ACM (2003) 78–85
[13] Turner, K.A.: Three-dimensional modeling with geoscientific information systems. In: NATO Advanced Research Workshop. (1990)
[14] Tryfona, N., Price, R., Jensen, C.S.: Conceptual models for spatio-temporal applications. In: Spatio-Temporal Databases: The CHOROCHRONOS Approach. (2003) 79–116
[15] Al-Taha, K.: Why time matters in cadastral systems. In Frank, A.U., Raper, J., Cheylan, J.P., eds.: Life and Motion of Socio-Economic Units. Taylor & Francis, London (2001)
[16] Al-Taha, K.: Temporal Reasoning in Cadastral Systems. PhD thesis, University of Maine (1992)
[17] van Oosterom, P., Lemmen, C., Ingvarsson, T., van der Molen, P., Ploeger, H., Quak, W., Stoter, J., al., e.: The core cadastral domain model. Computers, Environment and Urban Systems **30**(5) (2006) 627–660

[18] Stoter, J.E., van Oosterom, P.: Technological aspects of a full 3d cadastral registration. International Journal of Geographical Information Science **19**(6) (2005) 669–696

# Chapter 24
# *Working Group II – Acquisition – Position Paper:*
# Data collection and 3D reconstruction

Sisi Zlatanova

3D Geographical Information Systems need 3D representations of objects and, hence, 3D data acquisition and reconstructions methods. Developments in these two areas, however, are not compatible. While numerous operational sensors for 3D data acquisition are readily available on the market (optical, laser scanning, radar, thermal, acoustic, etc.), 3D reconstruction software offers predominantly manual and semi-automatic tools (e.g. Cyclone, Leica Photogrammetry Suite, PhotoModeler or Sketch-up). The ultimate 3D reconstruction algorithm is still a challenge and a subject of intensive research. Many 3D reconstruction approaches have been investigated, and they can be classified into two large groups, optical image-based and point cloud-based, with respect to the sensor used, which can be mount on different platforms.

Optical Image-based sensors produce sets of single or multiple images, which combined appropriately, can be used to create 3D polyhedronal models. This approach can deliver accurate, detailed, realistic 3D models, but many components of the process remain manual or semi-manual. It is a technique which has been well-studied and documented (see Manuals of Photogrammetry, 2004; Henricsson and Baltsavias, 1997; Tao and Hu, 2001).

Active scanning techniques, such as laser and acoustic methods, have been an enormous success in recent years because they can produce very dense and accurate 3D point clouds. Applications that need terrain or seabed surfaces regularly make use of the 2.5D grids obtained from airborne or acoustic points clouds. The integration of direct geo-referencing (using GPS and inertial systems) into laser scanning technologies has given a further boost to 3D modelling. Although extraction of height (depth) information is largely automated, complete 3D object reconstruction and textures (for visualisation) are often weak, and the amount of data to be processed is huge (Maas and Vosselman, 1999; Wang and Schenk 2000; Rottensteiner et al 2005).

---

Delft University of Technology, OTB, section GIS Technology,
Jaffalaan 9, 2628 BX the Netherlands
s.zlaranova@tudelft.nl

Hybrid approaches overcome the disadvantages mentioned above by using combinations of optical images, point cloud data and other data sources (e.g. existing maps or GIS/CAD databases) (Tao, 2006). The combination of images, laser scanning point clouds and existing GIS maps is considered to be the most successful approach to automatically create low resolution, photo-textured models. There are various promising studies and publications focused on hybrid methods (Schwalbe et al, 2005; Pu and Vosselman, 2006) and even on operational solutions (see van Essen, 2007). These approaches are generally more flexible, robust and successful but require additional data sources, which may influence the quality of the model.

In summary, 3D data acquisition has become ubiquitous, fast and relatively cheap over the last decade. However, the automation of 3D reconstruction remains a big challenge. There are various approaches for 3D reconstruction from a diverse array of data sources, and each of them has some limitations in producing fully automated detailed models. However, as the cost of sensors, platforms and processing hardware decreases, simultaneous and integrated 3D data collection using multiple sensing technologies should allow for more effective and efficient 3D object reconstruction.

Designing integrated sensor platforms, processing and integrating sensors measurements and developing algorithms for 3D reconstruction are among the topics which should be addressed in the near future. Besides these, I expect several more general issues to emerge:

1. Levels of Detail (LoD). Presently, a 3D reconstruction algorithm is often created for a given application (e.g. cadastre, navigation, visualisation, analysis, etc.), responding to specific requirements for detail and realism. Indeed, 3D reconstruction is closely related to the application that uses the model, but such a chaotic creation of 3D models may become a major bottleneck for mainstream use of 3D data in the very near future. Early attempts to specify LoD are already being done by the CityGML team, but this work must be further tested and refined (Döllner et al, 2006).
2. Standard outputs. Formalising and standardizing the outputs of the reconstruction processes with respect to formal models and schemas as defined by OGC is becoming increasingly important. Currently, most of the algorithms for 3D reconstruction result in proprietary formats and models, both with specific feature definitions, which frequently disturb import/export and often lead to loss of data (e.g. geometry detail or texture).
3. Integrated 3D data acquisition and 3D modelling, including subsurface objects such as geologic bodies, seabed, utilities and underground construction. Traditionally, the objects of interest for modelling in GIS have been visible, natural and man-made, usually above the ground. As the convergence of applications increases, various domains (e.g. civil engineering, emergency response, urban planning, cadastre, etc.) will look towards integrated 3D models. With advances in underground detection

technologies (e.g. sonic/acoustic, ground penetration radar), already developed algorithms can be re-applied to obtain models of underground objects.
4. Change detection. Detection of changes is going to play a crucial role in the maintenance and update of 3D models. Assuming that automated 3D acquisition mechanisms will be available, the initial high costs of acquiring multiple data sources can be balanced and justified. Changes can then be detected against existing data from previous periods or initial design models (e.g. CAD). In both cases, robust and efficient 3D computational geometry algorithms must be studied.
5. Monitoring dynamic processes. The focus of 3D reconstruction is still on static objects. Although most sensors produce 3D data, hardly any dynamic 3D reconstruction is presently being done. Most dynamic software relies on geovisualisation tools (e.g. flood monitoring; Jern, 2005) for analysis and decision making.

# References

Döllner, J., T. Kolbe, F. Liecke, T. Sgouros, Takis, K. Teichmann, 2006, The Virtual 3D City Model of Berlin - Managing, Integrating, and Communicating Complex Urban Information, In: Proceedings of the 25th Urban Data Management Symposium UDMS 2006 in Aalborg, Denmark, May 15-17. 2006.

van Essen, 2007, Maps Get Real: Digital Maps evolving from mathematical line graphs to virtual reality models. In this book '2nd International Workshop on 3D Geo-Information: Requirements, Acquisition, Modelling, Analysis, Visualisation, 12-14 December 2007, Delft, the Netherlands'.

Maas, H., Vosselman, G., 1999. Two algorithms for extracting building models from raw altimetry data, ISPRS JPRS, 54: 153-63. Manual of Photogrammetry, 2004, 5th Edition, Editor-in-Chief, Chris McGlone, Ed Mikhail, and Jim Bethel, ASPRS publisher, ISBN: 1-57083-071-1, 2004

Henricsson O., and E. Baltsavias, 1997. 3D building reconstruction with ARUBA: a qualitative and quantitative evaluation, Automatic Man-made Object Extraction from Aerial and Space Images (A. Grün, O. Kuebler, and P. Agouris, editors), Birkhaeuser Verlag, Basel, pp. 65-76.

Jern, M. 2005. Web based 3D visual user interface to flood forecasting system, in: Oosterom, Zlatanova&Fendel(eds.) Geo-information for Disaster Management Springer-Verlag, ISBN 3-540-24988-5, pp. 1021-1039

Pu, S., Vosselman, G., 2006, Automatic extraction of building features from terrestrial laser scanning International Archives of Photogrammetry, Remote Sensing and Spatial Information Sciences, vol. 36, part 5, Dresden, Germany, September 25-27, 5 p.

Rottensteiner, F., J. Trinder, S. Clode, K. Kubik, 2005, Automated delineation of roof plans from LIDAR data, ISPRS WG III/3, III/4, V/3 Workshop 'Laser scanning 2005', Enschede, the Netherlands, September 12-14, 2005, pp. 221-226.

Schwalbe, E., H.G. Maas, F. Seidel, 2005, 3D building model generation, from airborne laser scanner data using 2D GIS data and orthogonal point could projections, ISPRS WG III/3, III/4, V/3 Workshop 'Laser scanning 2005', Enschede, the Netherlands, September 12-14, 2005, pp209-214.

Tao, V. 2006, 3D data acquisition and object reconstruction for GIS and AEC, in: Zlatanova&Prosperi (Eds.) 3D Geo-DBMS, in '3D large scale data integration: challenges and opportunities, CRC Press, Taylor&Francis Group, pp. 39-56

Tao, C. V. and Y. Hu, 2001. A Comprehensive Study on The Rational Function Model For Photogrammetric Processing, Photogrammetric Engineering and Remote Sensing, Vol. 67, No. 12, pp. 1347-1358, 2001.

Wang, Z., Schenk, T., 2000. Building extraction and reconstruction from lidar data, IAPRS, 17-22 July, Amsterdam, vol. 33, part B3, pp. 958-964

# Chapter 25
# *Working Group III – Modelling – Position Paper:*
# Modelling 3D Geo-Information

Christopher Gold

3D geo-information can be thought of in several ways. At the simplest level it involves a 2D data structure with elevation attributes, as with remote sensing data such as LIDAR. The resulting structure forms a simple 2-manifold. At a slightly more advanced level we may recognise that the earth may not always be modelled by a planar graph, but requires bridges and tunnels. This 2-manifold of higher genus may still use the same data structure (e.g. a triangulation) but certain assumptions (e.g. a Delaunay triangulation) no longer hold. Finally, we may wish to model true volumes, in which case a triangulation might be replaced by a tetrahedralisation.

Each of these structures may be thought of as a graph - a set of nodes with connecting (topological) edges or links. Most workers in computational geometry, for example, would think in this way. However, because of the usual very large volume of geo-information the emphasis here has often been on (relational) data bases and their associated modelling techniques. More work is clearly needed on the integration of these two approaches. The discussion here uses the graph approach.

An example of a potential major application area is disaster management. This has become particularly relevant in the last few years, and the GIS response to this is very recent, as the 3D structures are not in place in commercial products. Latuada's (1998) paper on 3D structures for GIS and for architecture, engineering and construction (AEC) provides a solid summary of available structures and their different requirements. Briefly, there are surface or volumetric models and he suggests methods for combining 2D triangulations and 3D tetrahedralizations. Lee's (2001) PhD thesis correctly distinguished between the geometric and the (dual) topological structures necessary for building evacuation planning, but did not produce a unified

---

Department of Computing and Mathematics
University of Glamorgan
Pontypridd, Wales, United Kingdom
cmgold@glam.ac.uk

data structure. Meijers *et al.* (2005), Slingsby (2006) and Pu and Zlatanova (2005) discussed the structuring of the navigation graph (using the skeleton or dual of the geometric graph) and the classification of the building 'polygons' (temporary walls, doors etc.)

While this research is very new, a few things emerge. Firstly, both primal and dual graphs are required. Secondly these graphs need to be modifiable in real-time (and in a synchronized fashion) to take account of changing scenarios. This implies a joint data structure (not a hybrid) where the two are fully combined. Thirdly, the structure should not be restricted to buildings (which have relatively well-ordered floors) but should apply to overpasses, tunnels and other awkward objects. The same model would apply to queries about fire propagation and flammability, air duct locations and air flow, utility pipes and cables, flooding and other related issues, where data is available. The model would also apply to other 3D applications such as geology, since the algebraic system expresses all adjacency relationships for complex 3D objects. While it is always technically possible to calculate a dual from its primal graph, it must be emphasized that this is often not ideal. Coordinates and other attributes may be lost, and the navigation in the one space will be easy, while in the dual it will become complex. The integration of the primal and the dual within the one data structure simplifies the number of element types necessary, permits the development of an appropriate 'edge algebra' (as is the case of the Quad-Edge in 2D - see Guibas and Stolfi, 1985) allows verifiable navigation, and assignment of appropriate attributes. (For example, the question: 'How do I get from this room to the next?' directly becomes: 'Give me the properties of the dual of this relationship - of the intervening wall or door.')

GIS is the integrating discipline/system for geo-spatial data from many sources for many applications. It is the natural context for various types of disaster management, route diversion, and flood simulation problems. It is basically a 2D system. Traditionally static, it may permit route modelling, and often include terrain models (TINs). It is a natural 'hub' for the import of various geographically-distributed data types - roads, polygon data, property boundaries, rivers etc. A major emphasis is on querying the attribute and geographic information.

While a good foundation, it does not include proper 3D structures - only 2D terrain models with associated elevations. Full 3D structures are needed for bridges, tunnels, building interiors etc. (N.B. recent work on extending TINs - the Polyhedral Earth (Tse and Gold, 2004) - has allowed bridges and tunnels, but only to give an exterior surface representation - not building interiors. This has been extended in Gold *et al.* 2006.) Thus in the long run, in an operational setting, 3D structures would need to be integrated within a commercial GIS. Zlatanova and Prosperi (2006) discuss the ongoing convergence between GIS and AEC, including the need for topological structures, as do Zlatanova *et al.* (2004)

The core requirement for volumetric models is the development and implementation of an appropriate 3D data structure so that the application may be run in the GIS context. The objective, as given above, is to have a real-time modifiable 3D data structure that integrates the primal and dual graphs, along with their attributes. This should be mathematically verifiable (an algebra) and implementable.

We may classify 3D data models into: Constructive Solid Geometry (CSG); boundary-representations (b-rep); regular decomposition; irregular decomposition; and non-manifold structures (Ledoux and Gold, 2006). Of these, b-reps and irregular decomposition models are the most relevant. B-reps model the boundaries of individual 2-manifolds (surfaces) as connected triangles, rectangles etc. but do not model the interiors. Well known b-rep data structures are the half-edge (Mantyla, 1988); the DCEL (Muller and Preparata, 1978); the winged-edge (Baumgart, 1975) and the quad-edge (Guibas and Stolfi, 1985). The quad-edge is distinctive in that it directly models both the primal and the dual graph on the 2-manifold, and may be expressed as an algebra. (It is often used to model Voronoi and Delaunay cells in the plane.) Irregular decomposition models (e.g. for constructing 3D Delaunay tetrahedralizations) may be constructed with the half-face data structure (Lopes and Tavarez, 1997); G-Maps (Lienhardt, 1994) and the facet-edge data structure (Dobkin and Laszlo, 1989). Half-edges and G-maps do not directly reference the dual structure (a property we need), and the full facet-edge structure appears never to have been implemented. Ledoux and Gold (2006) have proposed the Augmented Quad Edge (AQE) as a navigational structure, but construction operators are not yet fully defined.

These are all graph storage structures from Computational Geometry. Within the GIS community most emphasis has been put on identifying feature elements and specifying their storage in a database. The actual topological connectivity would usually be established after their retrieval into memory (Zlatanova *et al.*, 2004). A possible approach to direct storage of graph structures is suggested in (Gold and Angel, 2006), where they use a form of Voronoi hierarchy to store edge structures in 2D, with the proposed extension to 3D.

# References

Baumgart, B. G. (1975). A polyhedron representation for computer vision. In National Computer Conference. AFIPS.

Dobkin, D. P. & Laszlo, M. J. (1989). Primitives for the manipulation of three-dimensional subdivisions. Algorithmica, v. 4, pp. 3-32.

Gold C. M. and Angel, P., 2006. Voronoi hierarchies. In Proceedings, GI-Science, Munster, Germany, 2006, LNCS 4197, Springer, pp. 99-111.

Gold, C. M., Tse, R. O. C. and Ledoux, H. (2006). Building reconstruction-outside and in. In Alias Abdul-Rahman, Sisi Zlatanova, and Volker Coors, (eds.), Innovations in 3D Geo Information Systems, Lecture Notes in Geoinformation and Cartography, Springer, pp. 355-369.

Guibas, L. J. & Stolfi, J. (1985). Primitives for the manipulation of general subdivisions and the computation of Voronoi diagrams. ACM Transactions on Graphics,v. 4, pp. 74-123.

Lattuada, R. (1998). A triangulation based approach to three dimensional geoscientific modelling. Ph.D. thesis, Department of Geography, Birkbeck College, University of London, London, UK.

Ledoux, H. and Gold, C. M., 2006. Simultaneous storage of primal and dual three-dimensional subdivisions. Computers, Environment and Urban Systems v. 4, pp. 393-408.

Lee, J. (2001) A 3-D Data Model for Representing Topological Relationships Between Spatial Entities in Built Environments. Ph.D. Thesis, University of North Carolina, Charlotte.

Lienhardt, P. (1994). N-dimensional generalized combinatorial maps and cellular quasi-manifolds. International Journal of Computational Geometry and Applications, v. 4, pp. 275-324.

Lopes, H. & Tavares, G. (1997). Structural operators for modeling 3-manifolds. In Proceedings 4th ACM Symposium on Solid Modeling and Applications, pp. 10-18. Atlanta, Georgia, USA.

Mantyla, M. (1988). An introduction to solid modeling. Computer Science Press, New York.

Meijers, M., Zlatanova, S. and Pfeifer, N. (2005). 3D Geo-information indoors: structuring for evacuation. In: Proceedings of Next generation 3D City Models, Bonn, 6p.

Muller, D. E. & Preparata, F. P. (1978). Finding the intersection of two convex polyhedra. Theoretical Computer Science, v. 7, pp. 217-236.

Pu, S. and Zlatanova, S. (2005). Evacuation route calculation of inner buildings, in: PJM van Oosterom, S Zlatanova & EM Fendel (Eds.), Geoinformation for disaster management, Springer Verlag, Heidelberg, pp. 1143-

1161.

Slingsby, A. (2006). Digital Mapping in Three Dimensional Space: Geometry, Features and Access. PhD thesis (unpublished), University College London.

Tse, R. O. C. and Gold, C. M. (2004). TIN Meets CAD - Extending the TIN Concept in GIS. Future Generation Computer systems (Geocomputation), v. 20, pp. 1171-1184, 2004.

Zlatanova, S. and Prosperi, D. (2006), 3D Large Scale data integration: challenges and opportunities, CRC Press, Taylor & Francis Group, ISBN 0-8493-9898-3.

Zlatanova, S., Rahman A. A. and Shi, W. (2004), Topological models and frameworks for 3D spatial objects. Journal of Computers & Geosciences, May, v. 30, pp. 419-428.

# Chapter 26
# *Working Group IV – Analysis – Position Paper:*
# Spatial Data Analysis in 3D GIS

Jiyeong Lee

One of major challenging tasks of 3D GIS is to support spatial analysis among different types of real 3D objects. The analysis functions in 3D require more complex algorithms than 2D functions, and have a considerable influence on the computational complexity. In order to maintain a good performance, not only are the algorithms implemented efficiently, but also the 3D spatial objects are represented by a suitable 3D data model. However, it is a difficult task to select an appropriate data structure designed for the characteristics of the applications, for example, objects of interest, resolution, required spatial analysis, etc. (Zlatanova *et al.* 2004). A model designed for 3D spatial analysis may not exhibit good performance on 3D visualization and navigation. In other words, different data models might be suitable for the execution of specific tasks but not others. In order to maximize efficiency and effectiveness in the provision of operations, Oosterom *et al.* (2002) proposed multiple topological models maintained in one database by describing the objects, rules and constraints of each model in a metadata table. Metric and position operations such as area or volume computations are realised on the geometric model, while spatial relationship operations such as 'meet' and 'overlap' are performed on the topological model. However, it is necessary to find out whether the developed 3D data models are designed for 3D spatial analysis.

3D Grid-based data models are used to support 3D volume computations for the applications of environmental models, such as 3D slope stability analysis and landslide hazard assessment. A shortest path algorithm is also implemented for an un-indexed three-dimensional voxel space using a cumulative distance cost approach. This approach produces a set of voxels, such that each voxel contains an attribute about the cost of traveling to that voxel from a specified start point, if there is uniform friction of movement throughout the representation. The three-dimensional shortest path algorithm moves

Department of Geoinformatics, University of Seoul,
90 Jeonnong-dong, Dongdaemun-gu. Seoul, 130-743,
South Korea jlee@uos.ac.kr

through the 'cost volume' along the steepest cost slope from target to origin using a 3 by 3 by 3 search kernel (Raper 2000). 3D topographic models combining 2.5D terrain models with 3D visualization systems are used for modeling noise (Stoter, *et al.* 2007) and odour contours, visibility analysis, line-of-sight analysis, and right-of-sunlight analysis in order to maintain a sustainable urban environment. The 3D city models in lower level-of-detail largely treat geographic features such as buildings as indivisible entities without internal partitions or subunits. Although the 3D topographic models have been developed and implemented for geo-science analyses and 3D visualization systems, they have some limitations with implementing 3D spatial analyses based on 3D topological relationships including adjacency, connectivity and containment.

The outputs of 3D topological analyses are in three forms: only retrieval of data for 3D visualization, the analytical querying of data once it has been structured in topological format, and the performances of spatial operations such as 3D route calculations, 3D proximity, etc. These topological analyses are relevant in applications where 3D models are extensively used such as earth sciences, geology, archaeology, chemistry, biology, medical sciences, cadastral and urban modeling, computer aided design and gaming.

In the applications, the analytical queries requiring identification of the topological relationships of adjacency and containment answer questions such as 'which regions are cut by a particular fault?', 'which Cambrian unconformities intersect Permian lime-stones?', which 3D buildings are in this 2D county boundary?', 'how many holes, or tunnels does the object have?', 'I am planning to build a tunnel of diameter X - what rock will the tunnel boring machine need to cut through?' and 'which 3D buildings will widening this road impact?' (Ellul and Kaklay 2006). In indoor location-based services to acquire indoor location information and to locate the position information into the 3D digital space using a map matching, the 3D topological queries are implemented to retrieve the context of a user's location to offer appropriate services.

The 3D spatial analysis based on 3D connectivity relationships among spatial entities within the urban modelling arena is performed to support emergency response systems. The applications require a network model through three-dimensional models of buildings for rapid determination of emergency exit paths. The network models present the topological relationships among 3D objects by drawing a dual graph interpreting the 'meet' relation between 3D and 3D objects. Such a structure might be quite powerful for calculations and visualization of 3D routing analysis and 3D topological queries (Lee, 2007). The 3D graph models are used for implementing 3D buffer operations in order to identify what is near features or within a given distance (Lee and Zlatanova, forthcoming). The 3D network models are used to define the network-based neighborhoods for 3D topological analyses in indoor space for analyzing human behaviors, such as an evaluation of neighborhood pedestrian accessibility.

Although many geospatial scientists have been interested in researching and implementing 3D spatial analysis in 3D GIS, a large amount of issues are still remained as challenging tasks in 3D geo-information analysis:

- Analytical 3D visualization to present knowledge on 3D geographic data;
- Analytical queries by identifying the topological relationships (adjacency, connectivity and containment) among combinations of 0, 1, 2, and 3D objects and between complex objects;
- Topological analytical functions including overlay and intersection analyses between two 3D, 2D, 1D and 0D combinations;
- 3D navigations through 3D indoor environments to support emergency response operations and urban modeling;
- 3D buffering and selections based on topological relationships among combinations of 0, 1, 2, and 3D objects and between complex objects.

# References

Ellul, C. and Haklay, M. 2006. Requirements for Topology in 3D GIS. Transactions in GIS, 10(2): 157-175

Lee, J. 2007. A 3D Navigable Data Model to Support Emergency Responses in Micro-Spatial Built-Environments. Annals of the Association of American Geographers, 97(3): 512-529

Lee, J and S. Zlatanova. forthcoming. A 3D Data Model and Topological Analyses for Emergency Response in Urban Areas. In Geo-Information technology for emergency response, eds. S. Zlatanova and J.Li, Bristol, PA: Taylor & Francis (in press).

Oosterom, P. v., Stoter, J., Quak, W., &Zlatanova, S. 2002. The balance between geometry and topology. In D. Richardson & P. Oosterom (eds), Advances in Spatial Data Handling, 10th International Symposium on Spatial Data Handling: 209-224. Berlin: Springer-Verlag.

Penninga, F, 2004, Oracle 10g Topology; Testing Oracle 10g Topology using cadastral data GISt Report No. 26, Delft, 2004, 48 p., available at : www.gdmc.nl/publications

Raper, J. 2000. Multidimensional Geographic Information Science. New York: Taylor & Francis.

Stoter, J., Kluijver, H. and Kurakula, 2007. Towards space-based modelling of continuous spatial phenomena: example of noise, 3D geoinfo 2007.

Zlatanova, S., A. Rahman, A. & Shi, W. 2004. Topological models and frameworks for 3D spatial objects, Journal of Computers & Geosciences, 30(4): 419-428.

# Chapter 27
# *Working Group V – Visualization – Position Paper:*
# 3D Geo-Visualization

Marc van Kreveld

Due to new data collection techniques, more data storage and more computing power, we can make and visualize 3D models of the world. However, the quality and interaction capabilities of these reconstructions are limited. It may be only a matter of time until high-quality visualizations with walkthrough and other interactive possibilities are developed. Some of the applications that can benefit from 3D geo-visualization include city architecture, landscape planning, soil analysis, geology, groundwater analysis, and meteorology.

3D visualization can be used for schematic representation of a geographic region in the style of a traditional map, but with an added 3rd dimension of space, for example, a topographic map draped over a digital elevation model. To get a visually pleasing 3D topographic visualization, 3D map objects are needed, such as 3D symbols and 3D labels.

Above-the-ground 3D data can be obtained by laser altimetry. Below-the-ground 3D data is obtained using bore holes and by several other techniques. Sometimes, data is sparse and errors may exist. For the advancement of 3D visualization, we must consider:

- how to visualize the original data,
- possible reconstructions into 3D models, and
- the uncertainty in these models.

A related question is how best to visualize data so that obvious errors can be detected and removed.

Other questions of interest to future research in 3D geo-visualization are the following:

- What is the role for animation and geo-exploration in 3D geo-visualization?

---

Department of Information and Computing Sciences
Utrecht University
the Netherlands
marc@cs.uu.nl

- What 3D visualizations are effective for various purposes, like geographical 3D data analysis, and illustrative summaries of 3D geographic data?
- How should we visualize patterns that are found by spatial data mining in 3D geo data sets?
- What computation is needed for 3D visualization, and how can we do it efficiently?
- Can ideas from the graph drawing research community be used in 3D geo-visualization?
- Can ideas from the visualization research community be used in 3D geo-visualization?
- How exactly should 3D dynamic/temporal visualization be done(processes, developments,...)?
- What is the best approach to the handling multiple levels of detail (e.g. smooth transition between levels)?
- Is the application of schematic abstract 3D representations possible (in a way similar to the 2D maps)? Note that this is quite different from a visualization that maximizes the realistic impression.
- How should 3D thematic maps be created (e.g. air quality, salinity of ocean, etc.)?

## Selected references

J. Döllner, K. Hinrichs, An object-oriented approach for integrating 3D visualization systems and GIS. Computers & Geosciences, 2000.

O. Kersting, J. D"ollner, Interactive 3D visualization of vector data in GIS. Proceedings of the 10th ACM international symposium on the advances in GIS, 2002.

N. Haala, C. Brenner, K.H. Anders, 3D urban GIS from laser altimeter and 2D map data. International Archives of Photogrammetry and Remote Sensing, 1998.

A. de la Losa, B. Cervelle, 3D Topological modeling and visualisation for 3D GIS. Computers and Graphics, 1999.

B. Huang, An integration of GIS, virtual reality and the Internet for visualization, analysis and exploration. International Journal of Geographical Information Science, 2001.

M.P. Kwan, J. Lee, Geovisualization of Human Activity Patterns Using 3D GIS: A Time-Geographic Approach. Spatially Integrated Social Science, 2004.

J. Stoter, S. Zlatanova, 3D GIS, where are we standing. ISPRS Joint Workshop on Spatial, Temporal and Multi-Dimensional Data Modeling, 2003.

A. Altmaier, T.H. Kolbe, Applications and Solutions for Interoperable 3d Geo-Visualization. Proceedings of the Photogrammetric Week, 2003.

S. Zlatanova, A.A. Rahman, M. Pilouk, Trends in 3D GIS Development. Journal of Geospatial Engineering, 2002.

S. Shumilov, A. Thomsen, A.B. Cremers, B. Koos, Management and visualization of large, complex and time-dependent 3D objects in distributed GIS. Proceedings of the 10th ACM international symposium on the advances in GIS, 2002.

S.S. Kim, S.H. Lee, K.H. Kim, J.H. Lee, A unified visualization framework for spatial and temporal analysis in 4D GIS. Geoscience and Remote Sensing Symposium, 2003.

Printing: Krips bv, Meppel, The Netherlands
Binding: Stürtz, Würzburg, Germany